GENERIC

GENERIC

THE **UNBRANDING**
OF MODERN MEDICINE

JEREMY A. GREENE

JOHNS HOPKINS UNIVERSITY PRESS | BALTIMORE

© 2014 Johns Hopkins University Press
All rights reserved. Published 2014
Printed in the United States of America on acid-free paper
9 8 7 6 5 4 3 2 1

Johns Hopkins University Press
2715 North Charles Street
Baltimore, Maryland 21218-4363
www.press.jhu.edu

Library of Congress Cataloging-in-Publication Data

Greene, Jeremy A., 1974– author.
    Generic : the unbranding of modern medicine / Jeremy A. Greene.
        p. ; cm.
    Includes bibliographical references and index.
    ISBN-13: 978-1-4214-1493-5 (hardcover : alk. paper)
    ISBN-10: 1-4214-1493-7 (hardcover : alk. paper)
    ISBN-13: 978-1-4214-1494-2 (electronic)
    ISBN-10: 1-4214-1494-5 (electronic)
    I. Title.
    [DNLM: 1. Drugs, Generic—history. 2. Drug Industry—history.
3. History, 20th Century. 4. History, 21st Century. QV 11.1]
    RM301
    615′.1—dc23   2013048851

A catalog record for this book is available from the British Library.

*Special discounts are available for bulk purchases of this book. For more information,
please contact Special Sales at 410-516-6936 or specialsales@press.jhu.edu.*

Johns Hopkins University Press uses environmentally friendly book materials, including recycled text paper that is composed of at least 30 percent post-consumer waste,
whenever possible.

To Allan

CONTENTS

ACKNOWLEDGMENTS

This book is dedicated to Allan Brandt, an incomparable mentor, teacher, colleague, and friend. It was Allan who introduced me to the vitality of the field of medical history and the joys of mapping a career at the intersection of clinical practice and historical scholarship. As I have researched and assembled the content of this volume, I have witnessed Allan continue to climb one Everest after another: he remains a uniquely self-actualized scholar and a singular human being.

In the decade or so that I have spent as a scholar of generics, I have accumulated many specific intellectual debts, not all of which can be contained in these acknowledgments. I would like to thank my editor at Johns Hopkins University Press, Jacqueline Wehmueller, who combines a graceful editorial style with remarkable efficiency, as well as the anonymous reader for the Press. Several individuals provided comments on chapter drafts, including G. Caleb Alexander, Chad Black, Bob Bohrer, Marilena Correa, Joseph Gabriel, David Herzberg, David Jones, Aaron Kesselheim, John Kuczwara, Ilana Löwy, Victor Manuel Garcia, Kris Peterson, Anne Pollock, Kaushik Sunder Rajan, Alisha Rankin, Charles Rosenberg, John Swann, Ulrike Thoms, Daniel Todes, and Nancy Tomes. Particular thanks to Rebecca Lemov, Deborah Levine, Scott Podolsky, and Sophia Roosth for repeated deep dives into the material. Thanks also to Carolyn Acker, Susan Bell, João Biehl, Alberto Cambrosio, Angela Creager, Christopher Crenner, Joe Dumit, Jonathan Erlen, Jean-Paul Gaudillicrc, Michael Gordin, Nathan Greenslit, Anita Hardon, Lynn Hsu, Margaret Humphreys, Evan Hepler-Smith, Greg Higby, Suzanne White Junod, Nick King, Howard Kushner, Richard Laing, Andrew Lakoff, Julie Livingston, Anne Kveim Lie, Neil and Lisa Martin, Bjarke Oxalund, Stephen Pemberton, Adriana Petryna, Joanna Radin, Peter Redfield, Susan Reverby, Samuel Roberts, Paul Sherman, Susan Strasser, Dominique Tobbell, Elly Truitt, Kiichiro Tsutani, John Harley Warner, Ayo Wahlberg, Keith Wailoo, Elizabeth Watkins, George Weisz, and Susan Reynolds Whyte for helpful conversations along the way. Early legwork on this project included a failed collaborative grant application with Cori Hayden; though we did not receive funding for our joint project, I gained a heightened appreciation for the value of articulating historical and ethnographic investigations of a mutual subject.

Early drafts of most chapters were presented at a number of conferences and colloquia, including the meetings of the American Association for the History of Medicine, the American Institute of the History of Pharmacy, the History of Science Society, the Organization of American Historians, and the American Studies Association, as well as invited lectures at Bowdoin College, California Western School of Law, Duke University, Emory University, Georgia Tech, Harvard Law School, Princeton University, the University of Albany, the University of Copenhagen, the University of Illinois–Chicago, the University of Oslo, the University of Pennsylvania, the University of Pittsburgh, the University of Rochester, and the University of Utrecht. Particular thanks go to the architects of the ESF Drugs Network conference series, especially Christian Bonah, Sophie Chauveau, Flurin Condrau, Jean-Paul Gaudilliere, Christoph Gradmann, Volker Hess, Ilana Lowy, Toine Pieters, Jonathan Simon, Ulrike Thoms, Carsten Timmerman, and Mick Worboys. An early sketch of the project was published in the *Journal of the History of Medicine and Allied Sciences*, and ancillary ideas were developed along the way in short publications in the *Lancet*, the *New England Journal of Medicine*, and *History & Technology*.

Early work on this project was supported by pilot funding from a Ewing Marion Kaufmann Junior Faculty Fellowship in Entrepreneurship Research and a Larry J. Hackman Research Residency from the New York State Archives. The larger research endeavor was made possible with key funding from a Scholars Award in Science and Technology Studies from the National Science Foundation. Subsequent support from the National Library of Medicine helped transform the research into a publishable book. I would like to thank Mateo Munoz, Emily Harrison, Oriana Walker, Kirsten Moore, Anita Dam, and Max Lloyd for their assistance with research. Thanks also to Bill Davis and Richard McCulley at the National Archives; to Stephen Greenberg and Mike Sappol at the National Library of Medicine; to Greg Higby and Elaine Stroud at the American Institute of the History of Pharmacy; to George Griffenhagen at the American Pharmaceutical Association; to John Swann and Suzanne Junod and the Freedom of Information Act officers at the Food and Drug Administration; to Marie Villemin at the World Health Organization Archives; and to Kathryn Baker and Jack Eckert at the Center for the History of Medicine at the Countway Medical Library, as well as the archivists at the American Medical Association, the University of California San Francisco, the University of California San Diego, the Chemical Heritage Foundation, and the New York State Archives. William Haddad, Alfred Engelberg, and Richard Burack provided access to personal archival materials and granted extensive interviews with the author.

This book owes its existence in no small part to Jerry Avorn, whose Division of Pharmacoepidemiology and Pharmacoeconomics (DoPE) at the Brigham and Women's Hospital has become a thriving center for interdisciplinary research on the role of pharmaceuticals in American society. Jerry's vision for the field communicates a deep and abiding sense that historical analysis can and should inform contemporary research into pharmaceutical policy and epidemiology, and I benefitted immeasurably from being an active member of his faculty. I particularly enjoyed the many long conversations and collaborations with fellow DoPErs Jennifer Polinski, Aaron Kesselheim, Niteesh Choudhry, Joshua Gagne, Michael Fischer, Sebastian Schneeweiss, and Will Shrank, and the research assistance of Mary Peterson and Uzaib Saya. In my new life at Johns Hopkins, I have likewise benefitted from conversations with fellow faculty members in the Center for Drug Safety and Effectiveness, especially G. Caleb Alexander, Kay Dickersin, Janet Holbrook, Tianjing Li, Brent Petty, Jodi Segal, and Sonal Singh. My work has also depended on close conversations with clinical colleagues, from Asaf Bitton, Andrew Ellner, Pracha Eamranond, Michael McWilliams, Jamie Redgrave, Gordy Schiff, Lori Tishler and the staff of the Brigham Internal Medical Associates, to Mike Albert, David Dowdy, and Sadie Peters and the staff of the East Baltimore Medical Center. Supportive and enlightened chairs of medicine, from Joseph Loscalzo at the Brigham & Women's Hospital to Myron Weisfeldt and Jeanne Clark at the Johns Hopkins School of Medicine, have likewise provided essential support for my continued career as a clinician-historian.

While all mistakes in this volume are clearly my own, some of the better ideas likely came from idle conversations with colleagues at Harvard and Johns Hopkins. I would like to particularly thank Mario Biagioli, Janet Browne, Steve Burt, Jimena Canales, Daniel Carpenter, Joyce Chaplin, Lizabeth Cohen, Paul Farmer, Peter Galison, Anne Harrington, Maya Jasanoff, Sheila Jasanoff, Andrew Jewett, Salmaan Keshavjee, David Korn, Shigehisa Kuriyama, Everett Mendelsohn, Ian Miller, Sarah Richardson, Dennis Ross-Degnan, Stephen Shapin, Adelheid Voskuhl, and Anita Wagner at the Department of History of Science and elsewhere at Harvard. Thanks as well to Sara Berry, Nathaniel Comfort, Veena Das, Mary Fissell, Yulia Frumer, Clara Han, Marta Hanson, Earle Havens, Bob Kargon, Sharon Kingsland, Bill Leslie, Graham Mooney, Randall Packard, Gianna Pomata, Maria Portuondo, Larry Principe, Christine Ruggere, Dan Todes, and especially Coraleeze Thompson for making me immediately welcome in the vibrant intellectual community of the Institute for the History of Medicine and elsewhere at Johns Hopkins University.

My family has supported this book project in innumerable ways, from the emergency childcare provided by Gene and Marilyn Dorosh to the writing weekends made possible by the generosity of my mother, Doren Greene, who along with my father and stepmother, Wayne and Patti Greene, also read drafts at the last minute. My grandparents Marilyn Freedman and Gerry Buter remain my final and most important reviewers. My overall well-being these past few years has been reliant on the nonstop rodeo that is my sisters and brother and sisters- and brothers-in-law and their children: David, Rachel, Ami, and Becca Greene, Jaime Wong, Emily, David, Amanda, Elena, and Sarah Asofsky, and Daniel, Jennie, Zoe and Berkeley Chamberlain. Calvin Greene is written into these pages as well, and we continue to remember him every day. I cannot count the many points that my wife and life partner, Elizabeth, has accumulated in helping me create the mental, temporal, and physical space to finish this work or how much joy my children Phoebe (seven) and Levi (four) have brought into my life as a scholar, helping me edit the manuscript even now, the night that I am sending it to the Press. "It's not a book for kids," Phoebe told me late this evening after reading the first page, shortly before falling asleep next to me on the couch, "but I can tell it's going to be good." I hope to live up to her expectations.

GENERIC

THE SAME
BUT NOT THE SAME

A name is a necessity. It distinguishes one thing from dissimilar things. A name also distinguishes one thing from other things that are similar but not the same.

—NELSON M. GAMPFER, NATIONAL PHARMACEUTICAL COUNCIL, 1961

The word *generic* suggests the ho-hum, the undistinguished, the easily forgotten. A generic argument is one that is not novel. A generic product is not noticeably different from its competitors. A generic response tells us something we already know.

So too, we might expect, with generic drugs.

Yet generic drugs have become objects of considerable interest. Over the past half century, generics have expanded from capturing less than 10 percent of the drug market to providing more than 80 percent of all prescriptions filled in the United States.[1] As the research-based pharmaceutical industry continues to suffer through a twenty-year crisis in innovation, with patents expiring on blockbuster drugs and fewer drugs emerging from the research pipeline to replace them, it now confronts the prospect of an increasingly generic future.[2] In turn, generic firms such as Teva and Mylan, which were small local companies in the 1970s, have mushroomed into multinational global players in their own right, ranking in the top ten vendors of prescription drugs by volume in the US market in 2010.[3]

Generic drugs are now omnipresent in clinics, drugstores, and households around the world. In most of the United States, a prescription written for a brand-name drug will automatically be filled with a cheaper, generic version in the pharmacy. We like to think of these tablets, capsules, patches, and ointments as interchangeable: why pay the brand-name price for Eli Lilly's Prozac when generic fluoxetine works just as well for a fraction of the price? Why pay more for the same? Nonetheless, as we fill our prescriptions at the drugstore, we are reminded materially that generics are never fully identical to the brand-name drug, or even to each other. When the same prescription is filled one month with green-and-yellow capsules, the next month with blue-and-orange, the next month with white-on-white, one is presented with a puzzle: is this unfamiliar object the same drug? On a conceptual, molecular level, all three products are the same: all contain 40 mg of fluoxetine and have passed stringent FDA bioequivalence protocols in human subjects. Yet, on a tangible, experiential level, all three products are not the same: they differ evidently in name and place of origin, in their color, shape, and size, and in their price tags.

An enormous amount of effort has gone into ensuring that, for the consumer of generic drugs, these superficial differences are trivial and the structural similarities are substantial. As a primary care physician in a busy neighborhood health center in the city of Baltimore, I fully expect most of the prescriptions I write to be filled with generic drugs. When, as I was drafting the manuscript for this book, my four-year-old son and I came down with strep throat, the antibiotics that we consumed were made by generic manufacturers. As a physician, parent, and patient, I regularly rely on the assumption that generic drugs are the same but cheaper. Indeed, the substitution of bioequivalent generic drugs for more expensive brand-name products can be considered one of the few success stories in a field of otherwise failed attempts to deliver equivalent value in health care for a lower price.

Yet this is a very recent history, and it has not been a gentle one. The generic drug that moves so easily in domestic and global commerce today was a hotly contested object only a few decades ago, and not an object at all just a few decades before that. Underneath their plain packaging, generic drugs have important stories to tell about the challenge of making things the same in modern medicine.

NO FREE LUNCH

On a balmy winter's day in late 1987, Marvin Seife sat down to a working lunch that would lead to the premature end of his career and almost his life. A physician who had entered service at the Food and Drug Administration (FDA) in 1965, Seife became the first head of the agency's Division of Generic Drugs in the 1970s. By most accounts his demeanor seemed to fit his position: thrifty, resourceful, described by a colleague as "a laconic Jewish New Englander who wears a Carter-type sweater with a hole in the elbow."[4] But when Seife sat down that day with an executive of a generic drug company called My-K Laboratories, he neglected his responsibility to report that a pharmaceutical firm paid the bill.

The $59.20 lunch, recorded on the American Express card of My-K's president, resurfaced a few years later in the hands of Representative John Dingell (D-MI). Generic drugs had by that time become a key symbol for bipartisanship in health policy under the administrations of Ronald Reagan and George H. W. Bush. The Price Competition and Patent Term Restoration Act of 1984—crafted by the liberal Henry Waxman (D-CA) and the conservative Orrin Hatch (R-UT)—showcased a well-regulated generic drug industry as a key mechanism for controlling costs in health care while maintaining an incentive for innovation. The Hatch-Waxman Act (or Waxman-Hatch, depending on which side of the aisle you saw it from) was praised by everyone from Ronald Reagan to Edward Kennedy as a "win-win" measure that balanced the advancement of public health and the virtues of the free market.[5] Yet by the late 1980s, Dingell began to hear complaints that generic firms were treated unequally by FDA reviewers. Over the next few years, Dingell's House Energy and Commerce Subcommittee on Oversight and Investigations would uncover a corruption scandal that led to multiple criminal convictions and the resignation of Frank Young as commissioner of the FDA.[6]

In the summer of 1989, two FDA officials testified in front of Dingell's subcommittee that they had accepted manila envelopes stuffed with several thousand dollars from the executive of the generics firm Superpharma. In exchange, regulators had orchestrated the slowing of reviews for competing generic products, burying key files in desk drawers while speeding others through the regulatory machinery.[7] Faced with this evidence of "arbitrary power of life and death over . . . the infant generic drug industry," Dingell vowed his investigation would provide "a large dose of the best disinfectant of all—sunshine."[8]

More sunshine, however, only exposed more rot. Unlike brand-name drugs, which the FDA approved for marketing on the basis of clinical trials demonstrating safety and efficacy in human subjects, generic drugs were approved on the basis of proof of similarity. If a generic drug could pass a set of tests in comparison to the brand-name version, it could be marketed as functionally the same. Dingell's investigation uncovered evidence that many crucial tests of similarity—especially the new protocol of in vivo bioequivalence testing—were performed not by generic manufacturers but by private testing companies. The outsourcing of bioequivalence testing enabled a rather audacious new form of fraud. Under pressure, the generic manufacturer Vitarine confessed that instead of sending its own generic copy of Smith, Kline & French's popular diuretic Dyazide for bioequivalence testing against the original brand, it had sent the testing company an unlabeled Dyazide capsule *as if it were Vitarine's own generic product.*[9] As Dingell noted with exasperation, Vitarine's approval to market a generic version of Dyazide "was based on a study in which Smith Kline's product was compared to itself," and Vitarine's product was never tested.[10]

When Dingell's committee subpoenaed the entire biosample library of the Baltimore bioequivalence testing firm PharmaKinetics, it found evidence of fraudulent biotesting in at least four other companies. In one particularly egregious case, the generic firm Bolar Pharmaceuticals had submitted an actual tablet of Sandoz's Mellaril with the Sandoz logo lightly sandpapered off and claimed that the pill was a Bolar generic product. Unfortunately for the generic firm, the pill artists at Bolar hadn't fully removed all evidence of the logo. FDA staffer Paul Vogel presented Dingell's committee with two microphotographs of pills submitted for bioequivalence testing—one "original," with the Sandoz logo of an *S* in a triangle clearly embossed on one side, and one "generic," which under close examination bore "very faintly the appearance of . . . a triangle and you can pick up some indication of an S." The FDA's approval of Bolar's generic product had been based on falsified proofs of similarity: proofs demonstrating only that Sandoz' brand-name product was equivalent to itself.[11]

Only five years after being praised by Ronald Reagan as a private sector solution to public health problems, the generic drug industry had metamorphosed into an illustration of new forms of graft and malfeasance that could occur at that intersection. Bolar withdrew its generic version of Dyazide in January of 1990, and the company (along with Vitarine) was expelled from the Generic Pharmaceutical Industry Association; several FDA officials were fined and sentenced to community service for accepting bribes; a number of generic industry executives received short jail terms.[12] As the director responsible for his division, Marvin Seife took the biggest fall. He reappeared in October of 1990, to face accusations that he had taken a bribe by accepting an undocumented lunch. As the credit card statements in Dingell's hands now proved, Seife had perjured himself when he lied under oath that he did not have meals with industry representatives.[13] He was sentenced to ten months in a prison near his home in San Antonio, Texas.[14]

When Seife arrived to begin his assigned time at a minimum-security federal corrections facility in Big Spring, Texas, the prison office had no record of him, nor did the nearby medium-security prison to which he was transferred. In accordance with standard policy for prisoners who arrive without specific documentation, he was placed in solitary isolation and issued standard prison clothing. Clinically speaking, however, Seife was not a standard prisoner: he was an insulin-dependent diabetic with avascular necrosis of both hips and feet two sizes larger than the largest prison-issue shoes. Locked up in solitary without his own clothing or medicines, he first developed foot ulcers, then a rapidly progressing infection that spread up his left leg, then gangrene that went untreated during nearly two weeks spent in solitary confinement before his paperwork was processed. When he finally collapsed and was brought to a nearby hospital, his leg was beyond saving and had to be amputated below the knee. Worse still, the infection had spread to the blood and caused septic shock, kidney failure, heart failure, pneumonia, and coma. After three weeks in an intensive care unit, including another overwhelming staphylococcal infection and the loss of a toe on his right foot, he somehow survived the ordeal and was transferred back to prison—this time in a private room with his own shoes.[15]

The relevant personalized information regarding Seife's life-threatening diabetes and the rest of his medical history had been lost along with the rest of his paperwork—including a letter from Seife's doctor noting his vulnerability to serious infections. As he recovered in his hospital room, Seife marveled at the gruesome irony of his own bodily suffering for the corruption of government officials and the generic industry.

What has happened to me is the most horrible thing on Earth. It almost killed me. I won all those awards, I was one of the most honored employees of the federal government, they wrote that I was "a breath of fresh air." And look at what they've done to me! . . . Out of nothing I was totally destroyed. Thirty years, I gave my whole life to public service! . . . It's cost me $250,000 to defend myself. I have to pay it myself

and those crooks in the generic drug industry get their legal bills paid
. . . And in Washington everybody is robbing the whole country blind.
I read in the newspaper about some of the things George Bush's sons
are doing—and they're walking around. And all I had—and it's not
even against the law—all I had was a lousy lunch!"[16]

Seife was the victim of an approach to health policy that posited the generic drug industry as a private sector solution to a public health problem.
By conceiving of the generic industry largely as an ally in the approval of
high-quality, affordable therapeutics, Seife underestimated the extent
to which generic firms could be just as vulnerable to the temptations
of graft, collusion, price fixing, and "pay-for-delay" as any other private
corporation engaged in the business of health care.

THE THYROID STORM

On the other side of the continent, as Seife was forgetting a lunch that
would prove to be far from free, another professional with pharmacological expertise was being drawn into a very different contract concerning generic drugs. As a professor of clinical pharmacy at the University
of California, San Francisco (UCSF), Betty Dong ran a laboratory that
pioneered new forms of research into the biological differences among
chemically equivalent versions of drugs, especially in the case of thyroid
hormones. Unlike drugs that work clinically once a minimal dosage has
been achieved, thyroid medicines only work when titrated to a precise
level in the bloodstream of a patient. Too much, and the patient will show
symptoms of hyperthyroidism; too little, and he or she will show symptoms of its opposite, hypothyroidism. Thyroid hormone also interacts
with a host of other entities en route from the stomach to its sites of
action.[17] Dong's research chronicled the many ways that similar versions
of the same drug could produce clinically significant differences.

It was not surprising, then, to Dong or the general counsel of UCSF,
that the British-based pharmaceutical firm Boots was interested in funding a larger study to compare the biological differences between their
brand of synthetic thyroid hormone, Synthroid, and a host of chemically
equivalent versions newly available for nearly half the price. Boots had
dominated the market for thyroid hormone ever since Synthroid was approved by the FDA in 1958; as a "hard to copy" drug, it remained without

generic competition for decades after its original patent had expired. When, in the mid-1980s, a generic competitor appeared on the US market, hospitals, health maintenance organizations (HMOs), and private and public insurers began to encourage the substitution of the cheaper levothyroxine at the pharmacy. Executives at Boots hoped to find proof of significant therapeutic differences between Synthroid and its new generic competitors, and Dong seemed just the right researcher for the job.

Because all drug products are slightly different, even when they contain the same chemical, Boots executive Carter Eckert reasoned that *some* proof of divergence between Synthroid and other levothyroxine products would emerge from a sufficiently large and sensitive study. After all, the product Synthroid was more than just the chemical levothyroxine. It was levothyroxine bound up in a secret blend of binders, stabilizers, fillers, and excipients, stamped under a precise degree of pressure and coated with a proprietary coating, all of which were protected as trade secrets. These trade secrets were part of the mystique that had made Synthroid a "hard to copy" drug in the first place. Eckert and others at Boots were confident that the right researcher could demonstrate the clinical significance of these differences in a clinical trial.

To Dong and Boots, the contract seemed another "win-win" affair. Dong would obtain funding to conduct a clinical study of bioequivalence on a scale hitherto impossible. Boots would obtain scientific proof of difference to defend its flagship product from competition. And so perhaps it is not surprising that when Dong signed a research contract with Boots in 1987, she took little note of the clauses granting ownership of data and publication rights to Boots. As stated in the contract, Boots conducted frequent site visits over the next few years as Dong's team recruited and followed its population of research subjects. Regular communication between sponsor and investigator suggested that her research would proceed uneventfully.

Once the data were analyzed, however, the results proved a rude surprise. All four versions of levothyroxine performed identically, according to the most sensitive techniques available to date. Instead of proof of difference, the UCSF team had found proof of similarity. As Dong's team prepared their findings for publication, executives at Boots scrambled to undermine the results of the study they had originally funded and actively monitored. After Dong's team submitted a manuscript to the *Journal of the American Medical Association* (*JAMA*) in January 1994, the VP of marketing at Boots marked up his copy "Actions—must review harshly—begin to get our ducks in order with the salesforce."[18] Dong was warned, in a letter from her sponsor, that "it will be necessary to use all legitimate, honest, honorable, and ethical means at our disposal to inform the scientific and regulatory community of the problems with this study if you choose to pursue publication of this manuscript," including litigation for breach of contract.[19] In January of 1995, just a week before her manuscript was scheduled to be published in *JAMA*, Dong received a call from UCSF's lawyers stating that they believed Boots would win such a lawsuit and they could neither protect her in court nor cover her legal expenses.

Dong pulled the manuscript. Instead of a *JAMA* study proving equivalence between Synthroid and its generic imitators, the medical literature in 1995 saw only the publication of a sixteen-page, one-sided critique of Dong's own unpublished data in a journal coedited by an executive of Boots.[20] That would have been the end of it, perhaps, had an investigative journalist for the *Wall Street Journal* not learned of the story through contacts at UCSF and broken it to the general public in a front-page feature story. The *Journal*'s analysis of the Boots/Dong affair ignited a firestorm over the brand-name pharmaceutical industry's ability to suppress negative results, leading eventually to Boots's reversal of its gag order and the final publication of Dong's original manuscript in 1997.[21]

The "Thyroid Storm" (as it came to be known) became the flash point for a new critique of drug industry involvement in clinical trials.[22] Yet, as Dong's ordeal came to stand more generally as a case study in the conflict of interests between academic researchers and their industrial sponsors, the telling and retelling of her story came to ignore the underlying issue at stake for both Dong and Boots: of scientific evidence of equivalence suppressed by a strong economic interest in perpetuating a belief in product difference.

The importance of these proofs of similarity and difference were not lost on the actors themselves, especially users of Synthroid and the public and private entities that paid for the drug. A class action lawsuit was launched in 1997 to recover some of money that had been overpaid for Synthroid based on false claims of its noninterchangeability with generic forms. Boots's new owner, BASF, ultimately settled for more than $100 million—a hefty sum but a mere fraction of the estimated cost gap of $800 million.[23] Dong's career would eventually recover but not before she—like Marvin Seife—lost years of her career as a casualty of public controversies over the similarity or difference between generic and brand-name drugs.

GENERIC HISTORIES

On a superficial level, both Seife and Dong faced problems unconnected to generic drugs: for Seife, the risk of a regulatory agency being captured and corrupted by the industry it regulates; for Dong, the fragility of authorship rights in a world where research is funded by powerful corporations. And yet Marvin Seife and Betty Dong can also be seen as parallel figures ensnared in different loops of the same web of interested claims of similarity and difference that relate brand-name and generic drugs. Seife, a decorated public servant, directed a new regulatory body responsible for ensuring that all drugs that claimed to be the same were, in fact, the same. Yet his bureaucratic division was riddled with corruption and favoritism, and some scientific claims of similarity were revealed to be fraudulent exercises that masked the underlying differences between generic and brand-name drugs. Dong, an emerging star in the field of clinical pharmacology, was contracted to perform a new scientific proof of differentiation between brand-name and generic drugs. Yet, when her research instead provided strong proof of similarity, it was suppressed by a powerful industry with a substantial interest in product differentiation. As both of their stories reveal, the science and politics of pharmaceutical equivalence can be a hazardous terrain to navigate.

Nobody comes off well in these stories of scandals—not the FDA, not academia, not the brand-name industry, not the generic industry. The sufferings of Seife and Dong remind us that there are few true heroes or villains to be found at the intersection of public health and private markets, that there is a good deal at stake, and that many of us suffer the consequences.[24] On the one hand, scores of patients cannot afford to purchase therapies that are protected by patent regimes, even though—had they contracted the same illness a few years later—the same therapeutic options might be generically available and thus affordable. On the other hand, many other patients continue to complain that for whatever reason, be it formulation, allergy, or placebo effect, they find that their medicines for depression, epilepsy, or diabetes just do not seem to work as well once they are switched from a familiar blue pill to an endlessly changing series of generic versions.

Is the generic a problem or a solution? It is manifestly both. To fully understand the controversies that have surrounded generic drugs over the past half century and to fully realize the promise that generic drugs might hold in years to come, we must come to terms both with Seife's problem (in which important differences between things are falsely erased by a seal of equivalence) and Dong's problem (in which effectively identical things are distinguished by a smokescreen of market-oriented claims of difference). Most stories we hear about generic drugs assume one or the other narrative to be true. To read the conflict between brand-name and generic drugs as something more than ideological, however, requires a willingness to engage with both sides of this argument even as we realize that both are fraught with powerful economic and political interests.[25]

This book chronicles the social, political, and cultural history of generic drugs in late twentieth- and early twenty-first-century America to examine what is at stake in calling two forms of health-care delivery the same. For every front-page chronicle of Seife and Dong, there are several equally important but less sensationalized conflicts about the development, circulation, and consumption of generic drugs. Tracing the historical debates over the value of generic drugs invokes powerful and unsettling questions as to the generalizability of medical knowledge, the role of science in public policy, and the increasing role of industry, marketing, and consumer logics in late twentieth-century and early twenty-first-century health care. Although these problems have become increasingly acute, we still lack a common language to talk about their role in medical practice, health policy, and everyday life. One of the goals of this book, then, is to help us stimulate through historical examination an explicit conversation in American medicine about the science and politics of similarity.

This is a recent history. There were no firms known specifically as generic drug manufacturers or anything clearly called a generic *drug* until the late twentieth century. Generic names can be traced back at least to the late nineteenth century; a pharmacy customer at the turn of the twentieth century might choose between Upjohn's morphine, Squibb's morphine, Parke, Davis's morphine, or Smith, Kline & French's morphine, but these were not "generic drugs." Certainly all of these companies attempted to market their own version as a superior product. But, as none of them had discovered morphine or held a patent on the drug, none of them could claim to sell the "innovator" product or the exclusive "brand-name" product. Only after the expansion of the innovative pharmaceutical sector—a brisk flow in the early decades of the twentieth century and then a flood in the years following the Second World War—did a pharmaceutical "life cycle" become naturalized in

North American and European markets in which a brand-name or innovative version of a new drug was protected by a seventeen-year (later a twenty-year) patent-based monopoly. The history of the generic drug begins once these monopolies started to end, and new kinds of markets for new kinds of pharmaceutical copies emerged.[26]

The expanding role of generic drugs in the late twentieth century is therefore a story of the risks and rewards of *unbranding* in modern medicine. Conflicts over the equivalence of generic drugs were at root conflicts over the extent to which an innovator company could continue to maintain a de facto monopoly through trademarks after its de jure monopoly through patents had expired. These battles were fought on many different registers: clinical, commercial, political, and legal. As late as the 1960s in many states—and up until the mid-1980s in a few—it was considered an act of adulteration, punishable by law, for pharmacists to fill a prescription for a brand-name drug with a generic version. Long after these anti-substitution laws were overturned, other efforts to promote generic drugs as functionally exchangeable with their brand-name counterparts in the 1960s, 1970s, and 1980s continued to be complicated by disputes over who defined what level of proof was "good enough" to determine equivalence: the manufacturer, the regulatory state, the insurer, the prescriber, or the patient who ultimately consumes the drug. Indeed, the question "when are two medicines the same?" has yielded different answers for different actors at different times and in different places. Even though this question is an old one, it continues to be central to the future of American health policy and medical practice.

THE SCIENCES OF SIMILARITY

Generic drugs have never been the same as their original counterparts in all respects. As pointed out at the beginning of this chapter, they are usually (but not always) cheaper. They tend to be different in color, shape, and size from the original drug. Claims of generic exchangeability are not based on generics being *identical* to the products they copy, but in being sufficiently *similar*—that is to say, the same in all ways that matter. Debates over whether generic drugs are or are not equivalent to brand-name versions can be understood, in this light, as debates over what aspects of a drug are crucial to the work it does inside the body. Is a pill just as good as the active chemical within? Or are there other parts of a drug that also do important therapeutic work? This question is not as easy to answer as it might first appear.

As the putative similarity or difference between generic and brand-name drugs became a site of increasing contestation in the second half of the twentieth century, proponents for both sides came up with a stunning variety of ways of proving their products were the same or different. Early generic manufacturers claimed that although the color, size, and shape of a given pill might be different from the original branded version, the drugs were identical if they contained the same active pharmaceutical ingredient at the same degree of potency and purity. But chemical similarity could, at times, mask biological difference. In the late twentieth century, a new scientific field of biopharmaceutics, financially supported by brand-name manufacturers, began to focus national attention on the many ways in which the same chemical might be differentially absorbed or circulated in the body, depending on the exact details of how a drug was made. Other new measures of difference and similarity followed, from industrial standards of quality assurance to regulatory standards of Good Manufacturing Practices (or GMPs), to the more recent debates over the biocomparability or biosimilarity of would-be generic biotech drugs.

These new forms of difference between otherwise similar therapeutic agents posed thorny problems for those who wanted to ground the practice of biomedicine with a set of universal laws of physiology, pathology, and pharmacology. In contrast to the physiological approach of earlier Hippocratic models of disease—in which every sick person became sick and received treatment in a unique way—biomedical sciences of the late nineteenth and early twentieth century posited an ontological understanding of diseases as specific entities knowable outside of individual cases. New forms of knowledge of specific pathogens (idealized in the bacteria grown in pure culture by Louis Pasteur and Robert Koch) allowed for a science of rational therapeutics (idealized in turn in the antisyphilitic "magic bullets" of Paul Ehrlich, the antibiotics of the mid-twentieth century or the rational drug design of the late twentieth century). If specific pathogens caused the same diseases in different patients, it followed that specific treatments could cure those same diseases in different patients. Part of the power of biomedical categories, both diagnostic and therapeutic, is their apparent universality: diseases and cures, abstracted from local context and understood on a microscopic level of molecular mechanism, should in theory work everywhere, should in theory be the same everywhere. This kind of biomedical knowledge itself claims to be universal. Biomedical objects can be found everywhere and are everywhere expected to perform the same way.[27]

Indeed, when an increasing number of scholars from the social sciences and humanities began in the 1970s to critique the field of biomedicine as overly reductionistic and deterministic and therefore dehumanizing, a central focus of these critiques was the apparently universal basis of biomedical knowledge. Sociologists and historians following Nicholas Jewson's account of the disappearance of the "sick man" from medical cosmology complained that modern medicine had made a Faustian bargain, gaining increased power to intervene on more and more precise levels—but losing its soul in the deal. Many medical anthropologists agreed with Arthur Kleinman that, among the world's healing systems, biomedicine was "the only one which had no room for the soul." Critical social scientists working in different sites around the world compared the apparent global knowledge of biomedicine with more "local biologies" produced by vastly different approaches to health and illness in other societies.[28]

More recently, voices of clinicians, scientists, and critics of medicine alike have come to realize that this apparent universality of biomedical objects like levothyroxine tablets is not a solid piece of external reality but a wishful, almost Platonic ideal. Comparative historians have revealed that biomedical practices took on very different shapes in different contexts: the world of medical specialties was divided differently in France than in the United Kingdom; early neonatologists used incubators in France in entirely different ways than their counterparts in the United States.[29] Comparative ethnographers moving between sub-Saharan Africa and North America have further fragmented our understanding of biomedical objects as singular: oncological chemotherapy in a biomedical cancer hospital in Botswana involves objects and practices that are similar, but different, from corresponding therapeutics in Boston, even if both are clearly marked as biomedicine. Within the same metropolitan academic medical center, a biomedical category like atherosclerosis can denote one kind of object in the cardiology clinic (symptoms, signs, and treatment strategies), another kind of object in the catheterization suite (images of coronary arteries, with demonstration of flows and blockages), and yet another in the clinical pathology laboratory (fibrofatty plaques in the intima arterial walls).[30]

Asking whether two versions of the same molecule can or cannot be treated as the same drug opens up a critical set of problems about the universality and generalizability of biomedical objects. Addressing these problems has required the formation of new scientific disciplines, new approaches to health policy, and new relations between medical industry and medical practice. Attention to generics likewise draws out the many forms of difference, or plurality, that exist in the seeming universality of biomedical objects. This book takes generic drugs as a starting point for examining these logics of similarity and difference in clinical medicine, public health, and the marketplace.

Initial claims that generic drugs were merely "the same but cheaper" than brand-name versions were grounded in logics of chemistry. These appeals to molecular equivalence invoked a field of scientific inquiry that from its alchemical origins had long concerned itself with arranging species of substances into tables based on similarity and difference. In late medieval Europe, significant conflicts erupted over whether artificial, or alchemical, gold was the same or not the same as natural mineral gold. Since gold was commonly used as a therapeutic agent, this question was of vital consequence: What to do if alchemical gold were indistinguishable from mineral gold by available tests yet caused physiological difference in the bodies of patients?[31] As the field of chemistry subsequently characterized the composition of more complex molecules, no small part of the work of the chemist lay in finding significant differences between seemingly similar objects.

Organic chemists in the nineteenth century began to realize that the numerical formula of how many carbon, hydrogen, oxygen, or nitrogen atoms were in a given molecule did not necessarily predict its physical properties. With a single Tinkertoy set containing a specified number of black, white, red, and green balls, one could build many different molecular structures, called geometrical isomers, with strikingly different boiling points, freezing points, smells, volatility, or tendency to burst into flame.[32] Alternately, two people, working with two Tinkertoy sets, could assemble two models of what appeared to be the same structure but were in fact mirror images of one another. As the field of pharmaceutical stereochemistry began to elaborate in the early twentieth century, left- and right-handed forms of the same basic structure (called optical isomers or enantiomers) had similar physical properties like boiling points and freezing points but had strikingly different pharmacological properties when they interacted with the physiology of animal systems.[33]

This illustrates a central theme of this book: any scientific claim of similarity also conceals alternative possible claims about difference. Both the right- and left-handed forms of the sedative thalidomide are equally effective as sleep aids, but only the left-handed form is metabolized into compounds that cause birth defects. Both the left- and right-handed forms of penicillamine are effective treatments for rheumatoid arthritis, but only the left-handed form causes inflammation of the optic nerve. In these cases, the difference between otherwise similar forms is clinically significant. In other cases, such as levalbuterol, the purified right-handed form of the airway-opening drug albuterol, putative clinical differences in stereochemistry turn out to be clinically trivial in most patients. Distinguishing which forms of molecular differences are significant and which are trivial is neither fully delineated nor answered by the field of chemistry alone.[34]

Rather, as this book will explore, questions of therapeutic similarity and difference are simultaneously questions of pharmacology and physiology, of economics and politics, of morality and belief. Pharmaceutical products are not merely molecular species; they are complex health technologies dependent on industrial and regulatory processes of standardization. Like the individual pieces that came together to form a Colt revolver or the interchangeable personnel involved in the process of mass production at Henry Ford's River Rouge automobile plant, the promise of generic drugs hinges on the hope that their interchangeability could help produce a more rational system of health care delivery itself. As with other industrial products, a complex set of manufacturing processes are involved in making different copies of the same drug: from the binders, fillers, and stabilizers added to fill a capsule down to the precise degree of pressure used to stamp a tablet. Over the past few decades, critics of generic drugs have repeatedly sought new forms of information to differentiate the manufacturing processes of generic drugs from their brand-name counterparts. As controversies over brands and generics have continued, the number of possible proofs of difference—and protocols for proving similarity—have continued to expand.[35]

UNBRANDING MEDICINES

The history of generic drugs and their brand-name counterparts brings these different ways of thinking about equivalence—chemical, biological, industrial—into contact with the logics of exchangeability that govern the relation of commodity to market. As the head of the Rugby-Darby generics conglomerate noted in 1980, continued state support of generic substitution "could easily turn generics into commodities. Do we really want our drugs to be commodities?"[36] In practice, the speaker was warning that generic manufacturers needed to be careful not to let their prices drop so low as to be unprofitable. This reflects a very specific usage of the term *commodity* to stand for certain kinds of goods that are bought and sold by volume and price alone. If pharmaceuticals were treated as bulk commodities—along the same market logics as, say, crude oil or grade 2 soybeans—then individual manufacturers would lose the ability to differentiate their products and charge more for that distinction. It is far from clear, however, that this is how generic drug markets work. On the one hand, the modern generic drug represents a kind of medicine that has been successfully *unbranded*; on the other hand, individual generic manufacturers have found ways to *rebrand* their particular generic products to differentiate them from competitors. From both perspectives, the generic becomes an instructive site from which to examine the role of branding and marketing in modern medicine and the many pathways through which medicines become commodities in the more general sense of the term.[37]

The narrative of the book is structured thematically and chronologically around three domains for understanding therapeutics as commodities: production, circulation, and consumption.[38] The first parts of the book are concerned with generic production and trace the origins of the international generic drug name, the generic manufacturing industry, and the generic drug itself from the mid-twentieth century onward. If marketing was based on differentiation, how did companies market products that claimed to be the same? Subsequent clusters of chapters are concerned with the generic circulation of lower-cost pharmaceuticals as inexpensive and interchangeable goods in an increasingly complex and costly American health-care system. What forms of proof allowed regulators to demonstrate that generic drugs were good enough to be swapped with brand-name drugs? What legal, financial, and logistical structures helped—or hindered—the circulation of generic medicines? The remainder of the book focuses on the problem of generic consumption and explores how debates over the existence or nonexistence of generic drugs changed markedly after a robust market for such commodities had taken shape. If the generic drug is functionally the same thing as the brand-name drug, why do brand-name drugs continue to exist at all after patent expiry?

The historical origins and possible futures of generic drugs have always been linked to broader global projects in medical science, industry, and public health. The first chapter of the book explores how the ideal of a single, universal generic name that would everywhere refer to the same drug emerged from a very specific set of negotiations between the World Health Organization (WHO) and the American Medical Association (AMA) in the 1940s and 1950s. The very last chapter traces the shifting geography of global generic drug production in the new millennium, as centers of generic production in North America and Western Europe become eclipsed by new "generic giants" such as Teva in Israel and Cipla, Ranbaxy, and Dr. Reddy's in India. In between, however, the book is concerned chiefly with the American context. As a private sector solution to a public health problem, generic drugs emerged locally to address problems of cost and access in health care far earlier in the United States than in many other places in the world.

Generic drugs are not innovative, we are told, but merely imitative. Nobody will win a Nobel Prize for discovering a generic drug, since the discovery of the "pioneer" product was already made by someone else, years before. No pharmaceutical firm will ever market a generic blockbuster, we think, because such a product is by definition no longer patent protected and therefore open to steep price competition. No incurable disease of today—be it cancer, diabetes, HIV/AIDS, or dengue—will finally be tamed with the advent of a new generic drug.

Or might they? Closer examination reveals the generic drug to be anything but boring or undifferentiated. This book explores the untold histories of generic drugs and what they reveal about the conflicts, the assumptions, the hopes and fears we hold out for biomedical innovation and the possible futures of our health care system.

PART I

WHAT'S IN A NAME?

CHAPTER 1

ORDERING THE WORLD
OF CURES

The Naming of Cats is a difficult matter,
It isn't just one of your holiday games
You may think at first I'm as mad as a hatter
When I tell you, a cat must have
THREE DIFFERENT NAMES.

—T. S. ELIOT, "THE NAMING OF CATS," 1939

As a caricature available to Lewis Carroll and T. S. Eliot alike, the Mad Hatter stood for scores of haberdashers, furriers, and dyers of the late nineteenth and early twentieth centuries who were thought to have been driven mad after years of close work with the synthetic chemicals essential to their trades. Those navigating the labyrinth of synthetic chemicals by the mid-twentieth century would find their names to be nearly as maddening. Chemicals had simply become more complicated things to think about. While the stuff of applied chemistry in the early nineteenth century could for the most part be described by simple monikers—quinine, morphine, calomel—the expansion of the petrochemical and dyestuff industries by the turn of the twentieth century had created a bestiary of new therapeutic compounds with correspondingly convoluted names.[1] Figuring out what to call these new therapeutics became a question of practical, economic, and political urgency.

Entire international conferences were dedicated to arguing about the names for new compounds coming out of the labs of organic chemists. To the extent that events like the Geneva conference of 1892 and the Liège Nomenclature of the International Union for Pure and Applied Chemistry (IUPAC) of 1930 succeeded in creating an international language of chemistry, however, they tended to sacrifice usability for specificity. Calling a synthetic cough remedy by its structurally accurate IUPAC name, (5α,6α)-7,8-didehydro-4,5-epoxy-17-methylmorphinan-3,6-diol diacetate, might help the analytical chemist but made life much harder for the physician trying to write a prescription or the patient or pharmacist trying to figure out how to fill it.[2]

The Swiss and German chemical houses that were the cradle of modern chemotherapeutics initially circumvented this problem by inventing memorable trademarks, or brand names, for their new products. The cough remedy referenced above could be more easily prescribed once Farbenfabriken Bayer glossed it with a short and memorable trademark, Heroin. Likewise, Bayer's popular fever-reducing drug, acetylsalicylic acid, could more easily be denoted by the optimistic and euphonious Aspirin. In the comparatively understaffed industrial research labs of American firms like Parke, Davis & Co. the synthetic hormone that would receive the IUPAC name (R)-4-(1-hydroxy-2-(methylamino)ethyl)benzene-1,2-diol could likewise be far more successfully marketed as Adrenalin. Pharmaceutical trademarks were short, euphonious, useful, and memorable, well suited for bridging the specific but unwieldy language of chemistry with the practical demands of the clinic and pharmacy.[3]

Yet some brand names for new drugs became *too* successful, in ways that began to threaten the interests of pharmaceutical manufacturers themselves. As these memorable little monikers circulated through a society itself learning to incorporate complex synthetic pharmaceuticals into its daily life, they often passed from private trademark into the realm of common language. To the chagrin of executives at Bayer and Parke, Davis, aspirin and adrenalin—like xerox and kleenex—quickly became the de facto common names for these substances in the daily life of physicians, pharmacists, and patients. This process is now known as "genericide," and from the standpoint of intellectual property there is one protection against it: creation of a public, nonproprietary name that protects the brand name as a form of private property.[4]

By the end of the nineteenth century, pharmaceutical manufacturers had learned the importance of generating a *third name* to publicly designate these new compounds. Between the intolerable specificity of (5α,6α)-7,8-didehydro-4,5-epoxy-17-methylmorphinan-3,6-diol diacetate and the singular brand of Heroin, one could produce one or more common names such as diacetylmorphine or morphine diacetate. After the 1917 Trading with the Enemy Act allowed the US government to seize

German pharmaceutical patents, the Federal Trade Commission (FTC) required that American versions of these drugs must be marketed using nonproprietary names. Bayer's barbituric acid derivative Luminal became phenobarbital; Veronal became barbital. In the 1920s, a few American pharmaceutical firms voluntarily coined nonproprietary names for their own new drug products, and the Federal Food, Drug, and Cosmetic Act of 1938 required that the label of all drugs should bear a "common or usual name." Yet these common names of drugs remained plural and poorly regulated. There was little sustained attention to *how* drugs obtained their names, *who* gave them their names, and *what* such names signified. The private and public qualities of drug names overlapped to form a Gordian knot of intellectual property claims: well into the 1940s a name like sulfamethazine could simultaneously be a common name in the United Kingdom and a trademark in France.[5]

This chapter describes an unusual moment in the 1940s and 1950s when a disparate group of actors attempted to order the world of cures such that each new drug possessed *one true name*. In the years following the Second World War, a group of reformers working within the World Health Organization, the American Medical Association, and a host of other institutions attempted to create an internationally recognized system of universal drug names. Coming from different industries, professional groups, and national backgrounds, these self-styled experts in pharmaceutical nomenclature sought a rational language of drug names that might in turn produce a more rational approach to drug therapy.

The relationship between names and things in global commerce and medical practice is a key starting point for this book. Before *generic drugs* could become things that could be treated as interchangeable in practice, *generic names* needed to be universally recognized as designating the same kind of thing. The naming and classifying of things into taxonomic systems always invokes a parallel set of genus/species decisions regarding what kinds of differences are significant and what kinds of variation are tolerable. To understand the significance of a given name, we must understand what kinds of shared characteristics it denoted among similarly named objects (generic qualities) while also observing what kinds of characteristics were still left for individuation and differentiation (specific qualities). Not all chicken bought in the grocery aisle is organic or free-ranging, grass-fed or antibiotics-free, but we presume that all products sold as chicken come from *Gallus gallus domesticus*: any violation of *this* assumption produces outrage, scandal, and calls for regulatory reform.[6]

Agreement on which shared features a generic name should denote and what kinds of variability it might still permit has been far less clear cut. It is precisely because the generic names of drugs are assumed to be part of a natural order—a simple (scientific) description of drugs "as they are"—that the social forces inherent in their naming are consequently unseen and therefore all the more surreptitious. The universal generic name emerged as a solution to a variety of problems erupting at new interfaces in the postwar political economy of health and medicine: as a means of defining national and international circulation of drugs as commodities, as a boundary between private and public modes of regulation in the marketplace of health, and a means of rationalizing and taming the emerging "therapeutic jungle" of expanding pharmaceutical chemotherapeutics in the mid-twentieth century.[7]

The universal generic drug name was itself the project of several decades' worth of work by scientists and practitioners who believed—in the same spirit of optimistic biomedical internationalism that underscored the founding of the WHO in the early postwar years—that the world of cures could be ordered rationally through a project of rational language. In some ways, this project was surprisingly successful. In other ways, it failed entirely.

THE NOMENCLATURA

In the fall of 1948, René Hazard, professor of pharmacology and materia medica at the Paris Faculty of Medicine, author of the influential *Précis de thérapeutique et de pharmacologie*, architect of the *French Codex*, and leading authority on pharmacy in the Francophone world, sent a series of memos to colleagues in Geneva. As Hazard complained, the ability of researchers, physicians, and pharmacists around the world to cooperate on therapeutic research was increasingly in peril because "every country selected a different common name designating the same prescription without reference to similar designations made by other countries." To further the progress of biomedical research and practice, Hazard proposed the creation of "a common international designation" that "should be internationally protected, and so should their translations into every language of the world."[8]

Hazard's dispatch was sent to the WHO—the specialized agency of the United Nations that had officially opened its doors only a few months earlier—and circulated to his fellow members on the WHO's Expert Committee on the Unification of Pharmacopoeias. The early WHO was the very definition of a technocracy; the bulk of its work was performed through carefully chosen transnational expert committees. The organization's charter and its first director general, Brock Chisholm, both clearly specified that the WHO's main purpose was not to be a political body or a supranational regulatory body but to function as a site for the technical intercalation of national health systems.[9]

A key project for the early WHO, therefore, was the task of universalizing biomedical knowledge so that an object might be understood to be the same object regardless of one's location or language. The WHO's international classification of disease (ICD) system, now on its eleventh revision, provided a comprehensive taxonomy of pathology now essential to all clinic visits and hospital stays in the United States. The WHO classification of tumors system allowed doctors and patients to compare prognosis against a unified registry of similar patients and know when to be watchful, when to be aggressive, and when to focus on comfort care. Likewise, the WHO's Pharmaceutical Department was tasked in the early postwar years with harmonizing the multiple and confusing systems by which pharmaceuticals were named into a single international scientific language.[10]

The quest for a rational language that might perfectly relate the world of things to the world of words evokes a set of Utopian projects reminiscent of the short stories of Jorge Luis Borges or the *Philosophical Language* of John Wilkins.[11] But the particular project inherited by the Expert Committee on the Unification of Pharmacopoeias could be traced directly back to the defunct League of Nations, which had attempted lamely since a conference in Brussels in 1925 to produce a universal *Pharmacopoeia Internationalis*. Initial plans called for the *Pharmacopoeia Internationalis* to be written in Latin, the "only language that can be adopted without reservation in the various languages." But Hazard conceded that by 1948 the mere Latinization of chemical names no longer sufficed to simplify communication. In the decades since the Brussels conference, different vernacular listings of drugs—chief among them, the *British Pharmacopoeia*, the *United States Pharmacopoeia,* and the *French Codex* often gave the same substance wildly different Latinate names or, alternately, gave the same Latinate name to dramatically different compounds. Hazard's American counterpart on the WHO expert committee, E. Fullerton Cook, complained that the discordance of drug names on French, British, and

American markets was trivial compared to the cacophony of their translation into Spanish, Dutch, Danish, Italian, Arabic, Chinese, Japanese, Russian, and other non-Romance languages.[12]

For older drugs, the international proliferation of official names was further confused by the intranational plurality of pharmaceutical synonyms. Even within a single country, many drugs had accrued a variety of common names in their long periods of use, all of which needed to be familiar to the pharmacist. The cardioactive drug digitalis might still be prescribed by an American physician in the 1940s as Withering's Tincture (named after its famous discoverer), foxglove (named for the plant from which it was extracted), or simply as digitalis. As an article in *Practitioner* complained in 1946, the nomenclature of new drugs was "a constant source of irritation and embarrassment: it may thus happen that a quite straightforward drug may have as many as a dozen names or more." As the expert committee prepared the first *Pharmacopoeia Internationalis*, they found that aspirin (acetylsalicylic acid) was now known by no less than sixty-six synonyms, some of which were common names in some countries and brand names in others, and vice versa.[13]

To complicate matters, newer scientific vernaculars were swiftly outstripping the utility of Latin as a universal language. Shortly after the preparation of the first *Pharmacopoeia Internationalis* in 1951, the directors of the *British Pharmacopoeia* declared they would be publishing all future volumes in English because it was becoming ridiculous to translate innovative therapeutic forms into Latin. "It would," they noted, "for example, have taxed the ingenuity of the most erudite scholar of the humanities to have rendered into tolerable Latin 'Concentrated Human Red Blood Corpuscles.'" Latinizing already unwieldy chemical names such as acetomenaphthone and sulphadimidine by adding extra syllables (to make acetomenephthon*um* and sulphadimidin*a*) did little to improve the lot of those who needed to pronounce and write such names in the clinic, hospital, or pharmacy.[14]

What language could take the place of Latin? By the end of the Second World War, French and German had clearly been surpassed by English as the dominant language of international scientific communication. The political economy of the immediate postwar period also ensured that the United States played a dominant role in the framing and funding of UN-based international bureaucracies—especially the scientific agencies such as the WHO. Few were surprised, then, when the WHO's new Expert Subcommittee on Nonproprietary Names looked toward the protocols of the *United States Pharmacopoeia*—which operated in close collaboration with the AMA's Council on Pharmacy and Chemistry—as its chief model for managing the complexity of the new world of drug names.[15]

POLITICS OF THE PHARMACOPOEIA

When the AMA Council on Pharmacy and Chemistry was formed in 1905, its publication entitled *New and Nonofficial Drugs* joined a bookshelf already filled with copies of the *National Formulary*, published by the American Pharmaceutical Association since 1888, and the much older *United States Pharmacopoeia*, published by the United States Pharmacopoeial Convention (USP) since 1820. Although these guides to the world of pharmacy were plural, privately managed, and occasionally conflicting, each claimed to link the worlds of pharmacist and physician by a set of common therapeutic standards. According the preface to its first edition, the *United States Pharmacopoeia* was devoted to eradicating the "evil of irregularity and uncertainty in the preparation of medicine" through rational nomenclature. "The essential properties of names," the original architects continued, "ought to be *expressiveness, brevity, and dissimilarity.*"[16]

In the late nineteenth and early twentieth century, these three compendia of official names had played a crucial part in the formation of the American ethical pharmaceutical industry. In contrast to the more secretive sale of patent medicines, these "ethical drugs" were sold by their nonproprietary or official (also known as "officinal") names clearly labeled on their packaging. Firms like Eli Lilly in Indianapolis; Parke, Davis in Detroit; Upjohn in Kalamazoo; Smith, Kline & French in Philadelphia; and Squibb in New York got their start by selling mass-market versions of the same compound—made according to the *United States Pharmacopoeia* or *National Formulary* specifications—and claiming that their own versions were particularly trustworthy because of their investment in analytic laboratories and quality control practices. Well into the first half of the twentieth century, many ethical drug manufacturers continued to sell different versions of the same thing: plainly labeled pharmaceutical agents, identical by the standards of the *United States Pharmacopoeia* or *National Formulary* but differing chiefly in terms of claims of purity, quality, uniformity, and dosage.[17]

Yet, as newly patentable compounds from adrenaline to insulin to the sulfa drugs began to emerge from the research laboratories of companies like Lilly, Parke-Davis, and Smith, Kline & French, these ethical firms began to sell more and more products under patent monopolies—with a tightly adherent product brand name. The Council on Pharmacy and Chemistry of the AMA instituted a Seal of Acceptance program for ethical pharmaceutical houses to vet their proposed brand names to ensure that the names of these new compounds did not beg the question of their efficacy. Effectively, a company could name its compound anything it wanted to, but if the council thought that the name of a drug reached a bit too far (for example, Radam's Microbe Killer, which could not be documented to kill any microbes), they could cut off access to the advertising pages of the *Journal of the American Medical Association*. Although the group was initially concerned with regulating brand names for ethical pharmaceutical houses, by the early 1950s, the AMA declared its intent to ensure the "adoption of an abbreviated scientific name for general use in prescribing, naming, and identifying agents with unwieldy chemical names."[18]

As the WHO cast its probing gaze toward the naming practices of the USP and AMA, the AMA and USP began to look warily at the intent and potential consequences of the WHO's interest in this matter. In 1952 Lloyd Miller, the new director of revision for the United States Pharmacopoeial Convention, returned from a set of meetings in Geneva greatly concerned about the project of universal drug naming and its implications for the USP and the American pharmaceutical industry. Though its first meeting had taken place in the US Senate chambers in 1820, the USP remained in its governance and accountability a private concern of the American medical and pharmacy professions and the pharmaceutical industry.[19] Miller was particularly worried that the WHO's project of universalizing generic names could trample on the trademark rights of the US drug industry.

To the director of the United States Pharmacopoeial Convention, naming was power, and that power was not a thing to be trifled with. By February of 1953, Miller had mobilized the American Drug Manufacturers Association, the American Pharmaceutical Manufacturers Association, the US Department of State, and the US Patent and Trademark Office to this cause, and organized an envoy to Geneva to present the American case at the World Health Assembly that May.[20] On the one hand, Miller joked to his contacts in the US drug industry not to overstate the risks of international law, noting in one memo that "the threat that the *Ph. I.* will displace the *U.S.P.* within our national borders must be ranked on a par with the likelihood that Congress will yield its legislative powers to the United Nations General Assembly. While the possibility exists, the risk seems slight indeed."[21] But Miller *was* worried that the WHO's efforts could have an immediate effect in places like Latin America where control over drug names could translate into control over drug markets.

In 1948, an aborted attempt by Latin American pharmacologists to create a *Pan-American Pharmacopoeia* had been characterized by Miller as a form of political resistance against pharmaceutical "Yankee imperialism." Thankfully, Miller concluded, this project had stalled, and Latin American pharmacopoeias remained "virtually ineffective, most of them being so far out of date as to be practically valueless as a national yardstick of drug potency and purity." So framed, this still-virgin landscape of Latin American pharmacy potentiated a turf war between the spheres of influence of the *United States Pharmacopoeia*, the *British Pharmacopoeia*, and the *French Codex*—as well as their corresponding national pharmaceutical industries, which Miller mapped as a table of pharmacopoeial empire (table 1).[22]

Miller characterized the *United States Pharmacopoeia*'s own system of drug names as a key weapon for American industry in its trade wars for overseas markets, and he criticized the US State Department for its weak protection of American drug names in Latin America. By way of contrast, he pointed at the French Embassy in Lima, which had recently sponsored a national conference of Peruvian pharmacists in which three hundred copies of the French pharmacopoeia were handed out as "a subsidy from the French government to use the *French Codex* in South America as a means of insuring a market for French drugs." Miller complained to contacts at the State Department that it was a "neat point of international relations that France receives direct financial aid from the United States" and uses American funds from the Marshall Plan to "jeopardize the foreign markets of the American pharmaceutical industry."[23]

Like the *French Codex*, an unchecked *Pharmacopoeia Internationalis* represented a potential threat to the *United States Pharmacopoeia* and whatever economic and geopolitical advantages it provided for American businesses. With proper stewardship, however, Miller saw the potential for American influence at the WHO to create a system of nonproprietary names that might be even more conducive to the globalization of American pharmaceutical markets.[24] Merck counsel John Horan, along with the US surgeon general Leonard Scheele, was sent with the US delegation

TABLE 1. Pharmacopoeial Spheres of Interest, c. 1953

	Date of national pharmacopoeia, if any	Other pharmacopoeias, official or used		
		USP	French	British
Argentina*	1943			
Bolivia*		use	use	
Brazil*	1926			
Chile*	1941			
Colombia		use	use	use
Costa Rica		use		
Cuba*		official		
Dominican Republic*		use	official	
Ecuador*		official	use	
El Salvador			use	
Guatemala*		official	official	
Haiti			use	
Honduras		official		
Jamaica				official
Mexico	1930	use	use	
Nicaragua*		use	use	
Panama*		official		
Paraguay	1945			
Peru*			use	
Puerto Rico		official		
Uruguay*			official	
Venezuela	1949			

Source: Adapted from Lloyd Miller, USP Board Committee on General Principles of International Cooperation, February 13, 1953, box 1, f 1, USPC, p. 9.

From information obtained January 1949 from Directors General of Public Health.

to Geneva to negotiate concessions from the WHO. Horan wanted to make sure that "to the extent that it promotes the use of a single generic name throughout the world," the WHO's decisions would involve substantial input from American industry. Writing back to his colleagues in the American pharmaceutical industry, Horan noted that "we found all of these men very anxious to obtain the cooperation of the American drug industry" and agreed to suspend the program until it had been satisfactorily reformed to meet the approval of the American medical profession and pharmaceutical industry.[25]

As a key part of the negotiation, the secretary of the AMA Council on Pharmacy and Chemistry, Robert Stormont, was appointed as an additional member of the Expert Subcommittee on Nonproprietary Names. Soon afterward, Stormont's own body at the AMA, the Council on Pharmacy and Chemistry, disbanded and jettisoned the Seal of Acceptance program and its responsibility for regulating pharmaceutical marketing in order to secure more funding from pharmaceutical advertisements in the pages of *JAMA*. No longer tasked with approving brand names, the AMA's newly renamed Council on Drugs refocused its efforts on the science of generic naming, working "in close cooperation with industry in the early selection and adoption of suitable convenient nonproprietary names for new drugs before they are marketed" to shape the common names of new drugs in the medical literature.[26]

A review of correspondence between Stormont's Council on Drugs and a number of pharmaceutical firms, however, speaks to some of the difficulties the AMA had in shifting from a body that approved brand names to a body that approved generic ones. In one example, an executive at Schering Pharmaceuticals wrote to the Council on Drugs in 1956 to pre-emptively "insure general acceptance of trade and generic names proposed for our products in the future." Attached to the letter was a long list of new drug names, including an anticholinergic drug Schering wished to call Miradon, with the proposed name methopenindione. Yet Schering refused to provide any further information about drug structure, noting, "For obvious reasons, we do not wish to reveal the exact chemical nomenclature and trust that the above classification will be sufficient." In asking the AMA to weigh in on a trade name, Schering was asking the Council on Drugs to do exactly the opposite of what it now claimed to do, while explicitly refusing to provide the information necessary to perform its stated task, the scientific evaluation of the relationship between generic name and chemical structure.[27]

Nor was Schering the only offender. Geigy Pharmaceuticals submitted the name of a new synthetic curare-like compound to the Council on Drugs in 1959, with the untenable chemical name of N, N, N′ N′, tetramethyl-N, N′-bis′(carbopropoxymethyl)-3, 14-dioxahexadecane-1, 16-diammonium bromide. Clearly this name would not do for clinical usage or purchase at the pharmacy, but Geigy proposed no generic name, only the trade name Prestonal. In replying to Geigy asking for more information about possible generic names, the Council on Drugs secretary could not refrain from commenting on the brand name, adding "I am not inclined to object to the trade name, Prestonal, although this does sound more like an antifreeze than a drug, since under our new program, we will list any and all trade names for drugs without attempting to determine their suitability."[28] Geigy executives seemed surprised that the AMA was more concerned with the generic name than the brand name; under duress they suggested the condensed name dioxahexadekanium. Though shorter than the chemical name, this eight-syllable mouthful still seemed too long to the AMA's Walter Wolman, who suggested instead the shorter prodekonium. While the latter suggested even less of the actual structure of the drug, Wolman noted that it at least contained some hint of it: (*pro*poxymethyl, dioxahexa*decane*, diammon*ium*). The rest of the Council on Drugs agreed that this formed a nice balance: memorability with a hint of structure.

Gradually, pharmaceutical marketers came to accept the AMA council as a liaison between the drug industry and a new global system of generic naming. Prospective names were solicited from industry, submitted for consideration by the Council on Drugs, circulated to the USP, APhA, WHO, British Pharmacopoeial Convention (and, in the case of biologicals and potentially addictive drugs, the National Institutes of Health as well), and then relayed back to the AMA for submission to Geneva. It helped, of course, that figures like Stormont and his successor, J. B. Jerome, either served simultaneously on or maintained close personal relations with the governing councils of *all* of the above-mentioned organizations.[29]

Even so, efforts to create a single universal drug name were not universally successful. In 1959, Warner-Lambert's new major tranquilizer, brand name Pacatal, was named "mepazine" by the Council on Drugs, but this name was refused by the WHO after the United Kingdom asked to call the drug pecazine instead. Pecazine was unacceptable to Warner-Lambert, which had already promoted several academic publications referring to the new drug as mepazine. Under pressure from Warner-Lambert and the AMA, the WHO's Paul Blanc agreed to withdraw pecazine as an international generic name, but the latter name remained in use in Great Britain and Commonwealth countries.[30] Similar schisms led to the geographical heterogeneity of the drug known by brand name as Tylenol in the United States and Panadol in the United Kingdom, which

generically named acetaminophen in North America but paracetamol
the United Kingdom and much of the rest of the world. Likewise, the
haled bronchodilator known by the brand name Ventolin is generi-
ly known as albuterol in the United States but as salbutamol on the
er side of the Mexican border. The process of defining and naming
..ugs has never reached quite the global universality that advocates in
the WHO initially hoped for. Yet, as these examples demonstrate, the
generic naming of drugs has since the 1940s been a cooperative—if often
competitive—project of physicians, scientists, and manufacturers across
many geographies.

INVENTING A RATIONAL LANGUAGE FOR PHARMACEUTICALS

As the WHO subcommittee sat down and tried to agree on the initial
building blocks of a new rational nomenclature that might "form a name
by the combination of syllables in such a way as to indicate the significant
chemical groupings of the compound and/or its pharmacological classification,"[31] members of the committee would have been aware of other
attempts at rational language production, such as Volaput, Interlingua,
and Esperanto. In postwar Europe, Esperanto in particular remained a
visible symbol of internationalism and had been formally proposed to be
the language of the United Nations Educational, Scientific, and Cultural
Organization (UNESCO) in 1948.[32]

Like Esperanto—itself the brainchild of a medical student, Ludwik
Zahmenoff—the kernel of the AMA-WHO drug naming system lay in taking a small number of common "stems" from many languages and using
simple prefixes and suffixes to add precision to the words themselves,
while restricting the alphabet to avoid sounds and spellings that did not
translate easily between tongues. The first attempt to parse this new
language of international drug names chart included fifteen stems that
could be translated easily between Latin, English, and French (table 2).

But this rational naming system also contained within it several of
the flaws that complicated the universal adoption of Esperanto. One of
the problems Esperanto has faced is that its stems and suffixes are not
really new—they all came from somewhere—and therefore carried some
baggage from their original language. So too with drug names. Looking
at table 2, we see that all stems listed have been chosen somewhat arbi-

TABLE 2. Initiation Proposal for an International System of Drug Nomenclature, WHO Expert Committee on the Unification of Pharmacopoeiae

Latin	English	French	
inum	ine	ine	for alkaloids and organic bases
inum	in	ine	for glycerides and neutral principles
olum	ol	ol	for glycerides and neutral principles
alum	al	al	for aldehydes
onum	one	one	for ketones and other substances containing the CO group
enum	ene	ene	for unsaturated hydrocarbons
anum	ane	ane	for saturated hydrocarbons
cainum	caine	caine	for local anesthetics
mer	mer	mer	for mercurial compounds
sulfonum	sulfone	sulfone	for sulfone derivatives
quinum	quine	quine	for antimalarial substances containing a quinoline group
crinum	crine	crine	for antimalarial substances containing an acridine group
sulfa	sulfa	sulfa	for derivatives of sulfanilamide
dionum	dione	dione	for antiepileptics derived from oxazolidnedione
toinum	toin	toine	for antiepileptics derived from hydantoin
stigminum	stigmine	stigmine	for anticholinesterases

Source: United States Pharmacopoeia, Circular 126, Appendix II, "General Principles in Devising International Non-Proprietary Names," box 251, f 6, USPC, pp. 423-24.

trarily from existing usage patterns. The supposedly comprehensive set of lexical rules governing the selection of stems is on further examination neither comprehensive nor consistent. Five of the categories refer to derivatives of specific compounds (mercury, hydantoin, sulfanilamide, oxazolinedione, sulfones), one refers to a pharmacological activity against a specific physiological receptor (anticholinesterases), four relate to the treatment of two specific diseases (malaria, epilepsy), one refers broadly to a therapeutic function (anesthesia), five refer to the basic structural elements of organic chemistry (alcohol, aldehyde, glyceride, ketones, hydrocarbons saturated and unsaturated). Meanwhile, the first category, alkaloids, serves as a wastebasket that manages to cover essentially the bulk of the materia medica extracted from plant matter existing up to the early twentieth century.[33]

The lumpiness of this proposed taxonomy created rifts within the AMA-WHO community of experts interested in producing universal rational drug names. To some on the committee, these relics of older, irrational drug names needed to be stamped out in favor of an entirely new, perfectly ordered naming system. To others, the solution to the inevitable multiplicity of meanings of drug names was to embrace the nonsensical. Only purely nonsensical words—words that were universally unintelligible and therefore not previously used for anything—could guarantee that novel synthetic agents would enjoy full international acceptance and freedom from intellectual property conflicts. A word with no meaning at all could be both specific and memorable—indeed, such a term would *only* refer to the drug and nothing else.

In the late 1950s, a new proposal for an "objective" system of naming was floated through the WHO, USP, and AMA after the chemist E. C. S. Little proposed a machine for naming synthetic chemicals, which he claimed utilized "a method whereby ten million new names could be easily coined to cope with the output of new chemicals." Little's Donomen system had as its chief virtue the fact that, unlike older systems of ordering the world from Linnaeus back to Aristotle, it came with no linguistic, social, or cultural baggage, no space for subjective valuation.

Instead, Little proposed an objective mechanism in which all names for new drugs would consist of four consonants (c) and three vowels (v) in a strict pattern of alternating repetition: c/v/c/v/c/v/c. After removing "confusing" letters (c, k, y) the English alphabet contained 17 consonants and 5 vowels, and $17 \times 17 \times 17 \times 17 \times 5 \times 5 \times 5$ combinations or 10.5 million. Little could boast that "there are thus enough names available between Bababab and Zuzuzuz to cope with the output of chemists virtually forever." Although he acknowledged that such names might appear ridiculous at first, they were no *more* ridiculous than many names already being used: "Suppose 'bababab' to be the name of a well known insecticide. After a number of repetitions of the word as part, say, of imaginary recommendations, it soon begins to lose its strangeness and becomes just another name—in style not so very different from DDT."[34] In this vision of nomenclature, it was precisely the meaninglessness of synthetic words— the babble of bababab—that gave them specific value in specifying new things without conflicting with other names already in use somewhere or other. Of the few dozen "Donomen words" already existing in English (e.g., general, janitor, seminar, venison), a number of them were already successfully used to denote industrial chemicals, such as dalapon (a weed killer), demiton (an insecticide), and fumarin (a rodenticide). The populace, it appeared, was already willing to accept these arbitrary clumps of synthetic-sounding syllables. Why not just make a machine that could produce them more systematically? Uniquely unsavory names (such as tisadud, which Little drolly suggested would be a nonstarter for any pharmaceutical manufacturer), could simply be weeded out and the entire problem of nomenclature mechanized.[35]

Little was not a crank. The *New Scientist* borrowed the computing power of the UK Atomic Energy Authority for a few hours to test out the Donomen system and concluded that the protocol did indeed work in producing nonsense words that could be specifically attached to new chemical substances. "These words," they determined, "have a music of their own, and would grace any pharmacist's label."[36] But the proposal as a whole still smacked of the Mad Hatter. "What could Lewis Carroll not have done with a vocabulary like this?" they concluded. Were a new cure to be developed for, say, a hangover, they surmised, "we may hear the Mad Chemist making his report":

> We sought a cure for mogosex
> Brought on by drinking gin.
> But mekitin, like palifex
> Still lets the mogins in;
> While lakubex, as one expects,
> Leaves blisters on the chin.
>
> We tried co-polymers of rope
> With femeral of tin—
> Then found an analogue of soap,
> And called it rifasin!
> As mogin dope it has not hope
> But it much improves the gin.[37]

The Donomen system circulated through prominent channels at the WHO, the AMA, and the USP and would later be described as the least subjective—and therefore "most objective"—system of naming during Estes Kefauver's Senate hearings into the naming and branding practices of the American pharmaceutical industry.[38] But it did not dislodge the AMA-WHO system that had already been set in motion, mostly because the rational grid Little attempted to impose upon the landscape of drug names could not easily be squared with existing uses of names already in practice. The AMA-WHO system, with its mongrelized set of categories, was equally arbitrary, perhaps, but far more acceptable to the medical profession and the pharmaceutical industry.

RATIONAL NOMENCLATURES, UNREASONABLE OUTCOMES

In a widely cited article in the September 1959 *Journal of Chemical Education*, Paul G. Stecher of the Merck Sharp & Dohme Research Laboratories credited the new popularity of generic names with two functions: the protection of trademarks and the identification of drugs by physicians and pharmacists. In spite of the WHO's efforts, generic names could still raise particularly thorny questions in international markets, in which syllabic meanings might prove mercurial. "An aerosol preparation of cortisone could well be called cortomist," Stecher warned, "sounding rather elegant in the U.S.A., but sales would drop off sharply in Germany where the word *Mist* means *dung*."[39]

Stecher's article is a window into the logistical and strategic problems that generic and brand names posed to multinational pharmaceutical firms at the close of the 1950s. Obviously, if Merck produced a new compound and was responsible for choosing both its generic and its Merck-specific names, they would want the Merck-specific name to be more memorable and the generic name less, so that Merck's market might be guaranteed for as long as possible. Savvy manufacturers, Stecher claimed, "are apt to choose awkward generic names, for example chlorphenpyridamine maleate," because "no medical doctor is going to write this on a prescription when there is an easy trademark such as Chlor-Trimeton®." Yet Stecher cautioned against designing a generic name that was too forgettable or too hard to use, as "non-use of a generic name may result in the loss of the proprietary nature of the trademark just as easily as the neglect to coin a generic name at all."[40]

In this chapter, we have explored the early politics of the generic drug name, as it became an object of increased interest in the new international order of the postwar period. In the 1940s and 1950s, an uneasy alliance had taken shape between the project of a technically oriented special agency of the United Nations in coining international names for modern chemotherapeutics and the project of the American medical profession and pharmaceutical industry in defining those therapeutics. As it attempted to coin new universal names for drugs moving in global commerce, the WHO was forced to look toward the AMA and USP for guidance. In turn, the AMA and USP took a far more nationally interested stance in the protocols of nonproprietary drug names once they perceived both threat and opportunity in the WHO's international naming project.

Only by exploring the local histories of the institutions involved in creating our present system of generic names can we understand how they were intended to function—and why and how they have both succeeded and failed. By looking to the *United States Pharmacopoeia* as a model for drug naming, the WHO became involved in proxy battles over the role of names in expanding markets for the American pharmaceutical industry. By looking to the AMA as a model for drug naming, the WHO unwittingly adopted a system originally designed to evaluate brand names of ethical drugs to evaluating generic names. Initially this seemed to be an unproblematic assertion: after all, wasn't structure, in chemistry and pharmacology, the same as function? And yet therapeutics as a field concerns far more than just pharmacology. The uses of a drug are also tied to the changing epidemiology of disease, the shifting awareness of practice patterns, and the vagaries of marketing. As a result, the AMA-WHO system of generic names presented a far less rational ordering of the world of drugs than its framers had hoped—and one that frequently substituted American national interest under a guise of international expertise.

Nonetheless, writing to the head of the British Pharmacopoeia in 1961, Miller reminded him that "the importance of [this] American viewpoint takes on emphasis when it is realized that at least half of the new drugs, and thus the new names, originate in the United States."[41] In spite of two decades of deliberation in Geneva, New York, Chicago, and Washington, by the close of the 1950s the generic drug name was far from a *res publica*. Generic names were created in multiple forms by manufacturers, in collusion with the medical profession, and were every bit as manipulable as marketing tools as were the brand names themselves.

CHAPTER 2

THE GENERIC AS CRITIQUE
OF THE BRAND

The anarchy that prevails in naming prescription drugs is epitomized in the fact that what are called generic names are not generic at all.
—GEORGE CLIFFORD, AIDE TO THE SENATE SUBCOMMITTEE
ON ANTITRUST AND MONOPOLY, 1961

In early May of 1960, American papers and television news programs covered the unusual spectacle of a roomful of senators interrogating physicians, pharmacists, consumer advocates, and executives from the pharmaceutical industry on whether the name of a drug affected its safety, efficacy, and utility in clinical practice. "We need to obtain answers to such questions," the chair of the Subcommittee on Antitrust and Monopoly Estes Kefauver, repeated. "What does the term 'generic name' really mean? How do products get their generic names? Is it true that consumers and other purchasing bodies can achieve substantial economies under generic name prescribing?"[1]

Answers to these seemingly simple questions would prove elusive. While the *United States Pharmacopoeia* had maintained a register of nonproprietary or "officinal" names of pharmaceuticals since 1820, the American Pharmaceutical Association (APhA) had not formally defined the term *generic name* until 1955, and as late as 1959 industry news sheets such as the *Di Cyan & Brown Monthly Bulletin* were still defining the concept of universal generic drug names to its readership as new "symbols for the identity of a drug . . . in a public domain."[2]

It was precisely the public quality of the generic name that Kefauver hoped to mobilize to counter the privacy and secrecy of the brand name. Having failed at two presidential bids by the late 1950s, Kefauver had nonetheless consolidated considerable power in the legislative theater. As chair of the Subcommittee on Antitrust and Monopoly, he interrogated several industries—from bread to automobiles—dominated by a relatively small number of large corporations able to effectively shape their own markets of "captive consumers" through "administered pricing" of effectively monopolistic brand-name goods. The pharmaceutical industry soon took on particular, almost monomaniacal interest in the eyes of the committee and its chair. Kefauver's research team, led by the economists John Blair and Irene Till, used the power of the subpoena to marshal documents and testimony related to the role that brand names played in creating de facto monopolies that persisted well after patents on innovative drug products expired. If brand-name prescribing had created a captive consumer, prescribing by generic names might set them free.[3]

The hearings were intended to compare drugs to other, more competitive consumer economies and understand why meaningful comparison shopping did not occur in the pharmaceutical realm. Witness after witness at these hearings made parallels between drugs and automobiles, washing machines, car tires, and canned goods. Working behind the scenes to aid Kefauver, Blair, and Till, consumer advocacy groups such as the Consumers Union (CU) suggested that widespread use of generic names might introduce a healthy dose of competition into the largely monopolistic world of brand-name drug consumption. If a consumer knew the generic name of the drug she was buying (for example, prednisone), she might be better equipped to shop around among competing manufacturers of prednisone products. As Kefauver's investigations revealed, though, many drugs were marketed under multiple generic names in the US alone, making it very difficult for such savvy shopping to take place at all.[4]

Kefauver and his subcommittee were surprised to learn that the same companies that coined a drug's brand name also coined its generic name. Charles O. Wilson, the dean of the school of pharmacy at Oregon State University and the founding editor of the *American Drug Index*, testified early in the Kefauver hearings of a systematic "confusion, misrepresentation, and lack of consistency which exists in the nomenclature" of American pharmaceuticals.[5] He pointed out that while the WHO and the AMA may have created a working relationship, it was still the legal right of any drug company to designate the final generic name for its product. Wilson waved the article by Merck's Paul Stecher (on the market strategies present in generic naming) as a confirmation that the pharmaceutical industry intentionally coined impracticable and unusable generic names to encourage the usage of brand-name drugs. He playfully suggested that that a manufacturer's guide to producing a generic name might read something like this:

RULE 1. No suggestion as to the chemical formula
RULE 2. No suggestion as to the use of the compound
RULE 3. No relationship to the brand name
RULE 4. Usually have name composed of several syllables
RULE 5. Best if name is long and awkward
RULE 6. Name should be reasonably difficult to pronounce
RULE 7. Name should not be as "catchy" as the registered
    proprietary name
RULE 8. Name should not be conducive to memorization
RULE 9. Spelling of the name should not be too easy
RULE 10. If similar molecules have a generic name, the generic
    name for this one should be different
RULE 11. Even if a common name for the substance already exists,
    a new generic name might be advantageous
RULE 12. When a generic name is available for a compound in the
    racemic or d-l form, a different generic name might be used for
    dextro or levo forms; for example, chlorpheniramine,
    amphetamine.[6]

Wilson's satirical rules of naming were quickly taken up by other reform-minded academic physicians and pharmacists who protested what they saw as the intentional production of "generic jawbreakers."[7] To many critical observers, this confirmed the failure of the AMA's stewardship over drugs, part of a "downward path [begun] some years ago when they abandoned all evaluation of therapy and dropped the Seal of Acceptance Program." Allowing drug companies to choose their own generic names "put generic terminology in the hands of the very people who have the most to gain by avoiding use of generic terms ... Thus in recent years new drugs have invariably had ridiculous names, in place of such arbitrary but perfectly usable old generic names like insulin, morphine, digitoxin, and the like." Even those academics who wanted to teach medical students using generic names found the practice very difficult.[8]

In the rare instances where a short and memorable generic name had been created, it could be muddled by the presence of many other less useful generic names. Acetaminophen had been proposed in 1957 as a reasonably memorable generic name for the analgesic Tylenol. Yet even in the United States, acetaminophen was but one of many common names for Tylenol. As late as 1960, marketing materials, textbooks, and journals actively referred to the substance as N-acetyl-p-aminophenol, p-acetylaminophenol, actetylaminophenol, N-p-hydroxyacetenilid, APAP (acetyl-p-aminophenol), or paracetamol—all of them equivalent common terms for the same substance. Likewise, the first semisynthetic penicillin in the United States was launched in 1957 by six different companies under three different generic names: Bristol's Syncill and Wyeth's Darcil both carried the generic name potassium penicillin-152; Schering's Alpen and Dramcillin-S, and Squibb's Chemipen all carried the generic name phenethecillin potassium; while Roerig's Maxipen carried the generic name alpha-phenoxyethyl penicillin potassium. All six brand names and three generic names referred to the same active molecular ingredient.[9] During cross-examination by Kefauver's lead economist, John Blair, Cornell's Walter Modell claimed that that in a given year some fifty new chemical entities were released in hundreds if not thousands of different brand-name formulations and combinations.[10]

Even when present, generic names were often printed in much smaller fonts (figure 1). Curt Weilberg, a principal agent in the drug industry's public relations body, the Medical and Pharmaceutical Information Bureau (MPIB), admitted under examination that MPIB had intentionally promoted popular publication of trade names in commissioned articles on new drugs in consumer periodicals from *Good Housekeeping* to the *Saturday Evening Post*, printing trade names of drugs in lowercase to confuse them with generic names. This allowed MPIB to use public relations materials to promote brand names instead of generic names in news articles about drugs. Wallace Laboratories, for example, the makers of the minor tranquilizer Soma, issued letters to the publicity offices of the MPIB emphasizing that the generic name carisoprodol was "not to be used in advertising promotion or publicity except when it is absolutely necessary. For example, it must be included in JAMA ads; it must *not* be included in other journal ads."[11]

Yet not all parties were convinced that the nonuse of generic names was due to the machinations of brand-name pharmaceutical marketers. Senator Roman Hruska (R-NE), the most senior advocate for the pharmaceutical industry on Kefauver's subcommittee, proceeded to challenge Wilson's critical claims in a series of exchanges of increasing absurdity.

FIGURE 1. Debates over the relative visibility of brand versus generic names of drug products repeatedly surfaced during the late 1950s and early 1960s. This advertisement for Ciba's Singoserp (syrosingopine) was used as evidence of that discrepancy in Senator Kefauver's drug industry antitrust act hearings in 1961–62. *Singoserp advertisement from JAMA, January 24, 1959, as reprinted in DIAA vol. 7, p. 3402.*

For one thing, Hruska thought Wilson's critique of the use of polysyllabic words was untenable, especially for a science-based industry:

> SENATOR HRUSKA: Is it possible to get very far in our English Language without getting into polysyllabic words? . . . So that when [the pharmaceutical firms] go into several syllable words, it is probably not because they choose to, but it is because the way our language is made, isn't it? Why should you poke fun at the manufacturers for using polysyllabic words when that is the way we do business? That is the way we talk . . . Would you want to deny the manufacturers the right to use something in a technical field that we use in every-day, ordinary conversation and usage?
> DR. WILSON: No. I am not denying them the use of syllables. That is kind of silly. But I am pointing out that they can be simplified.[12]

In return, Wilson repeated for Hruska his conviction that a *good* generic name had a utility independent from dosage form or finished product. The difference was as simple as the difference between chemistry and pharmacology: prednisone was a generic name for a pure chemical, while Meticorten was a Schering tablet containing prednisone, with a specific pharmaceutical dosage form whose exact composition was known only to the Schering Corp. "Any tablet of prednisone," Wilson noted, "is not equivalent to Meticorten tablets any more than just any beefsteak is equivalent to a Kansas City beefsteak." Yet it was only once the consumer had a concept of beefsteak that was different from, say, ground beef or even pork chops, that she would be able to meaningfully differentiate between Kansas City beefsteak and New York beefsteak as an informed consumer. "Or an Omaha beefsteak, Doctor," the senator from Nebraska grudgingly agreed.[13]

The generic name was meant to signify membership in a particular family of pharmaceutical products just as the name beefsteak signified a particular class of butchery. In contrast to a brand name, Walter Modell added, which served to mystify physicians into believing in the singularity of the brand-name drug, "the term 'generic' means what it means, it shows the genus of the drug, where it came from, what it is related to, what its heredity is, so to speak."[14]

UNSTABLE KINSHIPS

As Kefauver's hearings set the stage for sweeping legislative reform of pharmaceutical marketing, many within the AMA and USP became actively concerned that their generic naming system would soon become the subject of federal oversight. In an attempt to deflect further government involvement, Hugh Hussey of the AMA testified in the fall of 1961 that the AMA and USP had recently begun a new joint project to improve generic names, so that "the generic name should be as simple as possible and that chemically related drugs, used within a given therapeutic area, should preserve some relationship in their nonproprietary names." Hussey's counterpart at the USP, Lloyd Miller, likewise agreed that increased professional self-regulation of generic names through the USP and AMA was far preferable to any government intervention.[15]

Yet, shortly before the head of the USP was called before the Kefauver commission again in the fall of 1961, his deputy in charge of nomenclature, Walter Hartung, wrote him a letter noting that still "there are many aspects of generic nomenclature that make me unhappy." Hartung warned of the explosion of new stems that had taken place since the original listing of fourteen categories of drugs and drug names issued by the AMA and WHO in 1956. The fourteen lines listed in the WHO naming system of 1956 were now insufficient to handle the growing number of "radicals"—or potentially therapeutically significant chemical subunits—

still emerging from industrial chemistry labs. "I feel," Hartung confided to Miller, "as one who has already waded beyond his depth."[16]

Miller agreed: the WHO/AMA system had become like a system of file folders that had been initially useful to reduce clutter but had since become dangerously overstuffed. Even once a suitable and memorable generic name was derived for a new molecular entity, such as chlorothiazide for the diuretic Diuril, subsequent drugs received clunkier and clunkier names. The five-syllable "chlorothiazide" led logically to seven-syllable "hydrochlorothiazide," and later entries in the class like the eleven-syllable hydrochlorothiazide/triamterene" promised to bear still longer and more unwieldy names.[17]

In 1961 the AMA and the USP founded a joint committee on nomenclature and invited the WHO to a joint meeting in September of 1961—attended by more than seventy representatives from the pharmaceutical industry and additional parties—to attempt to deal with this issue. Most delegates agreed that the generic naming system needed to be more flexible, to expose more explicitly the genetic relationship between different drugs. As the AMA Council on Drug's Winsor Cutting concluded after the conference, "to be worth the effort of coining them," generic names required two things: to be used and to be useful. To be used, they needed to be short, easy to spell, "euphonious, and if possible with a little tune in them." To be useful, they "must tell a story." "The story," he confided to a colleague at the USP, "is to say what family they belong to (your syllables)—for with such a story comes automatically information about chemical nature, perhaps mechanism of action, very likely something about absorption and excretion, surely something of toxicity, and therapeutic use."[18] Against the backdrop of Kefauver's threat of external regulation, the 1961 conference of manufacturers, physicians, and pharmacists found common ground to adapt a more comprehensive private approach to drug regulation. Industry executives grumbled that "it seems rather clear that despite the announced purpose of this conference that some in attendance wished to make a case for generics vs. trademarks."[19]

Certainly, many others were willing to make the case for trademarks versus generics. Kefauver also received testimony from the National Pharmaceutical Council (NPC), a collaborative venture of pharmacists' associations and major pharmaceutical firms, which will be described in more detail in chapter 3. To properly understand the value of the brand, the NPC insisted, one needed to consider an alternate genealogy of the term *generic*, not from *genus* but from *Genesis*. Comparing the naming of drugs to the Edenic parable of Adam's naming of the animals, NPC president Nelson Gampfer emphasized the centrality of names in giving order and utility to the world of things. In Gampfer's terms, the difference between generic and brand names was like the difference between a crude tool and a precision instrument: "Drugs, like washing machines and automobiles, are the products of man's research and ingenuity and are fittingly named by their manufacturer with both common and brand names. The common name is usually a condensation of the full chemical name of a specific substance (frequently long, complex, and unpronounceable) and intended for the purpose of scientific classification in the compendia of standards and in the chemical reference lexicons. The brand name, on the other hand, provides a simpler, easier to remember, pronounce, and write name for the physician to use in ordering."[20]

Presented with a generic name for a drug, the physician knows merely what chemical it contains. Presented with a brand name, the physician knows at least three more aspects of the drug: the pharmaceutical makeup (e.g., ointment, tablet, capsule, elixir, etc.), the balance of ingredients, and the reliability of the manufacturer who stands behind that product. For one word to contain all of the above—and be memorable and easy to say and write to boot—suggested to Gampfer a much more parsimonious and efficient way to prescribe than the generic name. Far from being subjective, emotional, or irrational, the brand name was a more precise and therefore a more objective description of a drug product.

This argument was expanded in an NPC pamphlet—entitled *Twenty-Four Reasons Why Prescription Brand Names Are Important to You.* Here the NPC further specified the value of a brand name, which unlike a generic name guaranteed with a single pen stroke twenty-four basic pharmacological principles that could differ between pharmaceutical preparations of the same chemical: potency, compatibility, purity, sustained release formulations, enteric coating, disintegration time, solubility, particle size, choice of vehicle or base, quantity of active ingredient, allergic manifestations, irritation, pH, viscosity, caloric value, melting point, surface tension, viscosity, ease of application, flavor, quality control, and packaging. The brand name was not the problem, but rather the *solution* to the increased complexity of the modern pharmaceutical market.[21] APhA president Linwood Tice warned,

> It would be difficult to imagine the utter chaos which would result if, let us say, for the next week all prescriptions written in the United

States were written using generic names . . . [given that] physicians themselves are totally unfamiliar with these names and pharmacists, we suspect, do not know 1 out of 10 . . . Even when those who are given the task of coining generic names do it with complete objectivity and follow all of the standard rules for nomenclature such as those promulgated by the World Health Organization, they came up with names which are real tongue twisters and almost impossible for the average practitioner to spell.[22]

This critique of generic names turned the language of therapeutic rationalism on its head. Generic names were not the solution but the problem, one more twist in a series of attempts by out-of-touch academics to rationalize the vagaries of the health-care system. Defenders of the brand sought to portray the celebration of generic names by academics and international bureaucrats as genteel and out of touch with the pragmatics of the front-line practitioner. The classroom, not the clinic, would remain the most successful site for the uptake of generic nomenclature in the 1960s and 1970s.

GENERIC NAMES, SPECIFIC THINGS

Estes Kefauver had initially hoped to propose a bill to abolish the brand name in American pharmaceutical marketing and enforce generic prescribing by physicians. Yet the outcomes of his hearings into generic names overwhelmingly worked against this dream. Even though an ideal system of generic names might have helped to empower physicians and consumers to make better therapeutic decisions, the generic name as it existed at the moment in the AMA/WHO system was revealed to be far from ideal and could not clearly be relied upon to designate a set of equivalent therapeutics. Hruska directly challenged Charles Wilson after he suggested that prescribing by generic name could suffice to designate a desired therapeutic product:

SENATOR HRUSKA: Well now, Doctor, let's explore that a little bit. If it was a prescription for a given quantity of insulin, and that were taken to a pharmacist, would he get the same product in every instance?

DR. WILSON: Yes, sir. AU-40 is AU-40.

SENATOR HRUSKA: Isn't it true that there is some insulin made with a beef base and some insulin made from a pork base, and that some patients are allergic to pork-base insulin and some are not? And he would not get the same product if the prescription simply said insulin?

DR. WILSON: Well, you are talking about a finished pharmaceutical there, Senator.

SENATOR HRUSKA: It is a generic name by now; isn't it?

DR. WILSON: Yes, it is official.

SENATOR HRUSKA: It is a generic name, and if it was so many units of insulin, he would not get the same product, whereas if he said Lilly, he would get an insulin with a beef base, isn't it true?"[23]

To Hruska, prescribing by generic name alone simply was not kosher. Hruska pressed Wilson repeatedly to admit that he was claiming that all generically named drugs were equivalent, a point that Wilson repeatedly denied. "You see, Doctor," Hruska interjected in one particularly heated exchange, "a patient doesn't swallow a nomenclature. He swallows or has injected into him certain substances. The question is whether he would get the same substance."[24]

Wilson, and Kefauver as well, was forced to concede that there were clinically significant differences between pharmaceutical products bearing the same generic drug name. By the early 1960s, the generic name had become a necessary but not a sufficient step for establishing interchangeability among drug products. Some other step was always necessary to guarantee the quality of the drug dispensed: a skilled pharmacist, an active and rigorous hospital formulary, or the name of a trusted pharmaceutical firm. This came as somewhat of a surprise for Kefauver, who had thought that the wider use of generic names alone could bring about a more rational form of prescribing by physicians and greater agency for patients as consumers.

As one of Kefauver's staffers reviewed all of the testimony and evidence gathered at the end of the hearings in late 1960, the generic name still remained an ideal concept rather than a useful tool. "Perhaps the viewpoint of the public" he typed, "is expressed in the concluding remarks of a representative of a consumers' organization." Pasted below this line was a cutout section of the testimony of Mildred Brady of the CU, recasting the equivalency of generically named drugs as an ideal that the Food and Drug Administration should someday make into a reality:

By whatever means prove to be necessary, and in the shortest possible time, the Federal Government must create within the drug industry those circumstances under which any physician in the United States may prescribe with full confidence in its potency, effectiveness, and freedom from adulteration, any drug regardless of its price or its name, generic or otherwise, which he feels will most benefit his patient.[25]

Kefauver's bid to ban brand names and enforce generic prescribing failed outright. But the bid to have a single, useful, generic name for each drug was well received and was signed into law as part of the amendments bearing Kefauver's name in 1962. For the first time, it had become the responsibility of the federal government—largely through the FDA—to enforce the requirement that every drug have one deep and inscrutable, singular name.

With the passage of the Kefauver-Harris Amendments to the Food, Drug, and Cosmetics Act in 1962, the secretary of Health, Education, and Welfare (HEW) was granted the right to designate one true generic name for every drug product. The formerly private USP-AMA collaboration now had a public charter and a new name, the United States Adopted Names (USAN) committee. Speaking at an industry conference in late 1965, the FDA liaison to the USAN committee spoke optimistically that the "impractical and sometimes hazardous" field of generic drug naming could now be remolded into a rational process.[26]

The USAN has operated since 1964 as the single clearinghouse for generic names for the US drug market (and consequently one of the most significant arenas for the global pharmaceutical industry), and it remains so today.[27] The USAN now sets an upper limit on acceptable length of a generic name at four syllables, and it seeks to promote the use of a limited number of "stems" or "common, simple word element[s] . . . incorporated in the names of all members of a group of related drugs when pertinent, common characteristics can be identified." The original list of sixteen stems, however, has expanded to more than 240 categories, and the USAN continues to meet semiannually to discuss new stems, prefixes, and suffixes in the ever-expanding taxonomy of drugs. The volume that contains them, the *USP Dictionary of USAN and International Drug Names*, has since become a fat book with nearly 1400 Bible-thin pages at last publication.[28]

Many within the USP, FDA, AMA, and WHO hailed the new USAN as a more robust system for making generic names meaningful. Yet as with the initial AMA/WHO collaboration, several shortcomings of the USAN were soon apparent. Already by 1966, FDA commissioner James Goddard criticized the USAN as largely ineffectual in the reform of generic drug names. In a review of two hundred generic names recently approved by the FDA, he found only twenty-eight of them to be well suited to the needs of the prescribing physician.[29] Goddard announced this critique at a medical writers forum in Washington alongside his frequent adversary, PMA president (and former AMA counsel) C. Joseph Stetler. Stetler on this occasion agreed with Goddard. "I do not think," he added, "that doctors could prescribe generically if you forced them to do so today." As Goddard elaborated, although medical students learned generic names in the classroom, "by the time they are graduates, they have switched over to trade names. So, we have the strange system where we select names that please chemists and are not usable to the people who prescribe the drugs." Without further reform, "the present generic names," the FDA commissioner concluded, "are just worthless for the most part."[30]

Goddard's critique of generic names, however, would be reported by the trade newsletter *F-D-C Reports: Pink Sheet* as a statement that all *generic drugs* were worthless. Though the *Pink Sheet*'s error was apparently typographical, it foreshadowed a series of debates of increasing intensity over the relationship between names and things. Likewise in the immediate wake of the Kefauver hearings into generic naming, a group of entrepreneurs saw in the nascent interest in generic names a business model for a new sort of pharmaceutical company. Calling their venture Generic Name Pharmaceuticals, Inc., the principals wrote a letter to Lloyd Miller asking him exactly *what* a generic prescription might look like, so that they could refine their business plan to best capitalize on this emerging market.[31]

Along with other firms that would come to constitute the modern generic drug industry, these entrepreneurs saw possibility in the publicity surrounding Kefauver's hearings.

PART II

NO SUCH THING AS A GENERIC DRUG?

CHAPTER 3

DRUGS ANONYMOUS

Without this similitude, no industry is possible.
—GABRIEL TARDE, *PSYCHOLOGIE ECONOMIQUE*, 1902

To speak of generic drugs in 1960 was to speak of something that did not yet exist, at least not for most people. In the transcripts of Estes Kefauver's hearings into the pharmaceutical industry, one finds plenty of discussion of smaller producers and larger producers, generic names and brand names, but very little explicit mention of generic drugs, except in negation. By the end of the decade, however, the generic drug had become a self-evident object with economic and material substance. Its solidity was attested to by an increasing number of stakeholders interested in seeing therapeutic objects as exchangeable parts of a health-care system that could itself be rationalized, in much the same way that the assembly line of a factory could be rationalized by industrial engineers. When in that spirit Gaylord Nelson, the senior Democratic senator from Wisconsin, opened a new set of Senate hearings into competitive problems in the drug industry in 1967, he spent the bulk of his introductory speech describing the "generic drug" not as a concept but as a thing. "For a substantial number of the most widely used brand-name drugs," Nelson noted, "*there is a generic drug available at a substantially lower price.*"[1]

Like Kefauver before him, Nelson used congressional hearings as a highly visible stage for advancing the emerging politics of consumerism. Like Kefauver he instigated a broad public exposé of deceptive pharmaceutical marketing practices as a capstone to a broader critique of monopoly power in America. Unlike Kefauver in 1960, however, Nelson could point out in 1967 that a set of "quality generic drugs" already existed in the American marketplace, promoted by companies that now called themselves "generic drug manufacturers." As proof of this concept, Nelson pointed to the Department of Defense's practice of buying generic drugs for Walter Reed Hospital, where Lyndon Johnson himself received treatment. If generic drugs were good enough for the president, surely they would suffice for the average consumer?

But generic drugs had not yet become equally concrete in the eyes of everyone. Certainly not for Durward Hall (R-MO), former head of the military pharmaceutical procurement program, who retorted to Nelson's subcommittee that the president had never consumed generic drugs, nor had anyone else in the military, ever:

> I simply say to you that anyone suggesting that one drug firm is as good as another is a fool or naïve, or both ... The military ... restrict[s] its purchases to products of proven quality; not just generic drugs, but quality drugs made by companies they know, and know *well*. The very word "generic" implies anonymity, gentlemen; the program I have described to you is one emphasizing just the opposite. The result is when the President goes to Walter Reed or Bethesda, he does not get "generic" drugs, he gets drugs made by companies the defense department knows supremely well. That is as it should be![2]

Hall was one of the few physicians on Capitol Hill, a free-market ideologue who, as one columnist quipped, "prescribes strictly brand-drugs for his patients and John Birch politics for his constituents."[3] To Hall and his colleagues at the Pharmaceutical Manufacturers Association, the National Pharmaceutical Council, and the American Medical Association, the new popularity of genericism promoted a dangerous confusion between worlds ideal and real. The objects that Nelson and his army of "rabid consumerists" called generic drugs represented a Platonic ideal that had no material counterpart in the world of goods. All drugs, in other words, were made by *some* manufacturer and all therefore carried the reputation of that firm. Well into the 1970s, PMA and NPC pamphlets repeated the catechism that, "strictly speaking, there is no such thing as a 'generic drug.'"[4]

For the historian, however, the question "is there any such thing as a generic drug?" can be unpacked and contextualized to explore a more revealing set of problems at the intersection of science, technology, business, and medicine.[5] How did the generic drug become an economically and clinically relevant object? What were the historical conditions that made this new thing a possibility? Which economic interests, political alliances, and institutional actors encouraged its emergence as a stable entity in the world of goods? This chapter seeks to reconstruct the instability and fragility of pharmaceutical copies as social, political, and economic objects and trace the processes by which they became more concrete in all of these realms.

COUNTERFEIT HISTORIES: IMITATION AND ADULTERATION

Richard Burack first became aware of generic drugs in the mid-1960s in his practice as an internist at Cambridge City Hospital. Like other municipal hospitals that assumed local public health responsibility, Cambridge City strived to provide high-quality medical care to indigent populations. Burack, in charge of choosing which drugs should be stocked in the hospital's formulary, found that he could achieve cost savings with no apparent sacrifice in quality by purchasing these "generic-equivalent drugs": "No single patient of mine who has been treated with a generic-equivalent drug has experienced anything but the effect which could be expected. I have been unable to observe in patients any difference whatsoever between the efficacy of generics and that of brand-name drugs, nor has there ever been a suggestion that the former are any more likely to cause untoward side-effects."[6] Use of these new generic products bled gradually from his public indigent practice into his middle-class outpatient practice and finally into his own personal approach to pharmaceutical consumption when treating his own bodily ills.

Burack's positive experiences with generics motivated him to provide a practical guide to the use of these drugs for doctors and patients. When his *Handbook to Prescription Drugs* (discussed in more detail in chapter 10) was published in 1967, it was an instant best seller, undergoing multiple printings and yielding four subsequent editions. The book was praised by Senators Nelson and Edward Kennedy on the floor of the Senate and reviewed by the *Washington Post* as a work that "could rock the drug world as much as Ralph Nader's [*Unsafe at Any Speed*] shook Detroit."[7] But the *Post*'s review of Burack's *Handbook* also covered a second book on generic drugs published that year: Margaret Kreig's *Black Market Medicine*, a darkly lit thriller narrating the seedy underworld of pharmaceutical counterfeiters. Kreig's book would be reviewed in turn by the *New York Sunday News* as "a document which may be the most shocking survey of vice in America since Upton Sinclair's *The Jungle.*"[8] To Kreig, the generic drug was not a solution but a problem: part of a deep, dark, ugly secret reaching to the underbelly of modern medicine. Kreig linked the genealogy of generic manufacturers to a longer history of dangerous and illegitimate drug copies: the generic drug, in her account, was merely a smarter version of the counterfeit drug.[9]

Kreig's literary descent into the underworld of generic drug manufacture resonated with the public relations efforts of the Pharmaceutical Manufacturers Association, a highly visible lobbying group for brand-name pharmaceutical interests.[10] The PMA's own 1967 entrant to the book(let) trade on generic drugs, the glossy *Drugs Anonymous?* (figure 2), made this connection explicit: generic drugs were dangerous because of their anonymity. The lack of any mark associating them with a reputable manufacturer meant that any indication of their quality, purity, and efficacy was effectively a cipher. Generic manufactures used cheaper materials, cut corners, and purchased and packaged adulterated materials in dirty and uncontrolled environments. Because they were sold anonymously, there were no negative consequences for bad actors. Rather, the generic drug market encouraged a form of moral hazard where attention to quality was explicitly *not* rewarded.[11] Morally, economically, and physicochemically, generic drugs could be equated with adulterated goods.

Educating the public about the dangers of anonymous and adulterated pharmaceuticals had also been the chief function of the National Pharmaceutical Council since its founding in 1953. The organization grew rapidly in the early 1950s and realized a number of successes on the state level, as forty states in quick succession passed laws that prohibited pharmacists from dispensing any other version of a drug than the brand prescribed by a physician. The

FIGURE 2. *Drugs Anonymous?*, one of several pamphlets distributed by the Pharmaceutical Manufacturers of America (PMA) questioning the safety and efficacy of generic drugs. Drugs Anonymous? *(Washington, DC: National Pharmaceutical Council, 1967).*

structure and function of these antisubstitution laws will be explored in more detail in chapter 8; for now it will suffice to emphasize that the NPC was well practiced in the art of the educational scare campaign on state and national levels. NPC-produced pamphlets, letters, journals, and public relations materials portrayed generic drugs as a latter-day form of adulteration.[12]

The problem of adulterated medicines—and the emergence of municipal, state, and federal regulations as a solution to that problem—long predated the twentieth-century formation of the Food and Drug Administration. In 1820, the year that the first *United States Pharmacopoeia* was published, Frederick Accum authored a comprehensive *Treatise on Adulterations of Food and Culinary Poisons* that likewise warned consumers that cheap food and drugs often prove to be a costly bargain. In mid-nineteenth-century New York City, for example, public health reformers took up the cause of adulterated "swill milk" to argue for the need for a public health agency to regularly inspect food products. Likewise, widespread concerns of adulteration of suspicious foreign drug products helped speed the passage of the Drug Importation Act in 1848, which would remain the most significant federal drug law until the Pure Food and Drug Act of 1906. It is an enduring testimony to the political viability of adulteration that the FDA's primary tool in regulating the therapeutic claims of pharmaceuticals and medical devices remains the label—reminding us that the foundational function of this regulatory agency was to make sure that packaging accurately describes its contents and their therapeutic claims.[13]

And yet the creation of the FDA did not resolve concerns over the possible adulteration of drug products. Fears of the untrustworthiness of drug products continued to grow alongside expanding markets for mass-produced pharmaceutical products in the twentieth century, especially as medications were prepared at further and further remove from the ultimate consumer. As late as 1939, 75–80 percent of prescriptions filled in the United States had still been compounded locally by the community pharmacist himself in the back of his own independently owned pharmacy. But in the years surrounding the Second World War, the American pharmaceutical industry increasingly integrated backward into chemical supply and forward into drug formulation, selling more and more prepackaged drugs to pharmacists. In turn, the pharmacist changed from a highly specialized local manufacturer of finished drug products into a more mechanized automaton who counted pills from larger (stock) bottles into smaller (prescription) bottles. By the end of the 1950s, nearly 75–80 percent of all prescriptions were filled in this fashion.[14]

Although an earlier generation of pharmaceutical manufacturers had celebrated the greater standardization of drug products that came with the mass market, the production of medicines in increasingly distant industrial spaces instead of the local professional spaces of the community pharmacy likewise opened up new anxieties over drug counterfeiting as both a real problem and a moral panic.[15] As the CEO of Merck, John Horan, told the New Hampshire Pharmaceutical Association, by 1961 the menace of counterfeit pharmaceuticals had become a "big underworld business in the United States." Merck, a vertically integrated research-based pharmaceutical firm, had experienced its first brush with counterfeiters earlier that year: "Tablets bearing our symbol were traced to their source, an indescribably filthy warehouse loft in Hoboken, NJ. As physicians, you will readily understand the threat large-scale drug counterfeiting poses to the public and the health professions, in addition to the injustice it works on the legitimate industry."[16] In principle, counterfeiting and imitation were distinct problems: counterfeiters marketed an illegal mimic of a trade name drug, while imitators marketed legal products that merely claimed to have similar therapeutic properties as innovator drugs. But the NPC pamphlets in the 1950s consistently lumped "the imitators and counterfeiters" of prescription drugs into the same group. Attacks on counterfeit medicines in pharmaceutical and medical journals described in lurid detail "the fly-by-night factory of the maker of imitation," which did brisk business in short but lucrative stints of "illicit operations."[17]

Popular representations of drug counterfeiting and low-cost imitative products likewise blurred the two categories. When in 1960 *Parade* magazine published one of the first popular exposés of drug counterfeiting (figure 3)—documenting the role of the Mafia in a "zombie" drug market then estimated at $150 million, they tarred small drug manufacturers with the same brush. "Most zombie makers are small manufacturers," *Parade* claimed, "who use their legitimate businesses as a cover for pirating big-name drugs, or fly-by-night, hole-in-the-wall operators whose premises are often so filthy that their drugs are contaminated." Investigative journalists performed stakeouts and accompanied police on raids of counterfeit operations within small, generic-name pharmaceutical manufacturers such as the firm Merck had complained about: General Pharmacal of Hoboken, New Jersey. As one journalist described, Pharma-

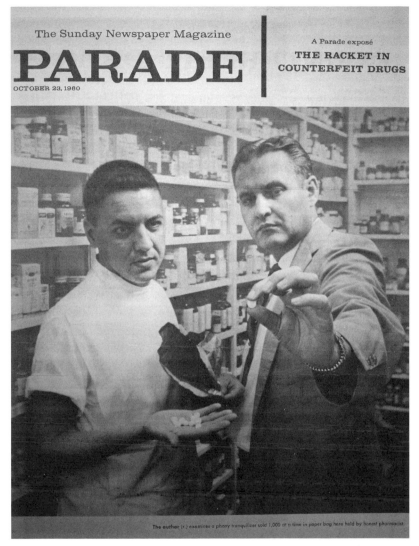

FIGURE 3. Popularizing the problem of counterfeit medication in the early 1960s.
Cover page from Parade *magazine, October 23, 1960.*

cal's laboratory consisted of "a dirt-crusted oven, stained counters, prim-itive sink, filthy compressing and compounding equipment—all crawling with cockroaches."[18] Readers wrote in to *Parade* with their own stories and fears of "savings squandered on phony drugs—from sawdust-filled 'vitamins' to fake 'wonder drugs' made of chalk, flour, and other fillers."[19] Whether the products were being passed off as brand-name versions or as generically named products—that is, whether they were counterfeits or merely imitators—the drugs could be considered tainted in either form.

Lucille Wendt, attorney to Kefauver's Subcommittee on Antitrust and Monopoly, was suspicious that firms like Merck were publicizing the counterfeit drug issue as a ploy to cast doubt on all copies of off-patent medicines, legitimate or not. "The thinking of members of the Subcommittee would be clarified if we would stop using the industry-coined term 'counterfeiting,'" she warned in a memo in early 1961, adding that "'Counterfeiting!' has now become a euphemistic term for getting government agencies to police the trademarks of the major drug manufacturers."[20] Nonetheless, when Estes Kefauver publicly called on the FDA a month later to create a registration program for all drug manufacturers, he fell back into exactly the same counterfeit imagery to depict a regulatory crisis that required a regulatory solution: "The idea of having a pharmaceutical manufacturer secure a license from the FDA before he started operating was to my mind . . . to keep the fly-by-nights out, the bathtub operators, and let the FDA know who they should inspect every two years. In other words, the situation that anybody without regard to their facilities can come into the business of making ethical drugs is untenable in my opinion."[21] Even with the implementation of Kefauver's registration system, the specter of counterfeiting seemed to grow only more menacing during the 1960s. As markets for prescription drugs continued to soar— especially for diet drugs, tranquilizers, and amphetamine-based "pep-pills"—opportunities for small- and large-scale syndicates of organized crime to encroach on such markets grew apace.

BLACK MARKET MEDICINES

Into this environment of heightened fear of adulteration and booming markets for imitative products strode the investigative journalist Margaret Kreig, embedded within a team of undercover FDA agents sporting wiretaps and sidearms. A former model, Kreig would become a beguiling character in her own exposé: the "Mata Hari of the drug sleuths."[22] *Black Market Medicine* provided a thrilling account of a Mafia-controlled underworld in which nothing was as it seemed. Beyond Kreig's white-hatted FDA agents, all forms of authority became suspect: chemists, pharmacists, manufacturers, physicians might all be on the take of the copycat pharmaceutical underworld.

As Kreig's wiretap tapes demonstrated, by the mid-1960s Mafia involvement in counterfeit drug production was undeniable, not merely in the distribution of drugs with recreational uses such as Dexedrine (amphetamine) and Seconal (barbiturate) but also for more workhorse drugs for the treatment of diabetes and hypertension such as Orinase (tolbutamide) and Diuril (chlorothiazide). In Kreig's account, a shadowy outfit known as "The Group," masterminded by an underworld boss called "The Chameleon," fronted a series of generic-name drug manufacturing firms, most of which were as filthy as those in the *Parade* coverage and which cranked out counterfeit imitations of drug brands behind closed doors in the off hours.

It was as if the twin targets of Estes Kefauver's two investigations of the 1950s—the profiteering prescription drug industry and the Mafia—had both changed shape, chameleon-like, into the form of his former darlings, the "small manufacturers" who sold drugs by their generic names. The Mafia overtones were only heightened by the observation that most generic drug manufacturers obtained raw materials from Italian pharmaceutical and chemical companies, who were free to produce patent-protected chemicals because Italy's intellectual property laws defended process patents, not product patents. Legitimate as this Italian-American trade may have been, the mere association could be trumped up as further evidence of a *cosa nostra* underworld coordinating the production of all generic drugs.

In Kreig's terms, a black market of overtly counterfeit products was nearly coterminous with a gray market of generic drug production. The generic drug industry was largely composed of shape-shifting "fly-by-night" companies that would set up shop in one location, produce a few batches of drugs of dubious quality, and then disband and reassemble in another locale. Even once the FDA learned about a new firm, it could take a few years before the first inspection actually took place. An astute fly-by-night operator could set up an apparently legitimate company and intentionally disband all operations before it was ever inspected.

In this darkly shaded perspective, the industry of off-patent drug production was not just a potential space for counterfeit production by a few bad actors but part of a larger pathological system that created incentives for further counterfeiting. "Everyone wants legitimate drugs at illegitimate prices," Kreig quoted trade journalist Irwin Di Cyan, "[and] that impossible demand creates the climate in which counterfeit drugs can flourish." Were there no such things as cheaper versions of brand-name drugs, there would be no market for counterfeits. Di Cyan described the connection between generic and counterfeit forms in converse: "How many know that, although rarely is a generic drug producer a counterfeiter, nearly every counterfeiter also produces generic name drugs?"[23]

Kreig's reporting meshed seamlessly with brand-name industry attempts to describe all generic manufacturers as schlock houses, fly-by-nights, or bathtub operators. Indeed, several correspondence files between the FDA and early generic drug firms document how quickly these terms seeped from manufacturers to pharmacies, from physicians to consumers. See, for example, the following letter from a Philadelphia pharmacist to the FDA in 1966 (typed on a pad promoting Wyeth's penicillin product Pen Vee K):

2 Sep 66

Dear Sirs;

I am a recently graduated Pharmacist and I work part time at a local pharmacy.

Today I had to refill an Rx for Prednisolone 5mg. Out of the 20 tablets, the 3 I am sending you were found. The reason I am so incensed about this "medication" is that my regular job is as a Quality Control Inspector for M.S.&D. and I know how hard we work to keep up the quality of all the products we turn out & I imagine the other large drug houses do the same. Then along comes this 'schlock' outfit & this is what we get.

This product is identified as

Prednisolone USP 5mg

Manufactured by

Aberdeen Pharmacals Corp.

Englewood, N.J.

Lot Number

3593

Thank you

KE

P.S.

I know my boss shouldn't have bought such cheap crap, but I figured I'd let you know anyhow[24]

The linkage of generic drugs with adulteration would continue to echo in industrial and professional attacks on generic drugs in the late 1960s and 1970s on broader registers. By buying drugs cheaply, consumers also purchased risk and danger and encouraged further risky entrepreneurs in a race to the bottom. By differentiating themselves on the market by their very anonymity, generic manufacturers invalidated the reputational basis that had been central to "ethical" drug marketing in the first half of the twentieth century (as defined in chapter 1): that a firm flourished or perished on the basis of its reputation for quality products.

In 1969, John Adriani, the head of the AMA's Council on Drugs, argued in a radio broadcast for the utility of prescribing by generic name so that physicians and consumers might be able to make rational choices in the pharmacy as they did in the grocery store, "just like oatmeal." "Oatmeal is oatmeal," Adriani continued, "and whether Kellogg's makes it or another cereal company makes no particular difference."[25] A few days later, in the midst of a torrent of angry retorts, he received a handwritten letter signed by "a patient" hotly contesting Adriani's credentials as a physician: " 'Just like . . . oatmeal.' This statement will result in good physicians using inferior drugs that might not even work to save your patient's lives. Oatmeal indeed—and for your information, I always buy 'Quaker' oats and I bet you do too. If not you deserve the mush you might get. Mush with bugs in it, with no nutritional value, that tastes bad, that will take three hours to cook, and mush your starving kid throws on the floor with delight."[26] This letter, like many others from consumers kept in the files of the AMA and the FDA, captures the power of equating anonymity with adulteration in the world of drugs and other consumer goods.

It is harder, perhaps, to understand why Adriani would have thought that comparing drugs with other consumer goods such as oatmeal, automobile tires, and dishwashers (to name a few of his comparisons), could have made generic names of drugs any more palatable to the increasingly brand-conscious American of the late 1960s. Yet others in medicine, pharmacy, and the pharmaceutical industry would soon join Adriani in asking: how might *generic* be reconstructed as a positive term?

CHAPTER 4

ORIGINS OF A SELF-EFFACING INDUSTRY

Generics are bustin' out all over and it looks like a long hot spring for the drug industry.

—*DRUG NEWS WEEKLY*, MARCH 13, 1967

In 1967, the new president of the American Pharmaceutical Association, William Apple, predicted that "the 'brand-name era' in pharmaceuticals is coming to an end."[1] With the rise of the modern generic drug industry, Apple told the assembled audience of senators, aides, and journalists attending Gaylord Nelson's hearings into competitive problems in the drug industry, interchangeable drugs would become the keystone in building a new and consumer-minded system of rational therapeutics. "*Generic* in itself has always had the connotation of being a bad word, an evil term, and I think we finally have come to an educational point where we recognize that what you call a drug is no criterion of its quality. The label, the nomenclature, goes on last. The quality is built in first and for this reason, I think prescribers and others, pharmacists certainly, are taking a keener interest in the use of generic drugs."[2]

Apple argued for the virtues of the generic. He recast the anonymity of the generic drug not as a risk (of adulteration) to be contained but as a value (of consumer empowerment) to be optimized. Along with other "genericists," such as the AMA's John Adriani and the Harvard internist and pharmacologist Richard Burack, Apple worked to invert the relationship between anonymity and adulteration by focusing not on the increased distance between producer and consumer but rather on the greater cost efficiency that generic drugs could provide as long as they delivered a certain baseline standard of safety and efficacy.

What were the conditions of possibility that allowed a group of pharmaceutical manufacturers to shake off the taint of adulteration, counterfeit, and copycat and embrace a new identity as a generic drug industry? A self-declared generic drug industry emerged at the intersection of three linked changes to the pharmaceutical market. These related, in turn, to the evolving intellectual property regime governing pharmaceutical monopolies, the changing role of the regulatory state in determining therapeutic efficacy, and the expanding role of the federal government in pharmaceutical delivery from the Kennedy administration to Lyndon Johnson's Great Society.

First, the very possibility of this industry was contingent on a broad shift in the pragmatics of pharmaceutical intellectual property taking place in the 1960s, as the patents on the postwar crop of "miracle drugs" launched in the 1940s and 1950s began to expire. In the later 1970s, as financial analysis began to dedicate analytic reports to the generic drug industry, they situated origins of its subject somewhat arbitrarily in 1967, the year in which all seventeen-year drug patents taken out prior to 1950 had expired. Each year thereafter brought a wave of new patent expiries. The precise date is immaterial: key to all these narratives is that, as the wonder drugs of the 1940s and 1950s went off patent in the 1960s, a new group of manufacturers began to see the potential for producing newly legal copies of these aging cash cows.[3]

If the passing of the patents on the first generation of postwar wonder drugs was a necessary precondition for the rise of the generic drug industry, its effects were amplified by the attention drawn to the field by Kefauver's Senate hearings. Take, for example, Herman C. Nolen, a former professor of marketing at Ohio State University who saw a business opportunity lurking in that publicity. Nolen was by that time the president of McKesson & Robbins, one of the nation's largest pharmaceutical wholesalers. Although McKesson & Robbins had not manufactured any of the drugs it sold for several decades, Nolen announced in 1961 that the firm would expand capacity to become a manufacturer of "generic pharmaceuticals" in 1961. This caught the attention of Kefauver's committee, and Nolen was brought in to testify after Kefauver proposed his sweeping drug industry reform legislation in December of 1961.

Kefauver in particular wanted to know more about the new generic drug program that McKesson was developing. Nolen answered that Kefauver's own hearings into drug pricing had produced a new demand for generically named drugs, and McKesson was simply responding to a new form of demand by opening up a new line of business—a generic drug business. "As we see it," he replied, "the decision to manufacture and distribute McKesson brand pharmaceuticals—to meet our customers' demand for generic drugs—is entirely in keeping with that concept of our business function."[4]

Second, if the publicity of Kefauver's hearings helped to make a market for generic drugs, the enactment of the 1962 Kefauver-Harris Amendments to the Food, Drug, and Cosmetic Act had the inadvertent effect of providing new federal aegis to copycat versions of these products on the American drug market. In 1966, the iconoclastic new head of the Food and Drug Administration, James Goddard, pledged to enact the new drug efficacy requirements not only prospectively for new drugs developed after 1962 but also retrospectively for all drugs previously approved by the FDA. This massive undertaking, conducted under the auspices of the National Academy of Sciences (NAS) and National Research Council (NRC), would have ramifications far beyond what its sponsors had originally intended.

Beginning in 1966 the Drug Efficacy Study (DES), and after 1969 its implementation (or DESI), reviewed the effectiveness of all drugs approved by the FDA between 1938 and 1962. For those drugs found ineffective and then removed from the market, the DESI program was the first step in a bitterly contested unmaking of vast segments of the pharmaceutical market. For those drugs found to be effective, however, DESI added robust academic and federal claims of therapeutic value to which any subsequent manufacturer could refer when marketing their own generically named product.[5] This was especially relevant as several pre-1962 DESI drugs were already off patent, and many more were going off patent each year.

After DESI, the FDA found itself in an unusual bind. To require all new copies of off-patent drugs to submit an extensive New Drug Application (NDA) merely to market a drug that had been approved and used for at least seventeen years seemed both economically wasteful and potentially unethical.[6] Instead, the FDA proposed in 1969 that manufacturers of newly off-patent drugs need only file an *Abbreviated* New Drug Application (ANDA) that showed proof of chemical equivalence according to standard pharmacopoeial compendia, in other words, that the active chemical was present at the dose specified without any contaminants. This regulatory action simultaneously lowered barriers to market entry for potential generic manufacturers and opened up a clearly sanctioned set of DESI-effective drugs as a new market space for would-be generic manufacturers.

Finally, it is impossible to understand the intensity of debates over generic drugs in clinical, popular, and policy literatures after 1965 without discussing the newly expanded role of federal and state governments as payors of pharmaceutical benefits under Medicare and Medicaid. The visibility of the public market for cheaper drugs drew companies into the generic drug business from many other sectors of the economy. "Whatever the future trend of generic-drug sales may be," the *Sunday Herald-Tribune* noted in January of 1966, "many companies are jumping into the swim. Only last week Cott Corp., chiefly a dispenser of soft drinks, announced it was forming a unit to sell 'a full line' of generic drugs. In an obvious attempt to capitalize on Medicare, the company dubbed its new division Medicure."[7] By the end of the 1960s, an estimated $500 million of federal funds was spent annually on prescription drugs. Advocates of generic substitution claimed that diverting resources from brand-name to generic drugs would save taxpayers more than $100 million annually.[8]

At the intersection of these congruent forces, a private sector "pull" and a public sector "push" combined to support a market and policy ecology favorable for the growth of generic drugs. Both genericist and antigenericist narratives agree that the rise of generic drugs in the 1960s was historically overdetermined by these broad economic and political forces. Surprisingly enough, however, few analysts have examined the role of the generic manufacturers themselves. The growth of the generic drug industry has in itself been assumed to be a natural economic cycle of the American pharmaceutical industry: as patents on innovative products expire, imitative products come to market. Yet the emergence of what we now know as the generic drug industry was not itself automatic. This history has largely not been told, because the companies themselves—like the products they marketed—did little to market themselves to the general public. Yet piece by piece, using letters preserved in the archives of the FDA, policy briefings, trade reports, popular periodicals, and marketing ephemera, it is possible to reconstruct a skeleton of the industry as it developed.

FROM ETHICAL TO GENERIC: PREMO PHARMACEUTICALS

When the president of Premo Pharmaceutical Laboratories, Seymour Blackman, was called to testify in front of Kefauver's subcommittee in 1960, he described his firm as a traditional "ethical drug company" that sold drugs under nonproprietary names, just as larger firms like Eli Lilly, Merck, Pfizer, and Squibb had done in the first half of the twentieth century.[9] Premo was less well known than these larger Pharmaceutical Manufacturers Association member firms, perhaps, but like its larger counterparts this small firm had nonetheless earned the commendation of the AMA's Council on Drugs and passed the stringent inspections of Durward Hall's Military Medical Supply Agency.[10]

Like many better-known PMA firms, Premo was founded by a pharmacist who sought larger and yet larger markets for his products. The firm's origins lay in the back of Theodore A. Blackman's pharmacy in the corner of Lexington Avenue and Ninety-seventh Street, where Blackman produced and sold drugs by nonproprietary names. Blackman was either a good pharmacist or a good salesman or both. As demand spread for his compounded drugs, this revenue soon afforded him a Park Place storefront. Business grew and continued to grow. When Blackman's son, Seymour, came of age, he joined the family business, moving the location to a large manufacturing plant occupying five storefronts on Broadway in 1933 under the name Premo Pharmaceuticals. Within a few decades the firm would expand to larger facilities in Hackensack, New Jersey.[11]

At a moment in which other ethical drug firms were investing in research laboratories and beginning to capitalize on patent-protected, innovative drug products, Premo invested instead in its production, distribution, warehousing, and quality control procedures.[12] By 1943 the company could boast of its state-of-the-art production facilities:

> Here we have acres of gleaming floor space, filled with the most modern machinery, tools and equipment for the production, warehousing, and distribution of the quality line of pharmaceutical products that bear the Premo label. All production is under the strict control of our analytical laboratory. Not a gram of raw material may be delivered to the production department without the seal of laboratory approval: and not a single package may be delivered to finished stock unless it bears the laboratory control number certifying that it has been tested and conforms to our exacting standards. This air-tight system of laboratory control is further supplemented by analyses and bio-assays from independent laboratories of national repute . . . Here our products must pass tests which are far more severe than any Government standards, before they are released for sale.[13]

FDA inspectors praised Premo's investment in hygiene and quality control, and Blackman believed that this investment set Premo apart from other, cheaper firms.[14] In his testimony in 1960, Blackman emphatically denied that any two generically named drug products were the same. Instead he warned Kefauver's subcommittee of the risks of adulteration in the field of generically named pharmaceuticals, observing that, unlike Premo, another company selling drugs by generic name "could very well be operating in a barn and meeting Federal specifications." Blackman was careful to point out that, even though his products were generically named, they were also branded with the Premo company name.[15] Nor did Premo conceive of itself as a generic manufacturer in a sense that would have precluded participation in research and development. The firm held several patents and was seeking to market an innovative new form of aerosolized penicillin, though none of its attempts to market innovative products had yet proven successful.[16]

Premo did not present itself as a *new* form of company (generic drug manufacturer) as much as an *older* form of company (ethical drug manufacturer) that was becoming increasingly incongruous with the trademarked, patent-oriented character of the postwar US pharmaceutical industry. Yet all of this would change after the Kefauver hearings: in their wake Premo found itself awash in a sea of favorable publicity. Medical and pharmacy directors of several hospitals wrote Premo and the FDA to ask more about purchasing "pharmaceuticals by generic name at substantial savings"; Premo began sending copies of transcripts of the Kefauver hearings to potential clients as promotional material.[17] Premo also began to enter more aggressively into the production of newly off-patent drugs, filing an early application to produce a copy of Upjohn's blockbuster antidiabetic pill, tolbutamide (Orinase), and others to follow. By 1966, the company's standard marketing letter to hospitals and pharmacies referred explicitly to their production of "generic products."[18]

In other words, Premo began expanding into private and public markets for generic products exactly as the market viability for such products was developing visibility. As the firm made bigger and bigger facilities to produce more of what they now called generic drugs at lower price, FDA inspections chronicled a backsliding of quality control.[19] As Premo became—to quote their corporate motto from the late 1970s—"Pioneers in Generics," they also began to bypass other FDA regulations they had previously followed, selling generic versions of bestselling drugs like Orinase, Librax, and Diabenese without regulatory approval because the market incentives were stronger than the risk of regulatory punishment.[20]

COOKING IN THE BATHTUB WITH BOB AND LARRY: BOLAR PHARMACEUTICALS

Premo's transition from the last of the ethical drug houses to the first of the generic firms must be understood as only one of many possible origin stories for the generic drug industry. The history of Bolar Pharmaceuticals, another firm that began to call itself a generic drug manufacturer in the late 1960s, sketches out a somewhat less "ethical" trajectory by which a small drug firm became a generic manufacturer. Unlike Premo, Bolar could claim no Progressive Era lineage, having been founded in 1958 by two business associates in Brooklyn with no prior pharmaceutical training, Robert Shulman and Larry Raisfeld. Indeed the name Bolar itself reflects the casual, collusive basis of their partnership—being a literal condensation of *Bo*b and *Lar*ry.

Unlike Premo's principals, Shulman and Raisfeld made no attempt to contact the FDA for approval prior to marketing their wares and were brought only grudgingly under regulation in the wake of a series of imposed FDA inspections between 1959 and 1963 that revealed cramped and inadequate production facilities. Initial FDA inspections of the Bolar plant recall the disturbing images from *Parade* magazine's exposé of Pharmacal: the conditions were filthy, and the firm was "operating in a crowded building with complete lack of manufacturing and laboratory controls." A court injunction was obtained, first in 1963 and again in 1966, to stop Bob and Larry from selling tablets of the popular minor tranquilizer meprobamate. The second injunction, it seems, finally prompted a move out of the dilapidated warehouse to a new building in Copiague, New York, which the FDA diffidently agreed could be considered an adequate facility for tableting drugs.[21] Around that time, President Larry Raisfeld asked politely in a terribly typed letter addressed simply to "FDA":

Gentlemen:

I manufacture PETN (Pentaerythritol Tetranitrate) tablets in the following strengths 10mg., 20mg., 10mg. with Phenobarbital 1/4 gr., and 20 mg. with Phenobarbital 1/4 gr.

Do I need a NDA for any or all of the above tablets? If so will you
please send me appropriate form. I amke [*sic*] no lable [*sic*] claims for
any of the tablets but sell it on a "Catuion [*sic*] Federal law prohibits
dispensing without a prescription legend," to Physician supply houses.

A prompt reply is hoped or as the above tablets are one of our principle
[*sic*] products.[22]

Read at face value, Raisfeld appeared to be claiming ignorance of the entire process of drug approval required to sell prescription drugs through
legitimate channels—though the possibility remains that he may have
been playing dumb to prolong his ability to operate outside of regulatory
purview.

By 1969, an intensified FDA inspection described Bolar as a generic
drug manufacturer whose product line consisted nearly entirely of amphetamines and vitamin tablets and a smattering of minor tranquilizers.[23] Only two of their drug products—both versions of the vasodilator
pentaerythritol tetranitrate—had even begun the process of formal
FDA approval through the ANDA process. Nor did the new facilities in
Copiague live up to their initial promise. Three years in the new facility
yielded three more recalls on Bolar products for failing tests of potency,
content uniformity, and disintegration. For much of the late 1960s, the
FDA waged a slow campaign against Bolar products as adulterated and
misbranded goods.[24]

It was doubly ironic, then, that when the FDA and NAS-NRC released
the first list of ineffective drugs published by the DESI project, one of the
drugs found ineffective was pentaerythritol tetranitrate. Faced with the
prospect of losing permission to market the only drug it was selling with
legitimate FDA approval, Bolar's CEO, Robert Shulman, wrote a frus-
trated plea to the head of the Bureau of Drugs, Henry Simmons. Would
the FDA please let Schulman and Bolar know which drugs the DESI review had deemed "effective"—and therefore a viable market—so that
Bolar might focus its efforts at regulatory approval in more productive
channels? Bob's typing was better than Larry's, his prose more fluid: "We
are a small manufacturer of generic drugs. In view of the recent NAS-
NRC studies and the Food and Drug Administration actions based on
these studies, we are very interested in re-evaluating our product line.
Accordingly, we are asking for a little guidance along these lines. Would
it be possible to obtain a list of drugs that have been evaluated and into

which category (effective, ineffective, etc.) that they fall. If we knew this information, we would be able to delete the ineffective items and replace them with effective drugs."[25]

However shady Bolar's origins might have been, the evolution of their correspondence with the FDA over the 1960s documents the means by which such firms could be recruited into new regulatory regimes— especially when led by the carrot of the new markets potentiated by the promulgation of federal standards of drug efficacy. The standardization of efficacy performed by ANDA and the NAS-NRC review was a double-edged sword: the same action that cut away profitable but DESI-ineffective items such as pentaeyrthritol tetranitrate could also stabilize markets for generic production of DESI-effective drugs. Further correspondence between Bolar and the FDA documents the transformation of this firm from a gray-market producer of illegitimate copies to an increasingly legitimate (if still occasionally intransigent) regulated producer of legitimate copies.[26] By 1973, FDA inspectors could note with some satisfaction that "as a result of the NAS/NRC studies they have discontinued manufacturing all amphetamine products which formerly constituted the bulk of their business." Bolar had shifted to a product line made up predominantly of "pending or approved ANDAs filed as the result of DESI announcements."[27]

From this perspective, Bolar's story begins to look like Premo's in reverse: a thoroughly unethical drug company with little regard for FDA protocol gradually becomes roped into a regulatory regime with a combination of carrot and stick. Generic markets and generic protocols gave Bolar reason to become a bit more "ethical" in its marketing.

REVIVING THE AGING WONDER DRUGS: ZENITH PHARMACEUTICAL CO.

The "patent cliff" has become a common metaphor in health and business journalism; it signifies the sudden drop in revenue that can occur when an innovative drug loses its patent exclusivity and faces generic competition. *Fortune* and the *Wall Street Journal*, not to mention *FDC Reports*, *Scrip*, *Pharmaceutical Executive*, and *Medical Marketing and Media*, are now filled with lists of drug products facing patent expiry with speculation as to what this loss of monopoly will spell for manufacturers and their investors. One of the earliest examples of this genre took form in

1965–66 as trade journals anxiously discussed the impending expiry of the patent for Parke-Davis's first blockbuster antibiotic, Chloromycetin (chloramphenicol).[28] Only three weeks after patent expiry, Zenith Pharmaceuticals—a firm based in Englewood, NJ, with a plant in the tax haven of the US Virgin Islands—launched its own generic version at less than five dollars per hundred pills—an 85 percent discount compared to Parke-Davis's original price of thirty dollars. Parke-Davis refused to lower its own price, and the industry held its breath.

Zenith had been founded in the late 1950s as a "private label" firm, packaging pharmaceuticals for other companies on demand. The company went public in 1962 and by 1965 announced it was interested in "moving into generic pharmaceuticals"[29] Unlike Bolar, Zenith swiftly developed the capacity to navigate FDA approvals and other regulatory affairs and began to articulate a voice for a new legitimate kind of generic manufacturer in state and local policy spheres. In a letter to the Oregon State Pharmaceutical Association in 1965, Zenith president Benjamin Wiener encouraged the State Board of Pharmacy to create a Seal of Approval program for quality generic drugs that would include Zenith products. "We are in a position to offer you a complete line of ethical as well as over-the-counter generics which are so timely today."[30] Though the Oregon proposal did not pass, Wiener continued to encourage other states to help shape markets for generic drugs. In an open letter to the state of Indiana, he urged legislators and pharmacists to "recognize their responsibility in assisting pharmacists to meet their obligation to dispense high quality drugs."[31]

An early lobbyist for the generic cause, Zenith worked to expand production capacity to deliver a full line of generic products. By 1967, the company offered a broad range of generic tablets and capsules in more than 250 formulations, ranging from amphetamines to sleeping pills, antibiotics to placebo pills. Zenith had filed a series of its own ANDAs and seemed reasonably successful at acquiring FDA approval for its generic products. Company reports proudly announced that generic drugs would account for 80 percent of Zenith's total volume in 1967 (or $2.5 million), compared with a mere 14 percent ($100,000) in 1963.[32]

The generic market that enabled the rise of Zenith, however, was no land of milk and honey. If the shifting terrain of FDA rule making presented entrepreneurial opportunities for an early generic drug firm like Zenith, it also brought hazards. In spite of a wave of FDA approvals for its new generic products, the second half of 1966 produced meager earnings for the company. As the *FDC Reports* noted, "problems related to the conformance with the new and more stringently enforced food and drug regs" lengthened time for NDA clearance and required the hiring of more laboratory staff, while at the same time dropping certain profitable drug products increasingly viewed as ineffective. By the end of the year, its profits had fallen to only $72,000—down from a high of $107,952 in 1965.[33]

Perhaps more than any other firm recorded in the FDA correspondence files I was able to obtain through Freedom of Information Act requests, the history of Zenith captures the sense of entrepreneurial risk and fragile development that characterized the early generic enterprise. Zenith's executives understood generic drugs as an emergent market space situated at the intersection of changes in pharmaceutical intellectual property, drug regulation, and the role of the state as pharmaceutical consumer. Zenith's early 1967 successes with chloramphenicol were followed in short order by the approval of an ANDA for its generic version of meprobamate—a popular minor tranquilizer sold under the brand names Miltown and Equanil—later that year and a series of antibiotics in its own "generic pipeline."

In 1965, Zenith had been one of the first generic firms to successfully challenge a patent on a drug (tetracycline) that it planned to copy *before* the patent expired. Zenith's executives saw a similar opportunity to challenge the patent on ampicillin, a broadly prescribed semisynthetic penicillin initially developed and marketed by the British firm Beecham and subsequently licensed to Bristol-Myers and other firms for sale in the United States. By 1968, Beecham and Bristol-Myers controlled the $70 million ampicillin market through a series of licenses and cross-licenses. In a confusing array of licensing deals, ampicillin was simultaneously marketed by Wyeth as Omnipen, by Squibb as Principen, by Parke, Davis as Amcill, and Polycillin by Bristol-Myers, while Beecham's own brand Penbriten was sold in the United States by Ayerst. After repeatedly petitioning Beecham and repeatedly being denied its own license to market a generic form of ampicillin, Zenith argued that Beecham's broad licensing of the drug had effectively invalidated the patent. The US Department of Justice supported Zenith's case, and the generic manufacturer prevailed again. Zenith priced its product at $14.50 per hundred, compared with the brand-name price of nearly $22 per hundred. Sales were brisk.[34]

Ampicillin can be seen as a prototypical generic drug of the late 1960s: it was a relatively easy chemical to purchase or produce, it was used in oral dosage forms and could easily be sold as a tablet or capsule, it was widely used in outpatient clinical settings as well as inpatient, and it was broadly indicated across a variety of conditions and patient populations. Any company that could prove their pill contained ampicillin of sufficient quality in the correct dosage could fill out an ANDA application and receive FDA approval to market their drug as a generic equivalent. By 1974 prescriptions for ampicillin comprised one out of every five generically written prescriptions.[35]

In 1975, the Department of Health, Education, and Welfare chose ampicillin as the index drug for its first attempt to establish a maximum allowable cost (MAC) for its payment for pharmaceuticals that "do not vary significantly in quality from one supplier to the other" (more on this in chapter 8). As late as 1976, ampicillin was still ranked number one in the top fifteen generic drugs (by prescription), followed by other antibiotic, cardiovascular, tranquilizer, antihypertensive, and endocrinological drugs that shared similar characteristics. By 1980, analysts predicted 70 percent of the top two hundred drugs would be available in generic form. Yet the popularity of easily manufactured big-market drugs like ampicillin among generic manufacturers can also be read as support of an inverse principle: not all drugs were attractive for generic production. For every drug like ampicillin that presented a large target for many generic manufacturers, there were a number of other, harder to make drugs that attracted the attention of only one or two manufacturers, if any. Early on in the history of the generic industry, a varied landscape was created in which some drugs would be available in excess—and some would be vulnerable to shortages.[36]

Zenith's aggressive stance toward marketing ampicillin also sheds light on new and different strategies opening up for generic drug manufacturers. Why wait for a patent to expire (as with chloramphenicol) when one could challenge that patent and market a generic drug even earlier? The growth of the generic ampicillin market also served as a wake-up call for larger, more established pharmaceutical companies to acknowledge a market space now increasingly reified as the "generics sector." By 1976 the US market in generic drugs was estimated at $2.3 billion—or 25 percent of the total US prescription drug market. By the end of the decade forecasters anticipated this percentage would rise to half of all prescriptions written, with a more than threefold increase in total generic sales in the course of a decade. Simply denying the existence of generic drugs was no longer a valid strategy for the brand-name industry.[37]

As a consumer good that claimed to be indistinguishable from its competitors, the generic drug presented analysts of the industry with a paradox: Even if such objects did exist, how could one market them? Marketing, if it was a science at all, was understood to be a science of differentiation, and generic products challenged this formulation. Yet, as the next chapter will reveal, there were many different ways of making things the same—and many different ways to market similarity.

CHAPTER 5

GENERIC SPECIFICITY

Don't take chances. Know who's making the generics you're dispensing. You can trust and buy Purepac generics for these important reasons.
—PUREPAC ADVERTISEMENT, *AMERICAN DRUGGIST*, 1978

As generic firms like Zenith, Bolar, and Premo sought to capitalize on the growing demand for drugs that were the same (but cheaper), they faced an unusual marketing dilemma: how to differentiate their products from the products of other companies claiming to market the same drugs? As an industry analyst later puzzled over this question in 1979, "The pharmaceutical industry, as a general rule, is very competitive. The generic pharmaceutical industry is even more competitive. This is a result of the lack of opportunities for creating product differentiation [and] large number of similar products due to lack of patent protection."[1] The marketing of generic drugs presented a paradox: while analysts struggled to understand how generic drugs could compete in any arena except price, somehow different generic firms were able to carve out distinct niches for themselves and their products.

Looking back through the precious few advertisements for generic drugs in medical journals of the 1960s and 1970s, we also might quickly conclude that generic manufacturers were uninterested in marketing. But we would be looking in the wrong place. Paging through pharmacy journals of the era, from *American Druggist* to *Pharmacy Times*, one finds a riot of advertisements for specific generic drugs.[2]

The marketing of generic drugs invoked many approaches to the challenge of selling the same thing, differently. Some of these advertisements stressed differences in quality between drugs that were merely inexpensive and drugs that were dangerously cheap. Others attempted to convince the pharmacist of the value of stocking their particular brand of generic products on the basis of reputation, service, and breadth of product line. Still others promised legal assistance to pharmacists newly liable for the clinical outcomes of their selection of a generic drug. Not all generic manufacturers were anonymous, it turned out, and many firms found ways to reinforce the specificity of their own generic product.[3]

CHEAP, BUT NOT RISKY CHEAP: PUREPAC

The early generic manufacturer Purepac became one of the most visible firms to promote its wares on this razor's edge between price and quality, unbranding and rebranding. Like Premo, Purepac had eked out an existence since the early twentieth century as a rather undifferentiated "ethical drug firm" (as defined in chapter 1) and had rebranded itself in the late 1960s as a generic manufacturer, adopting as its slogan "The Leading National Brand of Generics."[4]

Purepac advertisements would become omnipresent in pharmacy journals of the 1970s, often taking up full pages with large-font messages, such as "low priced . . . but not risky cheap" (figure 4). The slogan—like all of Purepac's marketing—differentiated their products along two margins. On the one hand, Purepac's quality was "every bit as good as SKF's, Lederle's, Pfizer's, or any other brand name manufacturer," whose drugs were by extension overpriced. On the other hand, Purepac's drugs were not *too* cheap. "If you're buying generics that are cheaper than Purepac," the copy continued, "chances are you're buying from a mail order house" or other purveyors of "risky cheap generics." What you don't know, these ads suggested, *can* hurt you. Purepac, by contrast, occupied a sweet spot of price and quality: reassuringly expensive, but still cost effective.

Other Purepac advertisements from the later 1970s added the promotion of convenience to the guarantee of quality. The breadth of their product line assured that Purepac carried *all* drugs relevant for generic substitution in the dynamic patchwork of state generic substitution laws (more on this in chapter 8). This promotional tactic allowed them to take advantage of the heterogeneity of regional and state-by-state approaches to generic prescribing and promoted to pharmacists a reliable, one-stop shopping source for all generic drugs, boasting that "Purepac has the generic line for every state substitution law." A pharmacy that stocked Purepac, this logic suggested, would be stocking an entire line of generics through one steeply discounted relationship.[5]

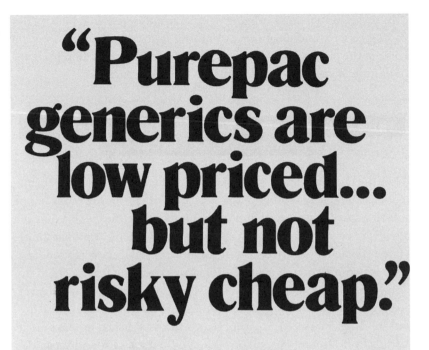

"Purepac generics are low priced... but not risky cheap."

Listen, if you're buying generics that are cheaper than Purepac, chances are you're buying from a mail order house and mail order houses are not manufacturers. Most likely you have never even heard of the manufacturers they buy from. And what you don't know, could hurt you. Look, mail order houses buy for price and price alone, from one manufacturer to the next. So, if there's a product liability suit and you can't pinpoint the manufacturer, you, Mr. Pharmacist, may be the only one liable for damages because you made the product selection and bought risky cheap generics.

Don't take chances. Know who's making the generics you're dispensing. You can tru and buy Purepac generics for these impo tant reasons.

Purepac's bioequivalence is every bit good as SKF's, Lederle's, Pfizer's or ar other brand name manufacturer who ha just recently started selling generics.

Pharmacists of America have dispense over one billion Purepac drug products i the last 46 years . . . and those Pharmacis have made us and are keeping us Numbe One! No other generic manufacturer ca match that record.

"Purepac generics are low priced . . . but no risky cheap."

PUREPAC
Elizabeth, NJ 07207

THE LEADING NATIONAL BRAND OF GENERIC
Available from your wholesaler.

FIGURE 4. Advertisement for Purepac, which called itself "The Leading National Brand of Generics."

American Druggist, *January 1976.*

In these and other advertising campaigns, Purepac stood out as the leading generic manufacturer promoting its brand of generics to the needs of the pharmacist. Purepac soon faced stiff competition from another form of generic competition: the "branded generic."

THE BRANDING OF GENERICS: SK-LINE, PFIPHARMECS, LEDERLE STANDARD PRODUCTS

While the Pharmaceutical Manufacturers Association, National Pharmaceutical Council, and other public relations organs of the brand-name industry spent a great deal of promotional and political effort denying the existence of generic drugs, a smaller faction of large firms like Smith, Kline & French, Pfizer, and Lederle Laboratories began to explore strategies for co-optation. If generic drugs—and generic firms—could not simply be waved away through denial, these firms could instead be vertically integrated into the general business of a large pharmaceutical firm and the drugs marketed as part of that firm's established product range. Already by 1974, only $180 million of the $2.41 billion generic drug market was sold by "commodity generic" firms such as Zenith, Purepac, Premo, and Bolar. The rest was captured by "branded generic" operations of PMA firms such as Eli Lilly, Wyeth, A. H. Robins, Squibb, Ayerst, Burroughs-Wellcome, Parke-Davis, Smith, Kline & French, Warner Chilcott, and Abbott Laboratories.[6]

One of the first prominent lines of branded generic drugs was the SK-Line brand of generic drugs sold by Smith, Kline & French (SKF). The first product in this line, SK-Line Ampicillin, was promoted at a price 25 percent less than the leading brand-name version of ampicillin yet at a price higher than the ampicillin sold by commodity generic manufacturers. If Purepac's ampicillin was not "risky cheap," SK-Line's ampicillin was even more reassuringly expensive.

Smith, Kline & French largely sold generic versions of products it had not originally held patents on itself. But unlike Purepac, SKF could use its well-trained team of sales representatives to expand its marketing from pharmacists to doctors. Starting in 1971, SKF dedicated a team of five hundred sales representatives to full-time duty promoting SKF's expanding line of generic drugs to prescribing physicians. The strategy of the pitch was explicit: physicians understood the importance of generic prescribing but needed to be able to obtain those cost reductions without risking reductions in quality. And what better vehicle than generic drugs vouchsafed by a well-regarded pharmaceutical brand?[7] By 1972 SK-Line accounted for $4 million in sales; by 1979 the SK-Line included thirty products and accounted for $15 million in sales; by 1982 SK-Line accounted for 15 percent of the company's total drug sales—though many of these branded generics were not produced in SKF factories.[8]

As other brand-name firms entered the branded generics market, including Lederle, Parke-Davis, and Pfizer, they coupled their names and reputations to new generic lines. Pfizer's branded generic division, Pfipharmecs inserted "Pfizer" into the name of each of its branded generic drug products, like Pfizercillin. Unlike SKF, Pfizer manufactured all of the generic drugs they sold—and so could differentiate its own branded generics from other branded generics on the basis of quality control of in-house production. Ad campaigns for other branded generics in turn highlighted the liability of the pharmacist—who was responsible for choosing *which* generic drug product was dispensed—in the event that a given drug did not work well or resulted in an adverse outcome. Lederle was among the companies that guaranteed to cover all legal costs and liability that might result from a negative outcome using their branded generic drugs in place of brand-name drugs.[9]

THE PERILS OF THE PRIVATE LABEL: MYLAN PHARMACEUTICALS

By the early 1980s the lion's share of the generic drug market had been captured by branded generics firms like Lederle, SK-Line, or Pfipharmecs, selling their products at prices as much as 50 percent more than commodity generics marketed by firms like Premo, Purepac, Zenith, or Bolar. This was all the more unconscionable—as Representative Al Gore (D-TN) asserted in a set of congressional hearings in the late 1970s and early 1980s—given that many of these products were produced in exactly the same factory.

Brand-name firms commonly purchased pills from generic manufacturers, put their label on them, and sold them as branded generics or even brand-name products. Wyeth, for example, outsourced all production of Amoxil, its brand-name version of amoxicillin, to Mylan Pharmaceuticals of Morgantown, West Virginia.[10] To be fair, this was not an entirely new practice. Kefauver's initial investigations into corticosteroid marketing had revealed a decade earlier that almost no drug companies actually made all of the drugs they sold.[11] A seemingly scandalized Gore asked Food and Drug Administration staffers: how was it that the mere presence of a Wyeth employee on the floor of the Mylan plant sufficed to allow Wyeth to sell drugs produced at that time at a markup as its own brand, while subsequent runs off the same manufacturing facility would be sold for a cheaper price as generic amoxicillin? How could the brand-name industry claim their products were superior to commodity generics if they were indeed produced by the same manufacturers?[12]

Wyeth was not the only brand-name firm selling drugs made by Mylan. In 1977, of the 144 drugs sold as Lederle Standard Products, only 6 were manufactured by Lederle itself. The rest were purchased from "private label" manufacturers like Mylan, which also supplied material for brand-name and branded generic lines to firms including Abbott Laboratories, Bristol, Mallinckrodt Laboratories, Parke-Davis, SKF, and Squibb. Although Lederle had to report to the FDA where each drug was manufactured, this information was considered a trade secret—and did not need to be related to pharmacist, physician, or consumer.[13]

Mylan was only one of several private label manufacturers whose wares could turn up on one shelf as a commodity generic, on another as a branded generic, and still another as a brand-name drug itself. Like Bolar, Mylan was founded in the 1960s in response to the growing market for low-price generic drugs. Like Bolar, the company was named after its founder, Milan Puskar. When the FDA first learned of the firm in 1962, Mylan was concerned chiefly with repackaging a dozen or so drug products—amphetamines, steroids, vitamins, and penicillins—and selling them to other firms to market under their own names. Over the course of the decade, the plant expanded its manufacturing capacities and built more plants in West Virginia, New Jersey, and the US Virgin Islands.[14]

Mylan's approach to marketing was literally self-effacing—erasing its own name from its products so that other brands, such as Wyeth, could stamp theirs in its stead. Concerns over this practice reached FDA officials in 1973, when it became apparent that Mylan was manufacturing the bulk of SKF's branded generics. Because advertisements for SK-Line generics claimed superior quality over commodity generics, the FDA argued that these claims were a form of misbranding. The "man-in-the-plant" model that Gore later uncovered had itself originated as a regulatory solution to this problem. Repackaged drugs from private-label firms could be sold as branded generics only if SKF maintained an inspector in Mylan plants at all times SK-Line products were running on the assembly line.[15]

After the Gore revelation, however, staking the future of the firm on the "man in the plant" model had become too risky. Mylan's increasing capital and capacity for mimicking brand-name drugs offered another possibility: for Mylan to produce its own versions of "hard to copy" generics like advanced diuretics and cardiovascular drugs. Years of learning how to make other companies' drugs for them on their own assembly lines gave Mylan a competitive edge in terms of pharmaceutical know-how necessary to make these hard to copy generics. "In the next few years," Milan Puskar claimed in 1982, "we anticipate adding to the roster of generic drugs some of the most widely prescribed pharmaceuticals still under patent."[16]

Like other generic firms that cut their teeth copying DESI-effective drugs the 1960s and 1970s, Mylan was strategizing a way to ensure a continuing flow of new generic products in the early 1980s. This would prove trickier than they initially thought.

WHEN DO NEW DRUGS BECOME OLD?

By the end of the 1970s, firms like Zenith, Premo, Purepac, Mylan, and Bolar, eager for growth, eyed hungrily the pool of drugs approved *after* 1962 whose patents were now beginning to expire. But there was a catch: the mechanism that had allowed the generic industry to market drugs based on proof of similarity alone—the Abbreviated New Drug Application—only applied to drugs approved before 1962 and declared effective by the Drug Efficacy Study implementation process. The copying of newer (but rapidly aging) drugs approved after 1962 remained a regulatory gray area. In 1975, Premo sent the FDA a wish list of forty drugs that were about to go off patent—from Atarax to Valium—asking how and when these formerly "new drugs" became old enough to market generically. Once a patent expired, did a generic manufacturer need to prove clinical efficacy (NDA) or only prove similarity (ANDA)? Or could they simply be "classified as old drugs, no ANDA required"?[17]

Zenith's counsel asked pointedly why the FDA required a full New Drug Application—including placebo-controlled clinical trials—to re-prove the efficacy of a drug that had already been proven effective nearly two decades earlier. Requiring proof of efficacy, when proof of similarity should suffice, was wasteful of time, money, and potentially in violation of human subject research ethics. As Zenith suggested in a series of increasingly forceful conversations with Health, Education, and Welfare lawyers, the FDA needed a protocol to reclassify aging "new drugs" into "old drugs," so that generic copies could be approved and marketed without the fuss of clinical trials. As a compromise, the FDA suggested that generic manufacturers could file a "paper NDA" including published trial materials for the original drug as part of their application in lieu of clinical trials on a generic drug. But this approach was confusing, generated large amounts of paperwork, and had no clear criteria for proof. Only a handful of generic drugs would be approved by paper NDA during the 1970s and early 1980s.[18]

All parties realized they had a problem on their hands. For Zenith and other generic manufacturers, unless these newer drugs could be made old, or at least more like the pre-1962 DESI drugs, the generic pipeline threatened to slow down to a viscous drip or stop altogether. For the FDA, this issue raised questions that would require a new legal and epistemological approach to the life cycle of a "new drug," questions they did not know how to answer in a politic fashion, even if they were sympathetic to their general principles. FDA counsel agreed to consider, however, that Zenith might "petition for old drug status"—that is, officially reclassify Valium from a new drug to an old drug. But the arsenal of the FDA's regulatory authority was bound to the concept of the "new drug." Truly transforming a new drug into an old drug would effectively take it out of the FDA's control—which was not an acceptable solution, either.[19]

With no answer forthcoming after three years of entreaties, Zenith VP J. Kevin Rooney began to use more pointed terms in 1978, asking why Valium was still considered new—and remained without generic competition—when so much wasteful spending was at stake:

> No doubt the Administration is well aware of these drugs, but we would point out that VALIUM® is the number one drug prescribed in the United States . . . Taking into consideration the length of time these drugs have been on the market, the billions of doses prescribed and utilized in treatment, and the numerous scientific articles published both in the United States and throughout the world, we seriously question whether these name brand drugs, at this point in time, as manufactured by the companies named supra should still be considered as new drugs.[20]

The FDA's hands were tied for most of the late 1970s by a seemingly imminent legislative reform of the FDA that never took place, even though all parties—the FDA, American Medical Association, PMA, and both Republican and Democratic leadership in the House and Senate assumed some form of bill passage would be inevitable. Several riders on the ill-fated Drug Reform Act of 1978—including, on the one hand, patent extension (inimical to the generic industry) and, on the other hand, a bill to establish ANDAs for post-1962 drugs (crucial for the generic industry)—were in or out of political favor seemingly on a weekly basis. Living for nearly four years with the uncertainty of looming reform greatly complicated the ability of the FDA to resolve the post-1962 generic drug problem through its own regulatory rule making. Instead, an inefficient system of paper NDAs continued, which had no clear set of guidelines for approval or dismissal and hopelessly delayed the approval process for generic drugs.[21]

Soon after its birth, the generic drug industry was facing a serious prospect of crib death. In protest, Zenith and other firms began to bring their generic products to market *without* FDA approval, flaunting these violative acts as a political statement. "The charges are true," Zenith's president Kenneth Larsen replied when accused of marketing an unapproved generic product, "and we believe that this is the occasion to set before you the horrendous and perhaps insurmountable problems that the entire ANDA mechanism, together with the (almost) 20-year history of the FDA's implementation of the 1962 Drug Amendments have caused."[22] In response, the head of the FDA's Bureau of Drugs, Richard Crout, asked for a meeting in May of 1980 to "develop a rapport between Zenith Laboratories and the FDA," along with a "few other responsible members of the generic drug industry."[23]

Larsen perceived an opportunity to present Zenith as the face of a legitimate generic industry. He recruited the heads of a few other firms to join him in Crout's office—such as Premo, Purepac, and Mylan. If the meeting did not lead immediately to a regulatory pathway for generic approval, it did lead to the founding of the Generic Pharmaceutical Industry of America (GPIA). With William Haddad, an enterprising journalist turned generic policy analyst, as its executive director, and a shrewd young patent lawyer, Alfred Engelberg, as its counsel, the GPIA positioned the generic drug industry as a key player in the American public health system. Unlike the more inclusive and polyvocal National Association of Pharmaceutical Manufacturers (NAPM)—filled with manufacturers of more secretive and proprietary medications as well—the membership of the GPIA was highly selective.

As the newly appointed head of the newly created generic industry lobby, Larsen wrote promptly to Representative Henry Waxman (D-CA) early in 1981 to offer the GPIA's services supporting new efforts to restart the stalled FDA reform legislation. Larsen testified in front of a congressional subcommittee to hammer home the need for a viable pathway to effectively manage new drugs once they had grown older. "Pre and Post 1962 drug products are spoken of as if this year—1962—had some special quality," Larsen complained, even though all present knew that "this year is no more than a date when the regulations were changed," rather than a date that described any material difference in the safety or efficacy in the drugs before or after.[24]

At the same time that Larsen and the GPIA were working to find faster strategies to make new drugs old, name-brand pharmaceutical firms were developing counterstrategies to keep old drugs "young." PMA firms argued in front of a series of congressional committees in the late 1970s that a "drug lag" caused by the longer FDA approval process effectively ate up years of meaningful patent life. The remedy, the PMA argued, was to extend patents by an extra three years—from seventeen to twenty—to make up for lost time imposed by regulatory delay. The GPIA's first series of policy actions, then, required simultaneously lobbying for a broader ANDA process (making new drugs old) while trying to forestall patent-extension legislation (keeping old drugs young).

In retrospect, it was the ability of the GPIA to help stall patent-extension legislation (claiming it would kill their young and beneficent industry) in 1983 that started the deal making that would ultimately produce a viable pathway for all new drugs to become copiable once their patent terms expired. In the immediate aftermath of the PMA's failed 1983 bid for patent extension, the GPIA executive director, William Haddad, received a summons to visit the office of Senator Orrin Hatch. The PMA, Hatch indicated, was interested in making a deal: if the generic industry *supported* patent extension for brand-name firms, the brand-name firms would support the extension of the ANDA approval pathway for all drugs.

The political alliances that produced this deal illustrate the heterogeneity that characterized both the generic and brand-name drug industries. For example, the GPIA did not include *all* generic houses, just those that had managed to secure a favorable reputation with the FDA. Likewise the PMA had to deal with a faction of its own members who opposed any deal that would recognize the legitimacy of generic drug products or the generic pharmaceutical industry. These "dissident companies" numbered eight major brand-name firms, including Schering-Plough, Squibb, Bristol-Myers, and Merck Sharp & Dohme, who protested the agreement. Of note, these firms represented the side of the PMA that had done the *least* to diversify into branded generics, unlike SKF and Pfizer, both of which were solid champions of the bill.[25]

Though it represented a remarkable political and economic compromise for brand-name and generic manufacturers alike, the proposed legislation did not represent the interests of all makers of all medicines. Rather, it emphasized the legitimate face of the generic industry and the side of the branded industry most savvy at engaging with the new realities of the generic market. On September 24, 1984, President Ronald Reagan signed the Drug Price Competition and Patent Term Restoration Act, flanked by the conservative Republican senator from Utah, Orrin Hatch, and a liberal Democratic representative from California, Henry Waxman. The Hatch-Waxman Act has since been held as one of the defining pieces of bipartisan policy making in living memory. The apparent unity of this bipartisanship, as we have seen, conceals a more complex intersection of branded generic and commodity generic investment in the future of the generic pharmaceutical market that had *already* existed for at least a decade.

The legislation effected a compromise not only between Republicans and Democrats but between the brand-name drug industry and the generic drug industry. Both sides came away with tangible results. Makers of brand-name drugs were granted longer periods of patent monopoly. The generic industry would now be protected by a streamlined ANDA application process and a 180-day market exclusivity awarded to the first firm to bring a generic drug to market. They also won a formal means to challenge drug patents (called a paragraph IV certification) and greater recognition of their exchangeability with brand-name drugs in the eyes of regulators, physicians, and the general public. When the bill passed in both houses, William Haddad—who had been an aide to Kefauver in the early 1960s—congratulated Congress on finally "completing what Estes Kefauver started thirty years ago."[26]

BIRTH OF THE GENERIC

The Hatch-Waxman Act of 1984 is often referred to as the birth of the generic drug industry. In the year following its passage, the volume of investment in generic firms skyrocketed. Companies like Bolar launched well-publicized initial public offerings, and generics were touted in some accounts as new "glamour stocks" on Wall Street.[27] As innovator firms gradually dropped their overt campaigns against the existence of generic drugs, the proportion of prescriptions filled generically grew from 18.6 percent in 1984 to 63 percent by 2007.[28]

Yet, as the chapters in this section have demonstrated, a single piece of legislation signed into law in 1984 did not create the modern generic drug industry. The architecture of this new economic sector was shaped by an entrepreneurial group of companies that emerged—and began to develop its own political organization—in the decades of the 1960s and 1970s. By the time the Hatch-Waxman Act was passed in 1984, the existence of such an industry was no longer really in question, as it had been in the beginning of the 1960s. True, Mylan's stock value doubled after its passage. But it had doubled already the year before, and it had doubled also the year before that. An industry that *did not exist at all* just two decades earlier had already become a robust sector of the marketplace well before passage of the 1984 act.

Reconstructing the otherwise hidden history of the generic drug industry helps us to shift our understanding of the role of the Hatch-Waxman legislation from the birth moment of the generic industry to merely an inflection point in its growth curve.[29] Debates over the existence of generic drugs were not legislated away as much as they were made irrelevant through changing market practices. As a financial analyst noted five years before Hatch-Waxman, the newfound solidity of the generic drug and the generic drug industry was already evident in 1979: "Just a few years ago, it was not uncommon to hear the generic drug industry being characterized as a 'passing-fad,' a 'me-too competition' with no place in the sophisticated drug business, etc. Things have definitely changed. With the introduction of generic product lines by five major trans-national pharmaceutical corporations and the tenacity, perseverance, and creativity in market development displayed by the smaller generic houses, the generic market segment has evolved into a major force to be reckoned with." As a kind of good, generic drugs were theoretically *unbranded*; in practice, a given generic product could still be effectively *branded* with recourse to a variety of marketing tactics.[30]

For reformers interested in rationalizing bloated and expanding American health-care expenditures, generic drugs were, to paraphrase the anthropologist Claude Lévi-Strauss, "good to think with."[31] For the same reason, generic drugs were immediately perceived to be threatening to the combined interests of the PMA and the AMA. As I have argued in this section, a polarized discourse on the existence or nonexistence of generic drugs in the 1960s and 1970s served to mask the emergence of a new economic form of life whose vitality was not immediately perceived by either side. Lamentably, most accounts of the role of generic drugs in US health policy have since replicated these polarized arguments: either generic drugs were always equivalent and the debate over their existence was economic interest posing as science, or generic drugs were never equivalent and the very premise of their existence was merely political interest posing as science. Taking the history and specificity of the generic drug industry seriously leads us to a very different conclusion: that a market for generic drugs emerged not with the resolution of these debates, but *in spite of their lack of resolution.*

For now it will suffice to note that the solution to the problem of the existence of generic drugs was not hatched in a laboratory but in the daily assumptions and practices of physicians and patients. Generic drugs became everyday objects through the formation of a generic market in which many actors—physicians, pharmacists, regulators, and consumers—found them useful objects and gave them an economic reality. And yet the increasing volume of trade in generics year after year was not sufficient to allay all fears that these products were equivalent and exchangeable in all ways with their brand-name counterparts. The market may have created generic drugs and the generic industry, but the question "what makes a medicine good enough?" could not be answered by the invisible hand alone.

Attempts to answer *this* question would require entirely new scientific approaches to the problem of equivalence. The strange, neglected history of these sciences is the subject of the next section.

PART III

THE SCIENCES OF SIMILARITY

CHAPTER 6

CONTESTS OF EQUIVALENCE

The desire to get the same therapeutic effect for less cost is a very reasonable one, but where can one find the data that would enable one to make this judgment?

—MAX SADOVE, "WHAT IS A GENERIC EQUIVALENT?" 1967

Faced with a broad shelf of nearly identical allergy remedies, how do you know which one to choose? Perhaps you choose the least expensive, thinking, if they are all the same, why pay more? Yet in the back of your head there is a lingering concern that buying a cut-rate drug will expose you to untold risks. Or perhaps you choose the most expensive, thinking, why take chances with my health? But this purchase, too, comes with the nagging concern that you are simply being fleeced. The generic drug provokes a paradox of similarity: we believe, and yet we do not believe, that similar things are the same. We believe, and yet we do not believe, that similar things are different. And we are not sure which forms of proof should convince us of their exchangeability.

When generic drugs were broadly introduced to the American public in Senator Gaylord Nelson's hearings on the competitive problems of the drug industry in the late 1960s, their interchangeability with brand-name drugs was assumed on the grounds of chemical identity. If a tablet of Parke-Davis's antibiotic Chloromycetin were ground up with mortar and pestle, subjected to mass spectroscopy alongside a similar tablet of generic chloramphenicol, and both were found to contain 250 mg of the active compound, should these two tablets not be fully exchangeable as therapeutics?

But chemistry is not the only relevant science of similarity. Two pills with the same amount of the same active therapeutic ingredient can cause different effects on the human body if they dissolve at different times in the stomach, if their active principles appear at different rates in the bloodstream, or if their binders, fillers, dyes, and shellac coatings influence their action in the human body in different ways. A drug is not reducible to a molecule. Even the simplest tablet needs to be understood as a complex technology for delivering a molecular agent from the outside world to a series of inner bodily sites necessary for pharmacologic action.

In the second half of the twentieth century, chemical claims of drug equivalence were complicated by a host of other scientific demonstrations of difference, from disciplines as disparate as physiology, epidemiology, economics, and marketing sciences. Entirely new fields of investigation emerged to document the differences between brand-name and generic drugs, and these fields took entirely different objects of study to demonstrate similarity: the physiology of absorption, the molecular biology of cell-surface drug receptors, the managerial sciences of quality assurance. These new sciences of difference could also be inverted to form sciences of similarity: the set of rules, laws, assays, and metrics used to prove when two objects were similar enough to be exchangeable.

In an insightful exploration of the role regulators play in the development of new scientific fields, Dominique Tobbell and Daniel Carpenter have narrated the pathway by which older physical and chemical proofs of similarity were displaced by more complex protocols for demonstrating biological equivalence, or "bioequivalence." They point to the Hatch-Waxman Act of 1984, which included bioavailability testing in healthy volunteers as a part of the Food and Drug Administration's Abbreviated New Drug Application pathway for the approval of all generic drugs, as proof of the arrival of biological over chemical modes of proving similarity.[1] But the problem of generic equivalence was not simply resolved by political bipartisanship nor by replacing a chemical proof of similarity with a biological proof of similarity. A few months after the passage of the Hatch-Waxman Act, Medicine in the Public Interest, a group headed by clinical pharmacologist Louis Lasagna, filed a Freedom of Information Act request demanding "greater transparency" into how the FDA actually determined the bioequivalence of the first three generic versions of Roche's Valium.[2] On the first anniversary of the passage of the bill that bore his name, Senator Orrin Hatch issued a denial that the science of bioequivalence had solved the problems of brand/generic difference: "It was never my understanding that the acceptance of the terms of last year's bill would somehow preclude this discussion. Some of the research-based companies accepted the generic drug title out of necessity because they thought it was worth it in order to obtain patent restoration; but not, as did Representative Waxman or the generics, because they believe that bioequivalence by FDA standards always means therapeutic equivalence."[3] According to Hatch, the "underlying premise" of the bill was not the ratification of bioequivalence as a gold standard but rather "the principle that generic drugs must be the same as the innovator *in all significant respects*."[4]

Generic drugs have never been identical in all respects to their original counterparts; rather, as Hatch suggests, their exchangeability depends on being similar in ways that are agreed to matter. Beyond bioequivalence, many forms of similarity and difference were still open for contest. Competing proofs of therapeutic difference could involve laboratory reagents, high sensitivity measuring scales, gas chromatography, the digestion of a tablet with simulated gastric juices in a glass beaker, spot checks on tablet shapes on assembly lines, bloodstream measurements of therapeutic biomarkers, or the epidemiology of patient experience with brand-generic switching. This section charts the emergence of these multiple sciences of therapeutic similarity as practices that constitute very different ways of proving that biomedical objects are, or are not, the same.

MAKING THINGS THE SAME

For centuries, pharmacists, physicians, manufacturers, and regulators have looked to pharmacopoeiae to adjudicate claims of therapeutic similarity and difference. In earlier chapters we explored the role of these compendia in ordering and (generically) naming the world of therapeutics. But the pharmacopoeia was always much more than just a book of official names. It was a technology of standardization. The listings under each entry linked words with things, describing protocols of identity, purity, and accuracy so that the reader could determine that the object one held in one's hands was indeed the therapeutic compound called for by the prescription.[5]

To take an early twentieth-century example, the 1900 (8th) edition of the *United States Pharmacopoeia* listed the following chemical proofs of identity as part of its entry for the narcotic morphine:

When heated slowly to about 200° C. (392° F.) it assumes a brown color, and when heated rapidly it melts at 254° C. (489.2° F.). Upon ignition, it is slowly consumed without leaving a residue.

Its aqueous solution shows an alkaline reaction to red litmus paper . . .

Sulphuric acid containing a crystal of potassium iodate gives with Morphine a dark brown color. (*Codeine* yields a moss-green color, changing to brown, and *narcotine* a cherry-red color) . . .

On adding 4 Cc. of potassium hydroxide T. S. to 0.2 Gm of Morphine, a clear solution, free from any undissolved residue, should result (absence of, and difference from, various *other alkaloids*), and no odor of ammonia should be noticeable (absence of *ammonium salts*).[6]

These proofs of therapeutic identity fell into five basic categories: identification, assay, weight variation, content uniformity, and purity. Of these five, identification tests ranked foremost. Some tests were qualitative, in which the addition of readily available reagents like sulphuric acid and potassium iodide should provoke a predictable response. Other tests, like the measurements of melting points or the chromatographic analysis of a drug in solution, were quantitative. Assays evaluated the quantity of drug present in a sample by physical or biological means. Weight variation tests measured pill-to-pill changes in size and therefore in dose. Content uniformity tests set forth acceptable limits of dose variation within and among drug samples. Purity tests functioned largely to identify substances like codeine, ammonium, or "other alkaloids" that were not supposed to be present in a sample of morphine. In 1950, a sixth standard was added, the disintegration test, which measured the ability of a tablet to dissolve in solution.[7]

The standards of the *USP* were adopted as the official protocols for proof of identity, purity, and uniformity by the Pure Food and Drug Act of 1906. Yet, even though the United States Pharmacopoeial Convention had a quasi-public status since its initial meeting in the US Senate chambers in 1820, it remained a private concern of physicians, pharmacists, and ethical drug manufacturers. Because the USP was private, its ability to articulate standards relied on cooperation and collaboration among the major drug firms, all of whom nonetheless sought to differentiate their own product lines from one another in their own marketing strategies.[8]

As discussed in chapter 1, until the middle of the twentieth century, the ethical drug industry largely sold versions of the same standard articles of the materia medica. All ethical firms benefitted from some minimal consumer confidence in pharmacopoeial standards, but the standards were understood to be a double-edged sword. The prominent placement of the mark of *U.S.P.* or *N.F.* on one's product separated the scientific marketing of ethical drugs from the proprietary drug manufacturer. But at the same time individual firms also sought to distinguish their own in-house assays and quality control procedures as somehow superior to the standard minimum. As historians of technology have noted in many other fields—from rifle making to electrical engineering—industrial standards can function to increase popular and professional confidence in products while still reserving to individual manufacturers some specific know-how, so that they can claim their own products are still somehow different in the very criteria of quality, uniformity, and identity that the overall standards are supposed to establish as the same.

Each tablet or spoonful of Squibb's morphine products were therefore comparable on one register with the basic standards described in the *United States Pharmacopoeia*, but only fully exchangeable with other versions of the product made by Squibb—or so Squibb's marketing department would have you believe. The distance between the two standards, public (USP) and private (Squibb), described the added value a physician, pharmacist, or consumer would derive from purchasing a Squibb-branded product. Yet Squibb's executives and scientific staff, along with their counterparts at competing firms like Eli Lilly, Parke-Davis, and Upjohn, were at the same time members of the USP Revision Committee and played a key role in the formation of the public USP standards against which they measured their own products. As in many other industries, the same people who created the private standards also created the public standards.[9]

Architects of drug standards built in a buffer between these public and private specifications, a vague space to be occupied by trade secret and know-how. For example, the specifications for digitalis—a prominent cardioactive agent used widely over the course of the twentieth century—did not include biological assays until 1916, even though firms like Parke-Davis had been using them for decades to claim their own digitalis was produced to higher specifications. When the first biological assay standard for digitalis was published in *United States Pharmacopoeia IX* (1916), it recommended that the drug be assayed by the "one-hour frog method," in which digitalis-poisoned frogs were compared with a control group; the details of the frog method were left intentionally vague. Only in 1939, after extensive critique of the variability of the *United States Pharmacopoeia Digitalis Reference Standard*, did ten collaborating laboratories pool the results of more than sixty thousand frogs to establish a clear standard of digitalis performance.[10] In the meantime, firms like Parke-Davis could continue to claim that their product conformed to higher standards.

In the middle decades of the twentieth century, the intersection of physiology and pharmacology produced new problems for the framers of public pharmacopoeial standards. Addressing the American Society of Hospital Pharmacists in August of 1960, a few months after the Kefauver hearings on generic names, Gerhard Levy of the University of Buffalo School of Pharmacy complained that pharmacists possessed a very limited set of tools to evaluate similarity and difference in pharmaceutical products. Levy pointed to the pioneering work conducted by researchers at the Canadian Food and Drug Laboratories in Ottawa—including J. A. Campbell and D. G. Chapman—who began in the early 1950s to study the absorption of micronutrients like riboflavin from different commercially available multivitamin products on the market and found that *none* of the tablets delivered an adequate dose of riboflavin into the bloodstream. Though all of the products contained adequate amounts of the chemical riboflavin in tablet form, the best of these tablets delivered only 80 percent of the expected dose to the bloodstream, while the worst of them delivered only 14 percent.[11]

The limited absorption of vitamins might have minimal public health significance, Levy pointed out, but the Canadian researchers had also found similar variability among long-acting preparations of the important antituberculosis drug p-aminosalicylate.[12] As the identity, purity, and dosage of the active agent could be verified in all products, the problem with p-aminosalicylate had not been anticipated by existing standards of similarity. The shellac coatings that manufacturers used to produce extended-release formulations varied widely in their application and affected the disintegration time of the capsule in the stomach and small intestine and therefore its absorption and overall clinical effectiveness.[13] As Chapman and colleagues investigated the situation further, they noted that pharmacopoeial proofs of equivalence had no criteria for measuring how and when these drugs were actually absorbed into the body.

Chapman and Campbell's work laid the foundations for a new science of "biopharmaceutics" that asked what kinds of knowledge besides the quality, purity, and dosage of the active ingredient might be necessary to make sure that two drugs were the same. Smith, Kline & French, for example, had by 1957 patented and trademarked their own shellacking technique—the Spansule—which consisted of a series of individually shellacked pellets contained within a gelatin capsule shell. When Chapman's group compared SKF's Dexedrine Spansule with seven other shellacked amphetamine products, they found the amount of drug excreted in the urine after consuming 15 mg extended-release capsules of amphetamine from different manufacturers varied considerably. In one preparation, only 5 mg of the dose was ever delivered; in another preparation, the full 15 mg was absorbed all at once.[14]

The problem was not limited to shellac. Almost any aspect of a drug's physical manifestation, it seemed, could affect absorption. In 1960 a Canadian manufacturer of the blood-thinning agent dicoumeral reported complaints by patients and physicians after they changed the shape of their tablet to make it easier to break in half. Patients on dicoumeral required repeated blood and urine tests to titrate a precise dose response: too much drug and the patient might bleed to death; too little drug and the patient risked a life-threatening blood clot. Yet patients on the new drug immediately experienced a drop in their blood levels, even though the new tablets scored identically on USP tests of purity, identity, content, and disintegration compared with the old tablets. Even after the company reformulated the shape of their tablets again with new attention to what they called "dissolution time," they received new complaints from physicians that the resultant tablet was now *too* potent. After advising physicians to retitrate their patients on the new formulation, the research laboratory published a letter to the editor in the *Canadian Medical Association Journal* noting that two lessons had been learned from this episode:

1. *In vitro* data cannot be used to interpret what may happen *in vivo*.
2. Different brands of products, although similarly composed with respect to active ingredient content, may not provide similar physiological responses. A brand name has implications beyond commercialism.[15]

As he addressed the community of hospital pharmacists in 1960, Gerhard Levy presented these and other new proofs of significant differences among drugs that met all compendial standards as evidence for the inadequacy of existing sciences of similarity. His plea was joined by several other physiologically minded pharmacologists such as John G.

Wagner, the Upjohn Professor of Clinical Pharmacology at the University of Michigan. Like Levy, Wagner became interested in the new problems posed by timed-release capsules and enteric-coated tablets. Wagner had created an animal model for drug absorption by training starved dogs to lie quietly on X-ray tables, feeding them enteric-coated tablets filled with radiopaque contrast agent, and then recording serial X-rays of their bellies to see how fast the contents of the tablets were actually released into the dogs' digestive tracts. Wagner's X-rays produced undeniable visual proof that the fate of seemingly identical pills could vary widely once they were inside the body.[16]

Chiding those physicians and pharmacists still willing to believe that therapeutic actions are "due only to the inherent activity of the *molecular structure* of the compound," Wagner urged the medical and pharmaceutical profession to demand further proofs of therapeutic similarity and difference. "Since dissolution, diffusion, absorption, transport, binding, distribution, adsorption on and transfer into cells, metabolism, and excretion are also intimately involved in drug action," he concluded, "the molecular structure, although vitally important, is only one factor in drug action."[17]

It is not surprising that Levy and Wagner's research found strong financial support from the research-based pharmaceutical industry. Wagner's connections to industry were evident in the title of his endowed Upjohn chair at the University of Michigan and in his cross-appointment at the Pharmacy Research Section of the Upjohn Corporation. At both Upjohn and the University of Michigan, Wagner explored the ways in which therapeutic activity might be influenced by the delivery of a drug to its target site. The new science of biopharmaceutics documented variations in the biological availability, or bioavailability, of a drug once consumed. For most orally consumed drugs, this meant studying the process by which the material inside a given capsule found its way out of the capsule and into the gut, how content in the lumen of the gut found its way across a series of membranes into the bloodstream, and how a pharmaceutical agent in the bloodstream made its way to its ultimate site of action.

Over the course of a prolific research career, Wagner documented in detail that the path from tablet to target often was not a straight line but an S-shaped curve: flat on both sides, but steep in the middle. Very low doses of drugs resulted in very low absorption, very high doses of drugs resulted in very high levels of absorption, but in the middle—especially in drugs that were relatively insoluble in water and had a tight margin between an insufficient dose and a toxic dose—therapeutic function could be dramatically altered by tiny changes in dissolvability or absorbability of otherwise molecularly equivalent drugs. Wagner's S-shaped curves formed a positive critique as well—expressible in a gathering series of laboratory practices, pharmacological pedagogy, and regulatory pathways—of how to create new in vitro and in vivo therapeutic tests.[18] One subset of this field concerned the measurement and modeling of how drugs circulated through different compartments of the body, which Wagner and University of California, San Francisco, pharmacologist Eino Nelson began to call pharmacokinetics. Though Wagner was not the first to coin the term, his work with Nelson on the biologically relevant differences of chemically identical drugs would play a key role in the spread of this new basic science of pharmacology.[19]

THE GUT IN THE LABORATORY: IN VITRO MODELS OF DRUG DISSOLUTION

As a positive science, pharmacokinetics could be used to document similarity as well as difference, and the immediate hope of many in the USP and FDA was to use pharmacokinetic modeling not to replace in vitro tests with costlier in vivo ones, but to build better in vitro models.[20] If the USP's laboratory standards of disintegration testing were insufficient to predict the absorption of pills in the guts of dogs and people, perhaps a better lab test might be constructed to predict absorption.

By the early 1960s, the USP's director, Lloyd Miller, was increasingly convinced that existing physicochemical techniques were insufficient for establishing the similarity of the small number of drugs that were relatively insoluble in water. Wagner had challenged the authors of the compendia to come up with a new test that would account for a capsule filled with fine granules of cement:

> A disintegration test merely measures the time required for the tablet to break up into a certain size range of granules which will drop through the certain size of a screen in the apparatus. The test tells us nothing about how rapidly the drug will be released from the small granules. Many substances, such as cement or metals, could be made into granules and compressed into tablets which would pass the U.S.P.

disintegration test; *in vivo* the granules would pass through the intestinal tract completely unchanged and be excreted in the feces.[21]

Miller tasked a group at the USP to create a better in vitro model for drug similarity—an improved "dissolution test"—that could be coupled with new pharmacokinetic data to account for the physiology of absorption and distribution in the human body. The Pharmaceutical Manufacturers Association likewise set up a Tablet Committee, headed by Charles M. Mitchell, a plant technical director for SKF. The committee tasked its member firms to collect data on sixty of their own products to look into possible problems of "tolerances for variation" and provide suggestions for new, pragmatic tests to make sure their products were what they said they were.[22]

By October of 1961 the PMA committee had drafted a new standard for the control of composition uniformity in tablets, which used the concept of tolerance as a metric of quality assurance that set a permissible upper and lower threshold on product variation. A tolerance standard that allowed for 10 percent variation would be measured as a 90–110 percent permissible range of acceptable values; a tolerance standard that allowed for 5 percent variation would be measured as a 95–105 percent range. Yet considerable debate ensued between large firms like Upjohn, Lilly, and SKF over exactly where one set the threshold for dissolution tolerance. Why 90–110 percent and not 99–101 percent or 80–120 percent or 75–125 percent? Was the task of the committee to set a minimum standard for pharmaceutical quality (i.e., 75–125 percent), against which other firms could market themselves as having superior standards (i.e., 95–105 percent)? Or an absolute standard to which all were accountable?[23] Mitchell appealed to PMA member companies to share their means of establishing how their own brand-name drugs controlled for batch-to-batch, tablet-to-tablet differences. "Unfortunately," he concluded, "none of the references or sources available to the Committee delineate a test method in sufficient detail to permit even a starting point."[24]

Some companies were more forthcoming that others. Upjohn's C. Leroy Graham—who worked closely with John Wagner and served on the *USP* Revision Committee as well—sent Mitchell a sketch of Upjohn's own approach toward operationalizing a dissolution test. Instead of simply watching a tablet disappear when shaken in a wire basket in a beaker of water (as instructed for the *United States Pharmacopoeia XVI* Disintegration Apparatus), the Upjohn technician had adapted a two-step process to more closely simulate the human gut. After an initial "gastric phase," in which the machine simulated the human stomach by bathing the tablet in a solution of warm hydrochloric acid, the test entered its "peptic phase," which used a buffered medium of sodium hydroxide and potassium phosphate to simulate the pH of the absorptive portion of the small intestines.[25] Other PMA and USP members murmured their approval and suggested modifications. The head of the *National Formulary*, Edward Feldmann, suggested adjusting the "standardized body temperature" to 37 degrees plus or minus 2 degrees Celsius to more closely simulate the inside of a human body.[26]

All parties wanted a way to replicate the human gut on a laboratory bench without needing to involve actual human guts in their product testing. Some of the attempts to simulate the human gut seem a bit baroque. One proposal from a Dr. Simons, forwarded to the Tablet Committee in 1963, promoted the Erweka AT-3 (figure 5), which moved beyond questions of acidity and temperature to visually and kinetically simulate the mechanics of the gut itself: "The Erweka consists of a hollow glass ring with uniformly spaced constrictions which serve to increase the internal surface of the ring. The constrictions, or ridges, interrupt the free flowing motion of the tablets within the rotating ring so that the tumbling and colliding action of the tablets against the ridges provides an eroding effect. The glass ring is rotated by a variable speed motor, and the test fluid is circulated by a peristaltic pump."[27]

Blown glass was to simulate the folds of tissue of the gastric rugae and intestinal plicae, respectively; a mechanical pump was to simulate the peristaltic squeezing of the gut wall smooth muscle. Committee members applauded the attempt to build a simulated gut within the laboratory as a biologically relevant in vitro version of an in vivo process.[28]

Yet several members of the PMA and USP committees voiced concern that the difference between disintegration and dissolution might have no real clinical implications or not *enough* implications to mandate a sweeping set of changes throughout analytical laboratories of hospitals, industry, and the FDA.[29] For one thing, the reagents were expensive. In addition to the machines themselves, laboratory reagents now specified the amount of enzymes and other biochemical reactants necessary for the vitro to sufficiently represent the vivo. By 1968 the PMA petitioned the USP to stop fussing so much over the inclusion of extra enzymes like pepsin (in *United States Pharmacopoeia* Simulated Gastric Fluids) and pancreatin (in *United States Pharmacopoeia* Simulated Intestinal Fluid),

FIGURE 5. Physiology of the tablet: the mechanical gut. The Erweka AT-3 machine, one of several mechanisms designed to simulate the absorption of a pill inside the human stomach and small intestine, was made of blown glass and a "peristaltic pump" that bathed a given tablet in artificial gastric and peptic fluids.

"Memorandum: Erweka Tester AT-3," September 9, 1963, box 43, f 25, USPC. Courtesy of Wisconsin Historical Society.

as they produced therapeutically trivial differences in drug testing. Although "enzymes were included in these fluids with the thought that the tablet disintegration test would then more closely approach the conditions encountered in the gastrointestinal tract," there had emerged "serious doubt that the disintegration test results are in any way affected by the presence of these enzymes."[30] Whether any of these in vitro models of drug absorption offered any true benefit to the practice of medicine or pharmacy was an open question by 1970, as the USP issued its first dissolution testing requirements for fourteen key drugs.[31]

INVESTING IN SCIENCES OF DIFFERENCE

These collaborative interests among pharmaceutical firms in creating public standards for biopharmaceutical similarity were complicated by the divergent interests of individual manufacturers to independently promote their own in-house standards of quality control as superior to the minimum standard. This logic was clearly articulated in the early 1960s by two Texas physicians—Jaime Delgado and Frank Cosgrove—in an article on the fallacies of generic equivalence.[32] The distance between a molecule and a medicine, in their account, was occupied by a terrain of "pharmaceutical know-how": the relative ability or inability of a firm to turn a mere pharmaceutical ingredient into a truly effective drug. The concept of pharmaceutical know-how illustrates the paradoxical process by which industrial standards intentionally leave out much of the relevant knowledge required to properly enact them.

As the historian of technology Amy Slaton has described in the case of electrical engineers, competing industrial parties will often agree on a set of technological standards to the extent that they require a common language to communicate and inspire broad confidence in the scientific grounding of their work and products. But a set of standards that defines *every* aspect of the job in precise language would threaten to erase the proprietary claim of a given company or the need for the expertise of the engineers themselves. Instead, Slaton argues, competing parties devise standards not to ensure all objects are the same but rather that they are "as near as practicable": explicit enough to enable many actors to work on a set of common objects but not so explicit as to render all objects equivalent and their own economic value obsolete.

Manufacturers, physicians, and academic pharmacologists echoed Delgado and Cosgrove's analytic of pharmaceutical know-how in separating reputable from disreputable firms. Reputable firms, the Harvard pharmacologist Dale Friend noted, "spend much effort in seeing to it that the best pharmaceutical 'know-how' is employed in compounding and that adequate controls are made and records kept." Products made by reputable firms simply could not be equated with those of companies "without adequate controls and with little pharmaceutical know-how and little effort to turn out a product of high caliber."[33] But Friend likewise lamented that "many of the differences between a branded product and its generic equivalent are manufacturing secrets of the originator of the drug product" and hoped that the new sciences of biopharmaceutics would likewise open up the black box of know-how so that its component parts might be better knowable and enforceable. "It is obvious that there is a definite need for the FDA and the USP to enlarge their interest in the formulation of active drugs so that agents will not only be identical in amounts of active drug but also compounded in a manner that ensures the full activity of the drug with the minimum of undesired effects."[34]

Friend's comments here suggest another dimension to the politics and economics of the sciences of similarity. Academic pioneers of biopharmaceutics and pharmacokinetics like Wagner, Levy, and Nelson were overwhelmingly funded by manufacturers like the Upjohn Company. Evidence of biopharmaceutic difference undermined standards of similarity and made claims of individual know-how all the more credible. These findings could be circulated by the public relations arms of the pharmaceutical industry to make broad arguments for the incommensurability of brand and generic drug forms. But at the same time new models of pharmacokinetics could be utilized by bodies like the USP and the FDA and in the process transform from sciences of difference into sciences of similarity. It is insufficient to merely dismiss these new sciences as economically interested. In studying the emergence of scientific fields (like pharmacokinetics) at the intersection of ideological opposition (like the brand/generic controversy), we find not just a polarized field but a dialectic: the emergence of a novel site of scientific activity at the intersection of competing interests.

By the end of the 1960s, negotiation with the USP over the biological bases of dissolution testing was perhaps the easiest of the pharmaceutical industry's many interfaces for articulating standards of pharmaceutical similarity and difference. Recall that on Capitol Hill in 1967, three separate bills to mandate the substitution of generic drugs in place of brand-name versions were proposed—although all were defeated. The principal architect of one bill, Senator Russell Long of Louisiana, claimed that substitution of chemically equivalent drugs would begin to save US taxpayers more than $100 million a year of the $500 million the US government was paying annually for prescription drugs. As we have examined, Gaylord Nelson's Senate hearings on the competitive problems in the drug industry also began that year with a volley of assertions about the substantial cost savings achievable through widespread replacement of brand-name drugs with their chemical equivalents. Nelson's hearings were orchestrated to substantiate—in a way that Estes Kefauver's had not—the burden of proof for pharmaceutical equivalence.[35]

Nelson, for his part, casually dismissed the scientific arguments of Chapman, Warner, and Levy as a smokescreen of insignificant differences loosely bound together by ample industry funding. Indeed, some of the strongest testimony Nelson elicited on the equivalence and interchange-ability of generic drugs came from the chief editors of the *National Formulary* and the *United States Pharmacopoeia*. The findings to date of the field of biopharmaceutics, they suggested, regarded largely theoretical concerns. Only a handful of cases of actual biological inequivalence in human subjects—fewer than five—had ever been documented among drugs that had met the *United States Pharmacopoeia* or *National Formulary* standards, and none had yet been clearly associated with a sustained difference in therapeutic outcome.[36]

If one insisted on thinking in terms of theoretical differences, Edward Feldmann of the *National Formulary* told Nelson, then brand-name drugs could not be assumed to be equivalent even to themselves. As the work of the PMA's Tablet Committee suggested, batches of the same brand-name drugs could differ among themselves by as much or more than the brand-name drug differed from the generic. Even if the FDA had the resources to measure all batches of all drugs produced for sale in all factories, it could not eliminate subtle pill-to-pill variations inevitable in all mass-produced objects. "From a technical standpoint," Feldmann continued, "there really is no such thing as complete 'drug equivalence.' "[37]

The goal of industrial standards, Feldmann argued, whether for pills, transistors, disc brakes, or chocolate bars, was not to prove that products were identical but to prove they were sufficiently similar in all important respects. No evidence existed to prove that the pharmacopoeial standards had yet posed an actual threat to pharmaceutical consumers. "Clinical deficiency among products which comply with the official compendia standards," he concluded, "is grossly exaggerated to haunt and confuse us . . . The problem is seen no longer as a giant foreboding phantom, but is reduced to a faint shadow, calling for objective, dispassionate, scientific investigation." The work of the pharmacopoeia was to carefully balance all observations of difference and determine which were significant and which were trivial.[38]

A few weeks later a quality control officer at SKF presented Nelson's committee with a list of 211 references and review articles that described further evidence of physiological effects of subtle differences in the formulation of tablets and capsules.[39] A Pfizer pamphlet would later critique Feldmann's dismissal of biopharmaceutics as a politically self-interested position. The question was not whether inequivalence was a problem but how often:

The question, clearly, is quantitative rather than qualitative . . . Let us say there are (for sake of argument) 100 classes of drugs, each containing several nominal equivalents. Let us then say that five of these groups have been studied for clinical equivalence, and that in each of them at least one of the generic drugs has shown significant failure of entry into the body. This could then be interpreted in two ways: (1) The problem of nonequivalence has so far appeared in 5 of 100 types of drug, or 5 per cent. *It is a faint shadow.* (2) The problem of nonequivalence has so far appeared in 5 of 5 types of drug studied, or 100 per cent. *It is a foreboding phantom.* The reader is invited to take his pick.[40]

Gaylord Nelson listened patiently—and at times impatiently—as academic pharmacologists and pharmaceutical executives explained the principles of biopharmaceutics and why this new laboratory field proved that existing compendial standards failed to guarantee therapeutic equivalence among chemically equivalent drugs. Yet when Henry DeBoest, vice president of Eli Lilly & Co., argued that brand-name drugs had certain intangible qualities that made them incommensurable with generics, Nelson exploded with frustration, asking DeBoest point blank whether he had any actual proof that an actual patient had ever actually been harmed by switching from a brand-name drug to its generic equivalent:

SENATOR NELSON: What you are saying is that the whole drug industry and the whole medical profession does not have any clinical evidence to demonstrate that two drugs that meet Pharmacopoeial standards are not therapeutically equivalent? Is that not your testimony? . . .

What I am asking you, sir, is whether any drug company in America has any evidence to prove the continually repeated assertion that even if two drugs meet USP standards, therapeutic equivalence may not exist. I have heard that time after time. I have asked each witness whether he has any clinical evidence at all from anyplace in the world. I have not received any yet.

MR. DEBOEST: No, sir; we do not have.[41]

Lilly's quibbles were largely theoretical, Nelson exclaimed, and had the rarefied air of high laboratory science but provided no actual evidence that generic drugs did not work as well as brand-name drugs in practice.

In the absence of proof of substantial difference, why shouldn't the taxpayer assume that cheaper generic drugs were just as good as the more expensive brand-name versions?

DeBoest may not have realized it, but the drama of his subpoenaed statement had been rehearsed well before his appearance on Capitol Hill. Nelson's team thought they had seen all of the evidence of brand/generic difference and that this evidence was fully dismissible. "You don't start a hearing," Nelson's staff economist and the key architect of the hearings would later recall, "unless you have the information at the very beginning": "The purpose for a hearing, in my judgment, is to create a demand for legislation. In other words, to move the public. It's an information thing—not a way to get information *for me*, but a way of getting the information *to the public*. I knew all the questions. I knew the answer to every question I ever asked. I wouldn't ask a question unless I knew the answer. The hearings were orchestrated to get the public to understand what's going on."[42] What was "going on" was a bid to create widespread support for the full exchangeability of chemically equivalent generic and brand-name drugs. Nelson and his team had witnessed the proliferation of new sciences of therapeutic difference described in this chapter, and they had attributed it to a form of industry interest. They had examined the cozy relationship between brand-name pharmaceutical manufacturers and the architects of pharmacopoeial standards. They had listened to the claims of experts from the new field of biopharmaceutics and considered that these findings were more theoretical than real, an industrially manufactured doubt where no real doubt need exist.[43]

But not even as well connected a man as Gaylord Nelson could have *all* the information from the very beginning. In October 1967, new data emerged to challenge Nelson's contention.

CHAPTER 7

THE SIGNIFICANCE
OF DIFFERENCES

Mere compliance with a U.S.P. monograph will not automatically
guarantee a pharmaceutical preparation of the highest quality . . .
any more than a literal following of a recipe in the finest cookbook
will guarantee a culinary triumph if the housewife is a bad cook,
or uses a dilapidated stove, unreliable measures, and poor utensils.
—SMITH, KLINE & FRENCH, "AN INDISPENSABLE BOOK IN PERSPECTIVE," 1961

Shortly before the expiry of Parke, Davis's patent on its blockbuster antibiotic Chloromycetin in 1966, the company's president, Austin Smith, met with the commissioner of the Food and Drug Administration, James Goddard, to discuss his concerns about his soon-to-be generic competitors. Smith contended that the FDA's process of licensing foreign plants that supplied raw materials to generic chloramphenicol manufacturers was overly lax and involved "processes, conditions, and raw materials" of production substantially different from those used in the Parke, Davis plant. "This subject," Smith concluded, "is highly complex and difficult to understand, much less express in writing." Perhaps the new generic competitors should be asked to submit their own clinical trials, compared against placebo, to prove that their versions of chloramphenicol worked as well as Chloromycetin?

Goddard replied somewhat brusquely that he understood the subject well enough without the need for a tutorial from a pharmaceutical executive. If the safety and efficacy of chloramphenicol as a drug had already been established by Parke, Davis's initial application for FDA approval, it would be unethical to require new clinical trials from other companies as a condition for approval. Furthermore, no batch of bulk antibiotic that met certification standards had yet been proven to be ineffective in the clinical realm. As long as the generic products were made with bulk chloramphenicol of sufficient purity, quality, identity, and strength, they were similar enough to be considered the same as Chloromycetin.[1]

Its arguments of difference defeated, Parke, Davis braced itself for major losses in the $70 million per year chloramphenicol market it had formerly controlled. But the fortunes of the drug firm would be saved at the last minute by the complaints of a West Coast tropical fish dealer. The gentleman in question, who had been in the habit of pouring a capsule of Chloromycetin into his fish tanks on a regular basis to treat infected fish, promptly purchased a bottle of McKesson & Robbins' generic chloramphenicol once it became available. When he emptied the first capsule into his tank, however, it refused to dissolve, leaving instead a white scum at the surface and a number of dead fish. The local Parke-Davis sales representative passed the anecdote upward through the ranks to Dr. Joseph Sadusk, the outgoing director of the FDA's Bureau of Drugs, who had recently joined Parke Davis's scientific management.[2] Sadusk's team ran a number of biopharmaceutic analyses comparing Chloromycetin to generic chloramphenicol products, which showed that the bloodstream absorption of the brand was different from that of the generic versions.

Sadusk presented these data to the FDA in April of 1967, and the agency agreed to conduct a comparative clinical trial of chloramphenicol absorption in human "volunteers" from a Florida military base. Utilizing techniques developed by John Wagner, the FDA studies documented that the "areas under the curve" of serum levels of chloramphenicol varied widely among generic formulations. Indeed, after some volunteers consumed the generic version sold by McKesson & Robbins, no trace of the drug could be found in the blood at all. Further biopharmaceutical analysis of chloramphenicol absorption suggested that, in addition to being relatively insoluble, the drug was only absorbed along a relatively tiny stretch of the alimentary tract, so that subtle difference in formulation could lead to significant changes in bloodstream absorption.[3]

As these studies were taking place, FDA commissioner Goddard gave an interview with Senator Russell Long (D-LA)—still at work trying to muster support for a bill to mandate generic substitution of brand-name pharmaceutical products—to document why the FDA had not been able to come out with a single list of generic equivalent products. The difficulty, Goddard noted, was due to a "rather marked change that has occurred in the sophistication of our drug supply" that apparently made small differences in absorbability more important than in the past, "because of minor differences in the method of manufacture, the size of the article, the excipient used, the kind of coating used on a tablet, the pressure on a tableting machine—I could go through a whole host of these things which have been written up in the scientific literature." From a regulatory standpoint, this presented new difficulties:

> Now admittedly this is not on a great many products, but it happens just enough of the time to give you concern and pause that no longer do you rely on the lab testing program *in toto*, which was once the main source of defense of FDA: did it meet USP standards. Take it in the labs and test dissolution or disintegration times and measure in beakers with hydrochloric acid which were found to be entirely accurate.
>
> We now know a good deal more about drug absorption. We know before the drug can get into the body, we know the particle size may alter its availability. The actual form of the drug, whether it is in an acid form or whether or not it has a different radical on it. The Defense Dept. had to learn this the hard way. It has had examples of it. So I am not trying to throw sand in your eyes. I am trying to be responsive, and at the same time help you to accomplish the objectives.[4]

Nobody within Goddard's FDA knew exactly how to account for the generic/brand chloramphenicol disparities, but they knew that they could not simply be brushed away. Chloramphenicol was a drug associated with substantial adverse effects; it was also a "life-or-death" drug ideally reserved for potentially fatal infections such as typhoid fever and overwhelming sepsis. The observation that one could swallow capsules of a generic version and absorb no drug whatsoever into the bloodstream called into question the competence of protocols used to claim the functional exchangeability of generic and brand-name drugs. When Leslie M. Lueck, Parke-Davis's director of quality control, testified at Nelson's hearings in late 1967 about generic chloramphenicol products, he cited no fewer than eight thousand analytic tests supporting thirteen bar charts and graphs comparing Chloromycetin's biopharmaceutical incommensurability with its three generic competitors (figure 6).

Attention soon turned to excipients, binders, and fillers, those substances *besides* the active chemical that were stuffed inside of tablets and capsules. Parke Davis let it be known that they used only lactose, or milk

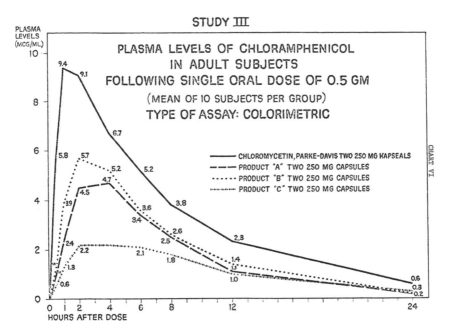

FIGURE 6. One of thirteen charts Parke-Davis presented before Congress in 1967 to document the differential serum availability of Chloromycetin and generic chloramphenicol formulations.

Series III, f "DESI PAC: Third Meeting," DES. Courtesy of the National Academy of Sciences.

sugar, where Zenith used magnesium stearate and talc and McKesson added starch and calcium sulfate. The insoluble starch of the McKesson tablets had no doubt been the cause of the white floaters in the fish fancier's tank; Parke, Davis began to issue samples of McKesson's generic chloramphenicol capsules to its sales representatives with instructions to pour them into a glass of water in physicians' offices, claiming, "See, the stuff doesn't dissolve."[5]

In response, the FDA suspended certification of all generic versions of chloramphenicol, effectively guaranteeing an additional year of marketing exclusivity for Parke, Davis's most profitable product.[6] Eventually the FDA came up with a new protocol for proving equivalence among generic versions of chloramphenicol, demonstrating in vivo evidence of serum biological availability in human volunteers by producing a graph like the one in figure 6 to document that the curve of bloodstream absorption was sufficiently similar to that of Chloromycetin.[7] When, in January of 1968, it finally allowed Rachelle Labs to bring a new version of generic

chloramphenicol to market, the FDA had taken the first step toward requiring in vivo scientific data to document proof of biological equivalence
of generic drug products.

Yet even after the demonstrations of Choromycetin/chloramphenicol
difference, Gaylord Nelson could still attempt to claim as late as March
1969 that "we have not been able to get any convincing evidence from the
manufacturers or anybody else that drugs that met USP standards are not
therapeutically equivalent." When confronted with the case of chloramphenicol, Nelson dismissed it as an outlier, replying brusquely that "that
is their only drug we are aware of and there is no proof yet." Since only fish
and prisoners had demonstrated this difference and no actual *patients*
taking generic chloramphenicol had been shown to have suffered a poor
treatment outcome, Nelson could still technically conclude that "there
is no proof that the different blood level achievement of Chloromycetin
versus the other chlorampenicols, indicates that one was any more efficacious, that there was any difference in therapeutic effect."[8]

Struggles over the burden of proof in the field of generic equivalence
continued to play out broadly in congressional hearings, promotional
pamphlets, and clinical journals. Did generic manufacturers need to find
new means to prove that their products had the same therapeutic effect
as the brand-name versions? Or was the burden of proof on the brand-
name industry to document that their largely theoretical concerns over
inequivalence translated into an actual public health problem?[9] Each
attempt to create a new standard for generic equivalence entered a politically charged field in which the standards—and burdens—of proof were
not equally shared.

MANUFACTURING DIFFERENCE

From the standpoint of other pharmaceutical manufacturers whose
blockbuster products faced patent expiry, Parke Davis's handling of the
Chloromycetin situation offered a model for preserving brand-name market share against new generic competition. In 1968, Upjohn deployed a
similar strategy pre-emptively, providing scientific data on the incommensurability of its popular antidiabetic agent Orinase (tolbutamide)
compared to generic versions—before any generic versions even existed.

The November issue of the *Journal of the American Medical Association* that year featured a lead article by Upjohn scientist Alan Varley,

entitled "The Generic Inequivalence of Drugs," that compared Orinase to an experimental generic version. Varley's team created a version of tolbutamide 500 mg tablets that contained half the amount of a proprietary disintegrating agent called VeeGum but otherwise met the *United States Pharmacopoeia / National Formulary* criteria for tolbutamide 500 mg. Ten nondiabetic volunteers were assigned to consume both pills in a double-blinded, crossover study. Levels of serum tolbutamide and blood sugar were followed over an eight-hour period. Varley then graphically presented the "area under the curve" for these two variables to form a visual demonstration (figure 7) of biological and therapeutic difference in two preparations that were technically equivalent by *United States Pharmacopoeia* standards.[10]

Like Wagner before him, Varley was a clinical pharmacologist associated with Upjohn who saw his responsibilities to the field of pharmacokinetics and to the firm that employed him to be unconflicted. His tolbutamide studies would feature prominently in the reopening of Nelson's drug probe after the senator's landslide re-election in 1968. In December of that year Nelson called William Bean, a leading national figure in medical ethics and clinical education, to discuss Varley's work as an example of industrial conflict of interest. Bean dismissed Varley's research as a form of laboratory tinkering, an elaborate attempt to dress industry propaganda in the guise of objective science. As Nelson's aide Ben Gordon concluded from Bean's testimony, "The only thing [Varley] shows is that it is possible for Upjohn Co. to manufacture a tablet which is not as good as another tablet they manufacture, of tolbutamide... Now how did he jump from that to his grand conclusion, when there is no such thing as generic tolbutamide, anyhow."[11]

John Adriani (of the American Medical Association's Council on Drugs) and Lloyd Miller (of the USP) were likewise upset that *JAMA* published such a study at all. Miller dispatched a curt note to *JAMA*'s editors complaining that "it is difficult to see what reasoning led to giving such prominence in this respected journal to a report on the execution of pharmaceutical legerdemain."[12] Adriani had especially sharp words for Varley after Upjohn refused to release its "experimental generic" tablet or the excipient VeeGum for other laboratories to test—claiming that the composition of both constituted a trade secret.[13] How could something essential to a drug's functioning be protected as a trade secret? "If it is crucial to the efficacy of the pill," Adriani asked, "shouldn't it be listed on the packaging?"[14] Adriani's question hinted at a far broader problem:

If the new sciences of biopharmaceutics demonstrated that there were things beyond the active molecule that contributed to the functioning of a drug—such as shellac, dispersants, binders, and fillers—shouldn't those aspects be copiable as well?

Varley's critique of *United States Pharmacopoeia* standards irked Lloyd Miller all the more because Upjohn scientists like C. Leroy Graham served on the *United States Pharmacopoeia / National Formulary* Joint Panel on Physiological Availability that drafted new dissolution test procedures that were about to be released in the 1970 *United States Pharmacopoeia XVIII*. As Miller fumed to Nelson's aide Ben Gordon, had Varley used the new *United States Pharmacopoeia* standards in his research, he would have found no difference—and would have had no front-page publication in *JAMA*. In a scathing letter to Varley, Miller pointed out that the irresponsibility of Upjohn's actions undermined the entire system of pharmacopoeial standards.[15]

In both cases, Chloromycetin and Orinase, industry scientists wielded techniques of biopharmaceutics as sciences of difference, demonstrating the incommensurability of brand-name and generic products. In both cases, standard-setting bodies such as the FDA and the USP were able to argue that the same biopharmaceutical tools used to demonstrate difference could be transformed into sciences of similarity. But here the parallels between the two cases end. Where Parke-Davis was able to use evidence of difference to extend Chloromycetin's monopoly against generic competition for another year, Varley's work on Orinase failed to convince a broader audience. Adriani, among others, remained skeptical of the generalizability of the problem and continued to publicly testify that the "paucity of convincing and well-documented data of clinical significance causes one to suspect that the situation has been grossly exaggerated."[16]

The chloramphenicol incident alerted parties in industry, pharmacy, medicine, and government that clinical differences in biological availability were not merely theoretical and demonstrated that the FDA could successfully demand in vivo proofs of "bio-equivalence" beyond existing pharmacopoeial standards. The flap around Varley's tolbutamide research indicated that drug manufacturers could manufacture the appearance of clinical differences in biological availability when no real clinical problem yet existed. Neither of these episodes, however, settled the more important question: When was it actually necessary to demonstrate

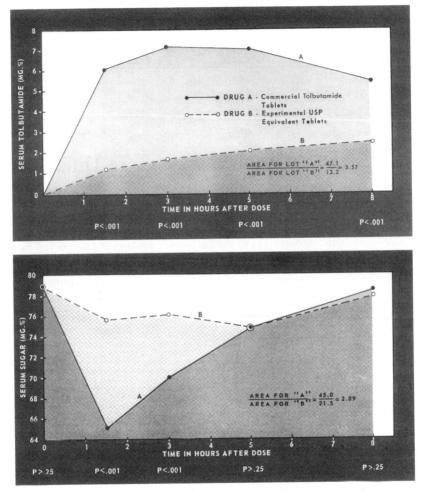

FIGURE 7. Alan Varley's area under the curve graphs of Orinase versus "experimental" generic tolbutamide demonstrating (*top*) biological inequivalence (differential drug absorption) and (*bottom*) therapeutic inequivalence (differential effects on blood sugar levels).

Alan B. Varley, "The Generic Inequivalence of Drugs" JAMA 206, no. 8 (1968): 1745–46. Copyright © 1968 American Medical Association. All rights reserved.

biological equivalence, and when would proof of chemical equivalence suffice? Attempts to answer this question carried the laboratory sciences of similarity into legislative and regulatory arenas, with far different burdens of proof at play.

BURDENS OF PROOF

In May of 1967, as Nelson opened his first hearings into generic drugs, Lyndon Johnson ordered his secretary of Health, Education, and Welfare, John Gardner, to establish a priority task force to investigate the costs of prescription drugs under Medicare. The work of the HEW Task Force on Prescription Drugs soon became deeply involved with the burden of scientific proofs of generic similarity or difference.[17] As Dr. Philip Lee, assistant secretary of HEW and head of the task force recalled, both genericists and antigenericists assumed that this fact-finding body must necessarily support their own position of generic equivalence or inequivalence: "On the one hand, most of the brand-name drug manufacturers, many representatives of organized medicine, and their adherents in Congress hailed the appointment of the Task Force and openly predicted that the group would inevitably support their stand, and find that 'generic nonequivalency' was so frequent and so dangerous that it represented a serious menace to health. On the other hand, proponents of generic prescribing were confident that the HEW task force would support *their* stand and find that 'generic nonequivalency' was so unimportant that it could safely be ignored."[18]

The example of Chloromycetin was cited by the task force as a crucial reminder of the fallibility of existing standards of chemical equivalence to establish the clinical equivalence of generic drugs. Yet the alternative of requiring full clinical trials for all copies of off-patent drugs seemed even more far-fetched. As the HEW task force considered the breakdown in the regulation of generic chloramphenicol, the intermediary proof of biological equivalence—as documented by the curve of bloodstream drug absorption in healthy subjects used to certify later generations of generic chloramphenicol—presented a reasonable middle path. Yet such testing was still likely to be expensive and wasteful in the case of *most* drugs, in which absorption problems did not really present a problem. Most drugs dissolved easily in water and had wide dose ranges at which they were therapeutically effective; for these agents proof of chemical equivalence might be enough.[19] Only a small handful of drugs was likely to pose the same problems as generic chloramphenicol, chiefly drugs that were (1) critical, or life-and-death, drugs available in (2) solid-dosage forms, with (3) relative insolubility, and (4) some anecdotal suggestion of absorption problems.

The assumption that most drugs had no bioequivalence problems

placed the burden of proof on inequivalence. The HEW task force identified only twenty-four compounds for initial evaluation, and, after four
years of study, significant differences in biological availability had been
found in only six products: four antibiotics (p-aminosalicylate, chloramphenicol, tetracycline, and oxytetracycline), one antiepileptic (phenytoin), and one anticoagulant (warfarin). In its final report in 1969, the
task force suggested that, for these and perhaps a few other products,
bioavailability testing along the lines of the FDA requirements for generic
chloramphenicol were called for. In general, however, they concluded that
"on the basis of available evidence, lack of clinical equivalency among
chemical equivalents meeting all official standards *has been grossly exaggerated* as a major hazard to the public health."[20]

If we lived in a world in which complex problems were simply solved
by expert committees, that would have been the end of that. Yet the
publication of the HEW task force report was not the resolution of
the brand/generic controversy but merely another step in its promulgation. At the same time that the experts on the task force were de-
liberating equivalence, a competing national committee of experts
on drug policy—the Drug Efficacy Study of the National Academy of
Sciences / National Research Council and the FDA—was also grappling
with similar problems. Like the task force, the DES team believed that
chemical equivalency sufficed for therapeutic equivalency in "the majority of cases," but the final report of the Drug Efficacy Study in 1969 was
far more vocal in advocating for the use of in vivo bioequivalence testing
as a regulatory practice.[21]

Because the same evidence of similarity and difference between brand-
name and generic drug products was used to different ends by the two
expert committees, a third committee, headed by Harvard professor of
political economy John Dunlop, was tasked by HEW to review the activity
of the task force and issue its own recommendations on their recommendations. Where the task force had placed the burden of proof on the
demonstration of difference, the Dunlop committee asked for proof of
similarity.[22] In a stunning demonstration of bureaucratic proliferation,
a *fourth* committee, run through the Office of Technology Assessment—
headed by the dean of Yale Medical School, Robert Berliner—was
appointed to resolve the discrepancies between the task force's recommendations and the Dunlop committee's recommendations. When issued in 1975, this report did little to settle the polarized field of generic
equivalence or inequivalence. As Berliner's report reshuffled the same

elements, balancing proof of equivalence against proof of difference, it still had difficulties settling the question of how to decide when biological proofs of similarity were really necessary in policy and practice:

> It is very important to point out, however, that two drugs may differ in availability, that is be bioinequivalent, but may still be therapeutically equivalent. It is entirely possible, and, in fact, frequently true, that the concentration in the blood produced by two products may differ substantially, yet, for both, concentration may fall well within the range between that required for the desirable therapeutic effect and that at which unacceptable toxic effects are produced. On the other hand, it is also true that in a very few instances, differences in bioavailability have led to well-documented therapeutic failures.[23]

Two issues had become evident as the question of bioequivalence testing bounced from committee to committee in the late 1960s and early 1970s. First, some protocol for demonstrating a convincing level of biological equivalence was desired by actors across many locations of the federal government—not merely the FDA, but also payor programs such as Medicare and Medicaid. Second, that the critical demonstration—and repeated invocation—of Chloromycetin had raised the status of biopharmaceutics from a domain of academic pharmacology to the status of an incipient regulatory science.

As FDA commissioner Alexander Schmidt noted in a talk at the USP on March 22, 1975, the constant public attention to the problems of generic equivalence was changing the "traditional" relationship of private and public regulatory science. Although from time immemorial, "USP has been the private developer and FDA has been the government enforcer of drug standards," Schmidt asserted that this older system was now giving way to the "bothersome" reality that "standard setting—once a quiet and almost private effort—has become a public issue. People are watching, what they once ignored. And they are watching with critical eyes."[24]

Schmidt still hoped that the USP's efforts to develop an in vitro test for bioequivalence might be successful, but he declared that the time had come for the FDA to be responsible for new pharmaceutical sciences of similarity. In a series of statements in the *Federal Register* over the course of the 1970s, the FDA gradually took the lead in defining a methodology of measuring therapeutic equivalence in terms of biological availability that worked to develop explicit protocols for bioequivalence.

THE PLURALITY OF EQUIVALENCE

The new science of biopharmaceutics entered the FDA via a revolving door of industry scientists. Trieste Vitti, the first chief of the FDA's Division of Clinical Research in the Bureau of Medicine came from Upjohn, but his tenure at the agency was brief, as was that of his next few successors. The agency worked closely with specialists—most notably John Wagner—from Upjohn and other firms but spent more time triaging drugs for bioequivalence testing than developing methods for conducting those tests. As of 1975, the FDA had commissioned five separate extramural contracts on bioavailability research, none of which agreed on a single method.[25] It wasn't until the mid-1970s that the FDA could announce the launch of a full-time Biopharmaceutics Laboratory that could focus its effort on developing its own standards of proof. Yet the tools to perform this translation were limited. Even the mathematical skill required to compare two absorption curves was lacking: as one director recalls, in actual practice the FDA staffers were tasked with plotting absorption curves on specially weighted pieces of graph paper, then cutting out the curves with scissors and literally weighing the different curves against one another on a laboratory scale.[26]

Bernard Cabana, who took over as head of the FDA's Division of Biopharmaceutics in 1975, had received a doctorate in pharmacology from the University of Buffalo and spent several years working on pharmacokinetics in the labs of Bristol Myers. Cabana quickly began setting new thresholds of tolerance that used mathematical modeling instead of weighing pieces of graph paper on scales. One of these early rules, the "75/75" rule, stipulated that as long as bloodstream availability of one drug was more than 75 percent but less than 125 percent of the other drug, in at least 75 percent of subjects studied, two drugs could be considered bioequivalent, provided there was no more than 20 percent variation in the maximal concentration and the total area under the curve. Cabana spoke optimistically of an in vitro model that would solve the bioequivalence problem and make costly in vivo tests unnecessary.

At the same time, Cabana's rulings swiftly increased the number of drugs requiring bioequivalence testing.[27] Several of his colleagues in the FDA thought he was out of touch with the actual pragmatics of regulatory science. FDA staffers sympathetic to generic manufacturers, such as drug monograph chief Marvin Seife (whose career-ending lunch was chronicled in the first pages of this book) found him confusing, condescending, and easily distracted by concerns of the brand-name industry. As Seife would write privately to Jean Callahan and David Langdon of the New York State Generic Drugs Investigation panel, "As long as the P.M.A., the A.M.A, and the F.D.A. hierarchy, seriously or otherwise, use Bernie Cabana's bioavailability pronouncements to their advantage, the whole bioavailability-bioequivalency problem (or madness) <u>will never</u> be settled. This include Bernie's present, new campaign regarding drug dissolution data—sometimes to be done using the 'paddle method,' sometimes the 'rotating basket' and other times no method. Even if a tablet dissolved in his hand in a matter of seconds, a dissolution study would be called for."[28]

Seife wrote a series of letters to consumer advocates outside the FDA—including William Haddad, who would become the executive director of the Generic Pharmaceutical Industry Association—bemoaning Cabana's undeserved power and the increasing influence of bioequivalence within FDA practice, which Seife thought to be overblown, overpriced, and merely another barrier intended to blunt generic competition for brand-name drugs going off patent:

> Note: be aware of vested interests in the biopharmaceutics game—the investors with their scientific front-men have found bioavailability testing to be a lucrative area. Schools of Pharmacy (university or non-university based) love this crap, since they can get government contracts to develop methodology or members of their faculty hired by FDA as consultants or have Bernie massage their egos with regular telephone calls for advice. Academicians can write papers for scientific journals re bio problems and be rewarded by large pharmaceutical firm grants or by being hired at higher salaries at the firms themselves. This game goes on and on *ad nauseam*. I can only stand by and watch the whole charade continue.[29]

Cabana's optimistic pronouncement that bioequivalence could be reduced to an in vitro test turned out to be a paper moon. Slowly, over the course of the 1970s, the FDA tamed the multiplicity of possible definitions of bioavailability testing (in vitro vs. in vivo, humans vs. animals, urine vs. blood) and largely stabilized into a more coherent set of protocols focusing on bloodstream availability in human subjects. By 1978, bioequivalence had become a more coherent object within the FDA, a regulatory science of similarity, born from earlier techniques of demonstrating difference.[30]

Where the FDA in the late 1970s had taken a largely reactive stance toward new sciences of similarity, by the early 1980s the FDA was taking a much more proactive approach to standardizing drug equivalence. As Shrikant Dighe, chief of the FDA Bioequivalency Review Division, told the Food and Drug Law Institute in 1982, the FDA had already begun to devise bioequivalence tests for drugs still under patent—such as ibuprofen, indomethacin, clonidine, and flurazepam—to create anticipatory standards of bioequivalence. By this time as well, the FDA had given up distinguishing "bioproblem" from "nonbioproblem" drugs and was merely drawing up in vivo bioequivalence standards for all new drugs. By the time the Hatch-Waxman Act in 1984 codified a single Abbreviated New Drug Application pathway for approval of all generic drugs, bioequivalence was required for all drugs as a commonplace regulatory concept, and the FDA was at the center of defining what bioequivalence meant.[31]

FRAGMENTED SCIENCES OF SIMILARITY

But if bioequivalence was coalescing into a more singular definition *within* the FDA, it still meant many different things to different members of the pharmaceutical industry. As described in the previous chapter, research capacity in biopharmaceutics was heavily concentrated in a small number of brand-name firms. Even among Pharmaceutical Manufacturers Association firms, more laboratory techniques for assessing bioavailability had been pioneered by the Upjohn Corporation than any other firm. As the FDA developed bioavailability policies, other prominent firms cried foul that the federal government was giving an unfair competitive advantage to Upjohn.

In 1974, for example, officials from Abbott Laboratories complained to Gaylord Nelson that Upjohn should not be able to use its bioavailability expertise to promote its version of the antibiotic erythromycin, complaining that "the main point that we wish to make is that Upjohn is using bioavailability studies for promotional purposes. To do this they are designing their bioavailability studies to be biased in favor of their product and negatively biased toward competitive products. Certainly we feel this is a prostitution of the science of bioavailability and does little credit to scientists who allow such distortion to occur."[32] The situation was far worse for generic firms, who now had to shoulder the additional costs of bioequivalence testing to bring their products to market. In the wake of the Hatch-Waxman Act, industry analyst Joseph Barrows worried that the cost of bioequivalence testing would "defeat the purpose of the Act as it will limit competition and your price for the drug may not be any different than that of the brand name company." Barrows cited the FDA's new standards for measuring bioequivalence for Valium (diazepam) tablets, which required nearly two thousand blood assays on thirty human subjects over sixteen days, with an estimated cost to a prospective generic manufacturer of $75,000–$125,000. The FDA's method, Barrows argued, was only one of many possible ways of testing bioequivalence. Why, for example, should these tests require human volunteers? Use of animal models for in vivo testing, for example, would be much cheaper, faster, and produce more standardizable answers. Of the many possible forms bioequivalence testing could take, he concluded, the FDA had chosen the one that would do the most to undermine the economic and social value of the nascent generics industry.[33]

While generic companies were complaining of the excessive burdens of the new bioequivalence regime, brand-name manufacturers and many physicians complained of their insufficiency. Bioavailability was a solution to the limitations of chemical equivalence, but it was not a simple, or even a singular, solution, and it could in turn be destabilized by alternate regimes of difference.[34] For example, if two drugs were the same for most people, who was to say they were the same for *everybody*? At a combined House/Senate hearing in July of 1983, Peter Lamy, professor of pharmacology at the University of Maryland, brought new research in pharmacokinetics to bear in a new critique of bioequivalence. Not only was it necessary to provide different tests of sameness for different types of drugs, Lamy pointed out, but also it was necessary to provide different tests of sameness for different kinds of bodies. Just because two drugs worked the same in most healthy adult bodies did not mean they would work the same in all kinds of bodies. Elderly bodies, for example, were known to have different pharmacokinetic properties than younger bodies, due to age-related changes in kidney function and muscle mass. Lamy warned that standards of bioavailability were vast generalizations that "aimed at 'average' and did not take into consideration the many factors revolving around the geriatric patient, which demand a much more individualized approach."[35]

The elderly were only one of many different kinds of people who could be segmented out as having a kind of body for whom standards of therapeutic equivalence needed to be considered differently. Another way to define kinds of bodies for whom a similar drug might not produce the same effects was by disease category. In January of 1986, Ayerst circulated a generic alert to all pharmacists in the United States that suggested that even if the new generic versions of its beta-blocker Inderal (propranolol) might pass the FDA's bioequivalence requirements, only Inderal was permitted to claim a new FDA indication for prevention of myocardial infarction in patients who had suffered their first heart attack.[36] For some conditions—like high blood pressure—a patient might do as well on generic propranolol. But for patients who had just suffered their first heart attack, only Inderal would do. Ayerst sent out other "Dear Pharmacist" letters suggesting that those who dispensed generic propranolol would be exposed to potential litigation by any patient who suffered a heart attack while on the generic pill.[37]

Other critiques of bioequivalence were made by statisticians—most notably Sharon Anderson and Walter Hauck of the University of California, San Francisco, who introduced in 1990 a new concept of "individual bioequivalence" to suggest that bioequivalence data based on populations (which sufficed to describe the initial "prescribability" of a drug to a patient) did not necessarily translate to equivalence data for each individual in a population (needed to prove the "switchability" of a drug a patient was already taking). In February 1993, as the FDA took up the subject of individual bioequivalence in its Generic Drugs Advisory Committee, Anderson testified that "the average person defines 'bioequivalence' to mean 'it doesn't matter which one I take'" but current FDA standards could not necessarily guarantee this statement was true. For the remainder of the 1990s, the generic drug review project of the FDA was convulsed over a series of proposals to mandate proof of individual bioequivalence to market generic drugs, including two draft guidances published in 1997 and 1999. Although these protocols for individual bioequivalence were abandoned in 2003 after both brand-name and generic companies came to oppose them as overly costly and the FDA conceded that individual bioequivalence was largely a "theoretical solution to a theoretical problem," the episode stands as another reminder that the development of new therapeutic sciences of similarity and difference did not merely stop with the embrace of bioequivalence by the Hatch-Waxman Act of 1984.[38]

THE ONE THE PATIENT TAKES IS NEVER TESTED

Bioequivalence was not the only battlefield. Other PMA firms rejected the entire project of the regulatory sciences of similarity. No matter how precisely the evidentiary basis of chemical or biological similarity was set for the approval of generic drugs, these tests and protocols were only as good as the social institutions that bound them to actual products circulating in actual markets. The FDA might be able to prove equivalence in the rarefied environment of the laboratory, these firms argued, but could they prove that equivalence was maintained outside of the schedule of FDA inspections?

This ethos was captured concisely in a new advertising campaign launched by Eli Lilly and Co. in 1976 (figure 8) that reminded physicians, pharmacists, and consumers in large type that "the one the patient takes is never tested." Supporting text was devoted to detailing the limitations of bioequivalence as a regulatory concept. Simply put, the dose of any drug actually consumed by a patient could never be verified by techniques of analytic chemistry or bioavailability, because these techniques were destructive and rendered each sample unfit for consumption. The Lilly ad speaks to another aspect of similarity far harder to quantify: that physician, patient, and pharmacist alike "depend on the manufacturer for assurance that the dose the patient takes is identical to the ones which have been tested."[39]

This firm-based guarantee of similarity was grounded not in chemical or biological sciences but on managerial sciences of tolerance, quality assurance, and quality control. "Quality" here became a potential space that could differentiate private forms of industrial know-how from public enforceable standards. To Lilly's marketing and public relations staff, the provenance of quality could not be regulated by any force external to the manufacturing firm itself: "At each step in the manufacture of a Lilly drug, test after test confirms the ingredients, formulation, purity, and accuracy—all the critical factors which assure that every Lilly medicine is just what the doctor ordered . . . And, of course, government standards alone do not assure the efficacy and consistency—the quality of each drug you dispense. As we at Eli Lilly and Company see it, the ultimate responsibility for quality is ours. For five generations we've been making medicines as if people's lives depended on them."[40]

The ironic distance in the last line may have been intended for professional audiences, but Lilly's appeals to pharmacists and patients were accompanied by more direct appeals to state legislators and public and private payors. In a briefing to the Indiana state legislative committee considering its new substitution law, Lilly argued that "even if there existed an infallible inspection program administered with unfailing diligence, significant quality variations would continue, for quality cannot

FIGURE 8. Eli Lilly advertisement: "The one the patient takes is never tested."
American Druggist, *September 1977.*

be inspected or regulated into a drug product. It must be built in by the manufacturer. Thus quality has in the past and will be in the future the result of the knowledge, skill, and motives of the manufacturer."[41]

Lilly's position—that only the private sphere could regulate quality—

willfully ignored a half century of the regulation of pharmaceutical quality at the FDA through public managerial sciences. As FDA historian John Swann has described, the protocols of manufacturing similarity known as Good Manufacturing Practices had their origins in a lesser-known tragedy of the early wonder-drug era. In 1941, more than three hundred people nationwide were killed or injured after consuming tainted batches of Winthrop's antibacterial sulfathiazole. Winthrop had been aware, as of late 1940, that some batches of its sulfathiazole tablets failed disintegration standards and had been contaminated by sharing a production line with tablets of the potent barbiturate Luminal, but the company had not notified the FDA of the contamination. The subsequent FDA investigation, tracking 1.5 million invoices and three thousand individual consumers, revealed the origins of the problem in poor manufacturing practices in the Winthrop factory. The machines used to compress sulfathiazole and Luminal tablets sat right next to one another in the same room, and there was no oversight of the personnel in charge of making the tablets.[42] After 1941, the FDA investigated all forty-seven firms manufacturing sulfathiazole explicitly to check their quality control procedures.

With the passage of the Kefauver-Harris Amendments of 1962, the FDA consolidated its experience with sulfathiazole and other forms of factory inspections into a formal set of protocols moving forward to ensure GMPs in the pharmaceutical industry, along with newly granted authority to prospectively inspect the factories of all companies submitting a New Drug Application to the agency. By the late 1960s, the FDA had expanded authority over drug quality assurance in four main areas: plant inspection, product surveillance, premarketing clearance, and certification of specific products (i.e., insulin products and antibiotics). While some degree of plant inspection had taken place since the 1906 law, the 1962 amendments broadened those inspections to specifically address eight core areas of good management: processing of dosage forms, upkeep of personnel and facilities, control of raw materials, comparison of formulas, manufacturing controls, sterile procedures, assays, and inventories. In addition, for a certain group of prespecified drugs—insulin and antibiotics—the FDA required evidence of batch certification in the interwar years. These powers were formally granted by acts of Congress in 1941 (for insulin products) and 1945 (initially for penicillin but extended to all antibiotics).[43]

Yet, as the case of chloramphenicol demonstrated, existing batch-certification techniques had been insufficient to pick up differences in bioavailability. In July of 1968—partly in response to the Chloromycetin case—these spot checks were extended into a system known as the Intensified Drug Inspection Program (IDIP), which required a more complete immersion of inspectors in manufacturing plants. Unlike prior inspections, the IDIP allowed for revisits to assess compliance (or lack of compliance) with recommendations made during the last visit. The FDA conducted IDIPs of 287 firms between 1968 and 1971. FDA field agents also collected dosage forms of all kinds from drugstores in their territories and send them back to the FDA laboratories for forensic evaluation (a practice named "pillistics" after the forensic science of ballistics; what worked for lead bullets should work for magic bullets as well). In the year 1967, the FDA conducted 43,388 domestic drug assays, which generated grounds for 631 recalls, 389 seizures, and 19 prosecutions. Estimates of FDA officials suggested that to perform optimal surveillance the FDA would need to increase its capacity by three to six times.[44]

In ensuing decades, FDA inspectors worked continuously to adapt quality control measures from industry into a regulatory framework. But these managerial sciences of quality control could still be used in different ways by manufacturers and regulators. Manufacturers claimed to have internal quality standards that exceeded the minimal FDA standards for quality. Regulators, in turn, claimed that any managerial sciences developed within a single firm could be generalized to a national regulation through extended policing and inspection by FDA field agents. In turn, manufacturers could claim new standards of difference, maintaining a distinction between the public use of quality control to demonstrate generic equivalence and the private use of quality control to differentiate the added value of a given firm's own brand-name products.

By the mid-1970s, the FDA expanded this approach of "national survey with automated technique" to thirty target drugs seen to be possible candidates for bioequivalence problems, including steroids, major and minor tranquilizers, anticoagulants, and oral contraceptives, among others.[45] FDA Bureau of Drugs director Richard Crout noted that his organization had "ample evidence to indicate that quality control problems can occur in any drug firm," not merely generic manufacturers.[46] Crout's sentiments here would be echoed by Marvin Seife, testifying in front of the Drug Formulary Commission of the Massachusetts Department of Public Health on September 8, 1977, that "a review of the recall lists for the past several years reveals the names of many major manufacturers as well as those that are not so well known. From this FDA is unable to conclude that there is any clear difference between large and small firms or between brand name and generic labeled drugs."[47]

But the FDA's recall data could also be used by epidemiologists at brand-name firms to tell different stories about recalls, brands, and generics. With a flurry of press releases and glossy displays, Lilly announced its own scientific review, entitled "FDA Enforcement Activities within the Pharmaceutical Industry: Analysis of Relative Incidence." Lilly researchers obtained a full series of the FDA's own court actions against pharmaceutical firms and compared the relative incidence of punitive regulatory action between brand-name and generic manufacturers by frequency and severity (figure 9). All in all, FDA court actions were taken against "nonresearch" companies 43 times more frequently than against research-based drug manufacturers. Generic firms were subjects of more recalls by a 7 to 1 ratio and had 50 percent more product problem reports as research-intensive firms. This work was widely promoted by the PMA, detailing the recalls and seizures of non-PMA compared to PMA firms.[48]

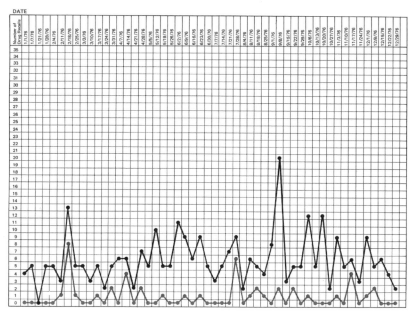

FIGURE 9. 1977 Eli Lilly study of FDA recalls of prescription and over-the-counter drugs from Kremers. The distance between the lines indicates the differential rate of recalls between generic and brand-name manufacturers. Darker line = total recalls from generic manufacturers; lighter line = twenty-seven brand-name manufacturers.

C 46 (p) I f 1, trademarks, tradenames, and product substitution—United States, KRF, University of Wisconsin School of Pharmacy.

These competing epidemiological studies of brand/generic difference illustrate the dynamic quality of the reputational sciences of similarity and difference. The divergence between the two curves of the Eli Lilly study accentuated the PMA's argument that, in spite of whatever testing was done to approve the similarity of a drug, the behavior of the firm itself varied by brand name. This graphical evidence was mobilized for legislative briefings on state and federal levels.[49] Generic drug advocates and FDA staffers shrugged off the Lilly analysis, arguing that the effect was due to a difference in denominators. Nonetheless, many state legislators heard these messages, as did physician groups such as those associated with the conservative journal *Private Practice,* which continued to promote the image of generic similarity as a form of willful deception that masked underlying differences (figure 10). These reputational arguments, which hinged on moral claims of reliability and deception, left a specter of nonexchangeability in practice and reputation that no amount of spot testing of bioavailability or regulatory gatekeeping could ever fully dispel.

BEYOND BIOEQUIVALENCE

While would-be therapeutic rationalists continue to dream of a therapeutic system composed of interchangeable parts that can be redesigned to optimize overall performance, brand-name manufacturers continue to argue that only the private standards of brand-name medicines can guarantee the high standards of American medicine. This debate is perpetuated by lingering contestation of therapeutic equivalence on multiple registers of theory, methodology, politics, and practice. To that effect, the standard of bioequivalence has not resolved but merely displaced earlier critiques of the insufficiency of pharmacopoeial standards of similarity. New sciences of similarity are still emerging. In June of 2013, as I was editing this chapter, the FDA held a public hearing on new forms of generic regulatory science funded by the Generic Drug User Fee Act of 2012 that might be recruited to prove new forms of equivalence where conventional bioequivalence was not enough. One of the exhibits presented—an artificial bronchial tree on which one could measure the deposition of inhaled drugs like corticosteroids and bronchodilators for the treatment of asthma and emphysema—bore a striking kinship to the failed Erweka AT-3 of a half century earlier.

We move through a world of goods whose positions within markets are predicated on claims of similarity and difference. We know implic-

FIGURE 10. Well into the late 1970s, conservative physician organizations such as the Congress of County Medical Societies continued to promote the image of generic similarity as a form of willful deception that masked underlying differences among pharmaceutical products.

"Generic Deceit: 'Look-Alike' Drugs and Your Patients," cover image, Private Practice, *May 1978.*

itly that one Hershey's bar on a store shelf can be considered the same as any other Hershey's bar, while at the same time we know this is not true: each of these Hershey bars *must* have a slightly different weight (if measured in micrograms) and a slightly different imprint (if examined under an electron microscope), must contain cocoa beans from a slightly different part of the world (depending on global fluctuations in commodities pricing), and may have been manufactured and packaged in any of a number of plants in the United States or elsewhere. And yet at the same time we are perfectly able to accept that all of these Hershey bars are similar enough to be functionally exchangeable. When a difference does erupt—when, say, the color is off or the chocolate is all melted into a clump or the stamp comes off funny, our sensibilities are offended. We may write to the Hershey Company demanding recompense; we may stop buying Hershey's products altogether. Assumptions of similarity, typically unseen, permeate our lives and allow us to function as consumers in a world of goods. We want to believe in Lipitor as we do in Hershey, even if each Lipitor tablet is constituted in a far more complex web of chemical supply and assembly plants stretched all over the globe. Our expectations of similarity are all the more important—and all the more difficult—when the good being consumed has vital consequences.

Generic drugs never were—and still are not—exactly the same as their brand-name counterparts, but they are similar in ways we deem to be important. The regimes of values that matter when we try to prove that two therapeutics are the same, however, have not been constant over time. Knowledge of which aspects of a given drug's materiality need to be attended to in order to demonstrate equivalence has proven to be a moving target. In this light, instead of situating this transition as a simple tale of progress (after decades of wandering in the wilderness, regulatory science finally found the right answer in bioequivalence) or as a mere shift from an old science of similarity (chemistry) to a new science of similarity (biology), the material in this chapter demonstrates that the brand/generic controversies of the 1960s and 1970s formed an active interface for the generation of a continually evolving series of sciences of similarity.

The sciences of similarity are multiple. They can be catalogued along many axes of potential difference: chemical, physical, serological, statistical, procedural, reputational. But they cannot merely be dismissed as interested nor elevated as disinterested. John Wagner's and Alan Varley's research on brand/generic difference was funded by the Upjohn Company in a context that cannot be separated from Upjohn's interest in extending its trademark monopolies long after patent expiry. Nor can Richard Crout's documentation of the efficacy of GMP inspections be separated from his own position as the Bureau of Drugs chief at the FDA. Yet, interested as these positions may be, the outcomes of these new forms of scientific engagement were never fully predictable in their origins. New techniques of proof and protocol cross over between industry and regulator and are fully captured by neither party. New sciences continue to emerge as the dynamic intersection of clinical practice, pharmaceutical marketing, and consumer regulation attaches new kinds of vital information to pharmaceutical products. If we accept Gregory Bateson's definition of information as "a difference that makes a difference" then the late twentieth-century proliferation of sciences of difference and similarity is another manifestation of the increased salience of drug-related information in American public life as we move into the early twenty-first century.[50]

These sciences of similarity became the rails on which the engines of generic substitution and therapeutic interchange now move. But these engines did not follow a predetermined course, and their passage—as the next chapter will describe—was rarely smooth, no matter how well laid the tracks.

PART IV

LAWS OF SUBSTITUTION

CHAPTER 8

SUBSTITUTION AS
VICE AND VIRTUE

*I am against drug substitution, but with equal vigor, I am also in
favor of drug substitution.*

—BRUCE CHADWICK, FEDERAL TRADE COMMISSION COUNSEL, 1976

Earl L. Casden was by all accounts a modest man, who ran a small community pharmacy in Michigan. One day in 1949 he was presented with a
prescription by a Mr. Robinson for Schering's Meticorten, one of several
versions of the synthetic steroid prednisolone available on the market.
Casden did not have any Meticorten in stock, and so he called Robinson's
physician and asked whether it was all right for him to give the patient
Upjohn's version of the drug instead. Even so, the Michigan State Board
of Pharmacy ruled that Casden was guilty of an act of substitution, a
violation of state law (M.S.A. § 14.741) and the ethics of the pharmaceutical profession. His penalty was the suspension of his license for a full
week—at great loss of income to his pharmacy.[1]

Casden sued for damages. In his defense, he argued that he was no
criminal, merely a pharmacist trying to do the best thing for the patient in
front of him. Would it not have been far worse for the patient to go without the drug he needed, just because the exact brand was not in stock?
Many companies made prednisolone products in 1949, and it was prohibi-
tively expensive for pharmacists like Casden to stock every single one just
to please the brand preferences of prescribing physicians—especially if
all products contained the same active pharmaceutical ingredient. The
circuit court judge agreed with Casden that since "chemically and by
assay the drugs were identical," Casden should be absolved of the crime
of substitution.[2]

To understand why Casden was caught in this particular bind, one needs to understand that, to the average pharmacist in 1949, substitution was a cardinal sin. For the pharmacist to dispense something other than exactly what was prescribed was considered a "gross immorality" that violated the ethics of the profession and community. Substitution was the pharmaceutical equivalent of a physician lying to a patient about a diagnosis or a surgeon accepting fees for a procedure not performed. As a "basic evil," drug law attorney Joseph H. Stamler would note a few years later, "we must consider how substitution breeds." Known by its proper name, Stamler continued, "the act of substituting has been defined as 'beguilement,' 'counterfeiting' 'connivance' 'misrepresentation' and by many other vividly apt and descriptive phrases."[3] Yet as the *Casden* case was appealed and reconsidered through the early 1950s, the state of Michigan became one of the first in the country to propose a law that would make Casden's actions legal and expressly define brand interchange of chemically equivalent products as a *virtue* instead of a vice.[4] Some forms of interchangeability, like switching the brand of a chemically equivalent drug for the good of a patient, might be understood as pragmatic, not deceptive.

Substitution is an action in which the politics of similarity intersect with the structures and contexts of product interchangeability. Substitution can be virtuous and rational if it enables a system to function under duress, like a public elementary school that needs to staff teachers every day even when a given teacher is ill. Alternately, substitution can be a vicious or deceptive practice if it constitutes "palming off" one good for another, like dispensing generic cola in place of Coca-Cola but charging the full price of the brand-name product.[5]

Substitution becomes all the more complex in the realm of prescription drugs, in which the interaction between consumer and product is mediated through a chain of state and professional judgments. If your physician prescribes a particular brand of drug for you—let's say the cholesterol-lowering drug Lipitor (atorvastatin)—would you be pleased or displeased to see a pharmacist fill your prescription bottle with a cheaper version of generic atorvastatin? Would you be more comfortable to learn that your pharmacist was required by law to give you the cheapest possible version of atorvastatin, or would that seem to you to be an unreasonable constraint? Depending on where you lived in the United States, your state legislature will have interpreted your own public interest quite differently: some states force substitution, others permit it, others—for many years—expressly prohibited it.

Earlier chapters discussed the problem of calling drugs the same, of making and marketing drugs as the same, and of proving that drugs are the same in ways that matter. But the generic project took on its greatest urgency in the policies and protocols of therapeutic interchange: the set of rules, laws, and incentives that allowed one drug to be substituted for another in practice. These politics of substitution were fought, for the most part, at the level of the states.[6]

"A GROWING EVIL": THE CRIMINALIZATION OF BRAND SUBSTITUTION

The *Casden* case exposed the pharmaceutical industry to the impending threat of generic-name substitution. Even though the proposed prosubstitution law in Michigan was defeated, firms like Upjohn and Schering alike began to worry that pharmacists might begin dispensing medication by generic name regardless of what physicians wrote on their prescriptions. In 1953 they were joined by ten other leading pharmaceutical manufacturers to form a collaborative organization to work with pharmacists to "promote drug standards and ethics of pharmacy and pharmaceutical manufacture" and to criminalize brand substitution. By 1960 twenty-two large firms composed the National Pharmaceutical Council, a political interest group engineered to create a favorable policy environment for the economic interest of pharmacists and large drug manufacturers. As the sociologists Neil Facchinetti and W. Michael Dickson would later argue, the NPC effectively engineered interest in brand-generic substitution as a "social problem" through artful management of professional and consumer concerns.[7]

The NPC used many strategies to educate the public on the evils of brand substitution and the need to enact a series of laws to prevent it. The organization trained employees of individual drug firms and state pharmacy boards to gather material evidence of brand substitution by sending company agents into pharmacies as fake patients armed with real prescriptions, to see which pharmacists would fill products with brand-name pharmaceuticals and which would offer generically named drugs as a substitute. Another NPC tactic encouraged the use of secret tracer molecules within brand-name pharmaceuticals so that they could be distinguished from "look-alike" counterfeits by chemical analysis. The NPC helped local organizations coordinate these efforts and translate them into meaningful policy briefings for state pharmacy boards and legislatures.[8]

The Pharmaceutical Manufacturers Association and American Pharmaceutical Association also jointly created public relations bureaus such as the Health News Institute and the Medical Publicity and Information Bureau, which placed positive articles on prescription drugs and pharmacists in popular magazines such as the *Saturday Evening Post* and *Good Housekeeping*.[9] One NPC publication on the importance of brand names, "Why All the Mystery in Prescriptions," was first published by MPIB author Donald Cooley in the June 1956 issue of *Better Homes and Gardens*, then packaged as a pamphlet and circulated to a broad network of physicians, pharmacists, policymakers, and consumers, with nearly a million copies distributed in total. Periodical and pamphleteering operations were accompanied by radio and television releases, speakers' bureaus, seminars for deans and faculties of pharmacy schools, and direct liaisons with state boards of pharmacy and state legislatures.[10]

Priority number one for the early NPC was the portrayal of brand-name substitution as "a growing evil," a moral crisis for pharmacists, physicians, and consumers alike.[11] In a move to pre-empt future losses like *Casden*, the NPC drafted a model legal definition of substitution as "the dispensing of a different drug or brand of drug in place of the drug or brand of drug ordered" and worked with state boards of pharmacy to place this language in all state laws governing pharmacists.[12] By early 1956 this language was written into state law in New Jersey, Oregon, and Pennsylvania; others soon followed.[13] The NPC also worked with contract writers to graft this new criminalization of brand-name substitution onto older legal and ethical codes of pharmacy.[14]

The NPC's attempt to sensationalize brand substitution represented a successful alliance of physicians, pharmacists, and the organized pharmaceutical industry. Even skeptical audiences, such as George F. Archambalt, the chief pharmacist for the US Public Health Service, noted the success of the NPC's campaign to redefine the "spreading evil" of substitution:[15] "Some 28 years ago when I started to practice pharmacy, 'substitution' meant one thing—the dispensing of a wrong chemical or drug, one different from that prescribed. Only occasionally did we hear 'substitution' then being applied to trade versus official substances. Then came

1955 and the NPC definition for substitution that enlarged the meaning to include a switching of brands without the expressed permission from the prescriber."[16] By 1959, the NPC could boast that forty-four states had introduced regulations to prohibit brand substitution.

The NPC tied together the mutual interests of pharmacists (APhA), American physicians (AMA), and brand-name pharmaceutical manufacturers (PMA). But by the mid-1960s this coalition was showing signs of wear.[17] The new president of the APhA, William Apple, began to articulate in 1966 that in the clinical arena knowledge about drugs was "more effectively utilized when the pharmacist has a greater role in the selection of the drug product to be dispensed than when that decision is preempted by the mechanism of a brand name." Apple represented a growing number of pharmacists for whom—as for Earl Casden—the criminalization of substitution had created unreasonable pragmatic and economic burdens.[18]

Apple's appointment as head of the APhA coincided with the enactment of Medicare and Medicaid and the ensuing expansion of federal and state government roles as payors for pharmaceuticals for the indigent.[19] The shifting role of the state in paying for drugs created both a new crisis and a new opportunity for pharmacists. Many were concerned about what would happen when the state as payor contradicted the state as regulator of professions regarding drug substitution. As APhA president Dean Linwood Tice noted in a letter to the chief research pharmacist at Smith, Kline & French, "some of our pharmacist members are being told by no less an a authority than the attorney general of a state" that substituting generics for brand-name drugs in the case of Medicaid patients "were in fact not a violation of those [antisubstitution] laws." Tice's contact at SKF brushed off this equation of substitution with poverty, asking why it was that "substitution is good for patients unable to pay for their own medicine, but perhaps substitution is not so good for those able to pay?"[20]

But the issue could not be brushed off so easily. At the May 1966 meetings of the APhA in Dallas, the Legislative Committee agreed that "the reason for the enactment of the antisubstitution laws in the states, even if valid at some prior time, no longer exist, particularly as these laws apply to governmentally-financed programs."[21] By the fall of that year, William Apple had created a special committee on substitution to address the problem, and the APhA House of Delegates recommended repeal of antisubstitution laws that no longer protected the interests of pharmacists or patients.[22] The *American Journal of Pharmacy* issued an editorial by Tice stating that antisubstitution laws could have no effect on the practicing ethics of pharmacists anyway, since such mandates violated the moral basis of pharmacy: "That day in American pharmacy when the profession is willing to sit quietly as a pawn of other vested interests is past. The profession has come of age. Its full services are needed if health care in this country is to achieve that level of competence that it must. Pharmacists are concerned that their role shall not become a purely mechanical one. They, too, can contribute toward improved patient care at a cost the nation can afford, and they are determined to do so."[23] Hundreds of practicing pharmacists were stirred by this call and wrote letters to the APhA and prominent journals of pharmacy in support. "It seems to me," A. E. Rothenberger, a Vermont community pharmacist, wrote, "that after five years in pharmacy school a pharmacist should be able to assume some responsibility rather than being a pill counter."[24]

In 1971—the same year that Missouri and Alaska became the last two states to formalize their antisubstitution laws—the APhA issued a controversial white paper calling for their repeal nationwide. This single action would shatter the foundations of the NPC, damage relations with the PMA and American Medical Association, and create a deep schism within the profession of pharmacy itself. Pharmacists working on or near the research and development process warned bluntly that this stance threatened continued support of the APhA by the pharmaceutical industry. Those who had risen to executive positions in the pharmaceutical industry wrote Apple with both civility and firmness. "Bill," one executive at Mead Johnson wrote to Apple, "you have a real facility for doing things once in a while that we do not think are good for your profession of pharmacy. You know how we took exception to your 'brand name' speech. Now we are confronted with your 'anti-substitution' program, which we believe is definitely a step backwards."[25]

The schism would not be healed. Later that year a large body of industry pharmacists seceded from the APhA to form the Academy of Pharmaceutical Sciences. Yet this growing divide between pharmacists and the pharmaceutical industry was occasionally generative of unusual transformations. In June of 1971, one pharmacist, Edward O. Leonard, wrote to terminate his membership in the APhA after learning of the organization's new prosubstitution stance. "I do not wish to lend my name or money," Leonard wrote, now that "the Association's support for the repeal of anti-substitution laws is completely alien to my own beliefs."[26] Yet after following the debate and watching carefully the increasing problems of stocking multiple versions of ampicillin and other drugs, Leonard found his position changing.[27] By October of the same year he wrote back to the APhA that he was prepared to "eat crow" after witnessing the "rapid proliferation of so called 'branded generics' ... Of course they are no more than 'me too' trade names, and if the only way to control their growth is through abolition of 'anti-substitution' statutes, then I must reluctantly support such effort."[28]

TECHNOLOGIES OF SUBSTITUTION

If research pharmacists working in industry or industry-funded academic laboratories had a particularly hard time accepting the virtues of substitution, this was not the case for those pharmacists who worked in hospitals. Even at the height of the antisubstitution era, brand substitution was always understood to be legal in the hospital. The principal tool used by hospital pharmacists to create and defend their own professional prerogative over substitution was a simple pen-and-ink technology called the formulary. This tool, a deceptively simple list of drugs, provided hospitals with a set of rules of substitution that linked prescribing physicians and dispensing pharmacists.[29]

"In its best sense," wrote the prominent hospital pharmacist Donald Francke in 1967, "a hospital formulary is therapeutic guide to the medical staff. Its purpose is to promote rational drug therapy." Like the pharmacopoeiae discussed in earlier chapters, formularies contained lists of drugs. But where pharmacopoeiae attempted to list *all* official remedies, formularies only listed medicines to be stocked at a given institution. Pharmacists might consult a pharmacopoeia to have a global sense of the knowledge of available therapeutics; they would consult a formulary to understand the medicines available—and increasingly interchangeable— at their own local institution. A given hospital's formulary formed a sort of Noah's ark of therapeutics, in which each kind of medicine had only one representative, to be obtained at the cheapest possible price at which quality standards could be guaranteed.[30]

The first hospital formulary in the United States, the *Pharmacopoeia Nosocomii Neo-eboracensis*, was created in 1816 to solve inventory problems of the New York Hospital, at the time one of only two voluntary hospitals in the country. At first the academic hospital formulary concerned itself mostly with practical methods for collecting and preserving herbs and producing basic therapeutic chemicals. But by the 1930s the formulary articulated a broader role in "the problems of selection of drugs for the rational management of disease." By the 1950s, the director of the formulary at New York Hospital could describe the tool as a "system of limitations" for "the generally accepted method of providing rational drug therapy in hospitals."[31]

The formulary system required four principles to function: first, the list of exchangeable drugs had to be approved by the medical administration of the hospital. Second, all physicians who wished to admit or treat patients at a hospital had to consent to the formulary. By 1959, all physicians who wished to admit patients to the New York Hospital had to sign a statement agreeing that "I am confident that the New York Hospital's formulary system, dating back 140 years, meets every possible moral, ethical, and legal requirement of sound medical and hospital practice." Third, it required an interdisciplinary expert panel of physicians and pharmacists, the Pharmacy and Therapeutics Committee (P&T Committee), that closely monitored the drug lists and quality of drug supplies the hospital maintained. Finally, it required acceptance of the use of generic names as a form of communication, prescription, and drug product selection. By 1961, the formulary director of the New York Hospital could boast that similar systems had now been put in place in 60 percent of all major hospitals in the country.[32]

Formularies spread quickly, and as they spread they took on greater talismanic and legal authority for defining a space for generic substitution.[33] Their successes in this arena did not escape the notice of the NPC, which in the late 1950s started but then aborted a campaign against hospital formularies as an "illegal and unethical" practice. Some individual pharmaceutical manufacturers continued to protest the power of the hospital pharmacists into the early 1970s. "I will challenge anybody anywhere to justify the assumption that other products bought at a discount price by some hospital pharmacist are 'similar to' Mead Johnson Products," one executive wrote hotly to the head of the APhA. "They may contain the same basic chemical, but this is where the similarity ends. It is presumptuous for any hospital pharmacist to label off-brand products as being 'similar to' our Mead Johnson products."[34] Yet the industry soon found that the hospital pharmacist's legal right to substitute was well defended by the structure of the formulary system—especially the signed consent of physicians associated with a given hospital. What had begun as an inventory tool had been reconceptualized as a tool for encouraging rational drug therapy, improving institutional cost effectiveness, and establishing rules of substitution.[35]

Although formularies were relatively uncontroversial in inpatient settings, the use of formularies rapidly became more controversial as it spread outward from the hospital and into the programs of state and local welfare agencies, private insurers, and the federal government. Earlier attempts to control the cost of drug benefits associated with public welfare programs in Baltimore and New York in the 1950s asked all participating physicians to agree to an outpatient formulary of drugs for indigent patients. In the late 1960s, as the Health, Education and Welfare Task Force on Prescription Drugs began to evaluate the feasibility of a nationwide prescription drug program for Medicaid, the prospect of an outpatient formulary became a central topic of interest. As they described, the outpatient formulary was a far more divisive topic than the hospital formulary. On the one hand, from the perspective of academic reformers and public payors, "the use of a formulary has been termed an absolute essential for rational drug therapy, a guide without which physicians cannot hope to keep abreast of new information on old drugs and the profusion of new drugs and their related advertising and promotion, a vital necessity for a financially sound insurance program and for economic hospital operations, an urgently required tabulation of drug specifications and standards, and a valuable guide to current prices."[36] On the other hand, from the perspective of many physicians and pharmaceutical manufacturers, "its use has been decried as a destructive infringement on the traditional right of the physician to prescribe according to the dictates of his professional judgment, a maneuver to require generic prescribing, an attack on brand-name drugs, dictatorship by the Federal Government (or a State government, a hospital administration, or a union), and a dangerous effort to control drug therapy by means of price rather than quality."[37] As the HEW task force looked abroad to understand how other countries used formularies to make decisions about which drugs were necessary for the populace, they found many different approaches to the formulary concept that had become national policy in places from Norway to New Zealand. Certain segments of the US population, especially soldiers and veterans, already received care through "restrictive" formularies, as did some employers and private insurers.[38]

The United Mine Workers of America Welfare and Retirement Fund, one of the oldest employee-based health insurance systems in the country, had instituted a flexible formulary of drugs for treatment of chronic diseases by the late 1960s. The Kaiser Foundation health plan had likewise implemented a plan of regional "voluntary formularies" that did not restrict physicians' prescriptions but limited which brands would be stocked and dispensed in their outpatient pharmacies, clinics, and health centers. The Group Health Cooperative of Puget Sound also used a formulary of about two thousand items indexed by generic and brand name. Like hospital formularies, these private community formularies created a structure that simplified inventories and allowed for bulk purchase of the least expensive chemical equivalents. By 1977 the United Mine Workers of America health insurance plans had created a basic drug list that encouraged generic drug use.[39]

Private insurance programs were private affairs, perhaps. But when public payors began to establish outpatient formularies, the process invited far more political theater. Public drug benefits spread outward from municipal programs (such as those in Baltimore and New York City) to statewide programs like California's MediCal system. At its inception in 1957, this program initially guaranteed that indigent Californians could access *any* drug on the market; but within a year drugs comprised nearly 50 percent of the program's expenditures, with projections for the drug benefit increasing. In 1958 California initiated a restrictive formulary of 350 drugs, which cut the total costs per beneficiary by roughly 50 percent; after 1967 Medicaid programs encouraged this process in other states. By the late 1960s, experimental formularies for outpatient drugs were also built into the expanding system of neighborhood health centers funded by the federal Office of Economic Opportunity as part of President Johnson's War on Poverty. Of the thirty-one states whose Medicaid systems included drug benefits by 1967, thirteen had established welfare systems with drug benefits under listed formularies.[40]

Some of these states hired advisory committees of physicians and pharmacists to produce their own formularies. Other states merely adopted existing lists, such as the inclusive but not particularly selective *Physicians' Desk Reference*. As a result, the size of formularies varied state to state, from nearly twenty-four thousand chemical entities allowed in Pennsylvania to only one hundred allowed in Kentucky. Some states, like Louisiana and West Virginia, avoided naming drugs altogether and simply listed a formulary of "specified serious and chronic illnesses" covered by the program. Louisiana covered pharmaceutical costs for only forty-three diseases. Other states kept lists of nonreimbursable drugs—a kind of anti-formulary or therapeutic blacklist. These lists include over-the-counter items, basic analgesics, antacids, antiobesity drugs, and multivitamins—anything not seen to be an essential medicine for treating a serious or life-threatening disease.[41]

Nor did all states use formularies to encourage generic substitution—some used them entirely to the contrary. The West Virginia public drug fund as of 1967 explicitly stated that public funds should be used to reimburse brand-name instead of generic products, stipulating that "payment from the Medical Services Fund is limited to drugs marketed by pharmaceutical companies involved in basic medical research, product development, and marketing." The majority of states, however, like California and Maryland, used formularies to exempt pharmacists from antisubstitution laws and developed preprinted prescriptions for patients receiving state assistance that expressly permitted pharmacists to substitute drugs despite the antisubstitution laws still on the rolls.[42]

As the politics of the formulary played out differently in different states, a lumpy map of brand substitution was traced across the United States. By the close of the 1960s, some states, such as Utah and Minnesota, had dropped their formulary experiments entirely after direct opposition by physician groups. Other states, like California, Maryland, and Massachusetts, had in turn become bold laboratories for experiments in rational drug utilization and cost-effective public medical care. The unevenness of this map of drug availability would be further accentuated over the course of the 1970s, as state legislatures began to replace the antisubstitution laws with new laws favoring generic interchange.

LEGALIZING SUBSTITUTION

In 1971, the profession of pharmacy made a bold move to seek the repeal of antisubstitution laws. With a declaration by the APhA House of Delegates and the publication of a "White Paper on Drug Product Selection," organized pharmacy took a strong stand against both the AMA and the PMA and aligned itself with two new partners: the American Association of Retired Persons (AARP) and organized labor (the AFL-CIO). This realignment would significantly alter the field of pharmaceutical policy and link generic substitution to the growing consumerist movement in powerful new ways.[43]

The APhA's new partners were no strangers to state politics. To AARP, access to health care represented one of the few universally galvanizing issues the organization could count on to rally and unify the otherwise disparate political interests of retired persons. AARP's Washington offices quickly mobilized grassroots organizers to work on state and local levels to design and pass drug substitution laws across the country. The organization produced a "generic drug kit" containing a pamphlet entitled *A Short History of the Rise and Fall of State Antisubstitution Laws*, an AARP model state drug product substitution law, pamphlets refuting arguments of generic inequivalence, lists of local experts who could speak in favor of repealing antisubstitution laws, and scripts for patients to use when speaking with their physicians. As the AARP's Fred Wegner recalled, "[The scripts] concerned what to say to physicians who had refused to prescribe generically. The letter helped patients who wanted to 'talk back' to doctors who failed to take into account that cost should be a factor in the prescribing of medicine. It also pointed out that most of those who refused to prescribe generically were following a double standard with regard to patients seen in their offices and patients in hospitals."[44] While the AARP mobilized consumers, the APhA moved to educate its rank and file on the importance of overturning antisubstitution laws and teaching the principles of drug product selection in schools of pharmacy. Pharmacists, not physicians, should be considered to have the most appropriate training to discriminate between generic versions of the same pharmaceutical agent.[45] As one California hospital pharmacist would testify before Congress in 1976—as the call to repeal antisubstitution laws was at its loudest—-physician claims to superior knowledge of drug brands was a convenient fabrication:

> In the University of California hospitals we have physicians who start as students, go through their internship, residency, and they go out and practice, and I can tell you, through their entire training period, most of them have no idea of what brand of ampicillin, penicillin, tetracycline the hospital pharmacies dispense.
>
> So my question is, where is this knowledge acquired? Where, after the date the physician leaves his residency at a teaching hospital and he opens up on Fifth Avenue, does he become an authority on brand names of drugs? It is just not the way it works.[46]

The APhA-AARP alliance worked directly with state legislatures in the 1970s (much as the NPC had in the 1950s) to develop and pass new laws about brand substitution. As more and more states began to propose substitution laws, the PMA hired a lawyer explicitly to monitor the progress of state bills; by 1971, he was keeping track of 1,600 separate drug-related bills pending in state legislatures. In 1972, Kentucky became the first state to repeal its antisubstitution law. Two more states followed suit in 1973, and between 1975 and 1979 another thirty-five states followed suit; by 1984 antisubstitution laws had been repealed and replaced with prosubstitution laws in all fifty states, Washington, DC, Puerto Rico, and Guam (figure 11). But as substitution was legalized in different ways in different places, these new legal frameworks were sewn piece by piece into a patchwork quilt of generic drug policy that persists to this day.[47]

State substitution laws disagreed over how pharmacists should be allowed to use the technology of the formulary to make substitution work. Some states—including Kentucky, Massachusetts, New Hampshire, New Jersey, and New York—created positive formularies, or lists of drugs that

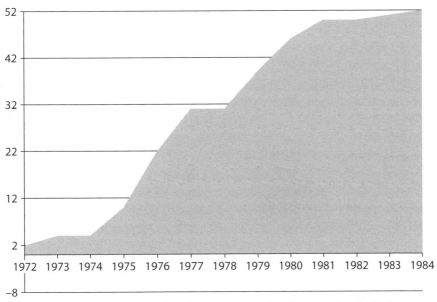

FIGURE 11. Number of states/provinces with laws favoring generic substitution. Adoption of substitution between 1972 (first law) and 1984 (last law) in fifty-two relevant jurisdictions, including Puerto Rico and Guam.

Adapted from Theodore Goldberg, Carolee A DeVitro, Ira E Raskin, eds. Generic Drug Laws: A Decade of Trial—a Prescription for Progress *(Washington, DC: US Department of Health and Human Services, 1986), p. 131.*

were generically interchangeable. Other states, such as Arkansas, California, Delaware, Florida, Iowa, and Washington, relied instead on negative formularies, which specified a group of drugs that were *not* substitutable and assumed all others were. Legislative battles over positive and negative formularies carried the conflict over the burden of proof of similarity discussed in chapters 6 and 7 into a new terrain. States with positive formularies asked for proof of equivalence; states with negative formularies asked for proof of difference.

Even within an overarching concept like a negative formulary, the actual contours of state substitution laws could vary greatly. Some states, like Arkansas, used as its negative formulary an FDA publication of 193 potential bioproblem drugs (which posed potential problems of bioequivalence). Florida's substitution law called for local physicians and pharmacists to create "a statewide negative formulary for drug products based on evidence of biological or therapeutic inequivalence," which produced a list of only fourteen problem drugs when first published in 1977. California's 1975 law stated, "*if anyone could document* for the state's Department of Health Services Food and Drug Section that a drug was bioinequivalent, that drug would be placed on a list (negative formulary) and could not be substituted." But the standard of proof of inequivalence was so high that APhA president Robert Johnson could state a decade later that "in the years since the implementation of the California law, no documentation of bioinequivalence has been presented to the Food and Drug Section and, consequently, no such negative formulary exists."[48]

States that had chosen to use positive formularies found the prospects of their development even more daunting, because they placed a greater burden of determining product equivalence on generic manufacturers and state agencies. Wisconsin, for example, issued its first formulary for generic substitution in August of 1976—a list containing only seven drugs allowable for generic substitution. Kentucky's positive formulary contained only forty-nine drugs by 1976; Rhode Island listed only thirty-two drugs by 1978. The small size of these lists of exchangeable drugs greatly minimized their potential economic and public health benefits. Indeed the expense of this process led Michigan to abandon its plans for a positive formulary, while Massachusetts complained that the cost of producing its formulary (through creation of the State Drug Equivalency Commission) might outweigh the cost benefits of drug product substitution.[49]

RULES OF SUBSTITUTION, LOCAL AND GLOBAL

This chapter has narrated the rise and fall of antisubstitution laws passed in most states in the 1940s and 1950s, which made it a crime for a pharmacist to dispense a generic form in place of a brand-name drug. Already by the late 1960s, state governments newly responsible for purchasing pharmaceuticals on a large scale began to search for ways to use the power of the purse to encourage substitution of the cheapest possible medicine in cases where all else was equivalent—even when this practice contradicted existing laws. By the early 1970s, a few states began to repeal their antisubstitution laws and then replace them with laws that encouraged or required pharmacists to use cheaper generic forms when possible. Public payors were joined in this project by private insurers, consumer advocates, and would-be therapeutic reformers from within academic medicine and pharmacy who sought to rationalize pharmaceutical use and medical expenditures through the large-scale substitution of generic for brand-name drugs.

Arrayed against these genericists (a term coined in opposition), the manufacturers of brand-name pharmaceuticals insisted that substitution was dangerous and based on a series of logical overreaches. In tandem, the organized medical profession portrayed generic substitution as a violation of the doctor's autonomy in choosing therapeutics for his or her patients.[50] In this parlance, substitution became a byword for the encroachment of state and federal governments on the practice of medicine and pharmacy—one of several intrusions on the sanctity of the doctor-patient relationship in the 1960s and 1970s. These forms of resistance to substitution will be explored in more detail in chapters 10 and 11.

Laws, regulations, and reimbursement structures favoring substitution were assembled in different ways in different places, torn down, patched together, and assembled again. As states began to overturn their antisubstitution laws and replace them with laws promoting substitution, they built formularies—a powerful set of paper technologies for promoting the interchangeability of therapeutic substances—to buttress and support new logics of therapeutic interchange. But this process did not take place evenly or in an orderly fashion, and it was opposed in different ways in different places. These local efforts to establish substitution programs were gentle at first, but as they took on increasing mandates for pharmacists, physicians, and patients, they attracted greater and greater controversy over whose choice mattered in the selection of drug products: the patient, the physician, the pharmacist, the insurer, or the state.

In 1973, HEW secretary Caspar Weinberger extended these debates over the rules of substitution from the level of the states to the federal government. In a series of announcements, Weinberger unveiled a new federal standard for Medicaid reimbursements "at the lowest cost at which the drug is generally available unless there is a demonstrated difference in therapeutic effect." This scheme, known as maximum allowable cost, or MAC, was typically set to the price of the generic drug. The MAC program constituted a nationwide positive formulary of drugs considered to be interchangeable and therefore reimbursable at the lowest cost. MAC reimbursement schemes were initially planned for thirty-two MAC-substitutable drugs; HEW projected $48 million in savings within the first year.[51]

MAC placed the burden of proof on those who would assert differences between brands and generics. But, as problems of bioequivalence and bioinequivalence of generic drugs became increasingly voiced at the FDA in this period (as described in chapter 7), Weinberger's program soon came under fire from those who would place the burden of proof on the side of similarity. Weinberger was forced first to declare that the MAC would not represent a "restrictive formulary," then to refrain from publishing any MAC recommendations until after the finalization of the ongoing Office of Technology Assessment inquiry into generic bioequivalence. When published in 1975, the Office of Technology Assessment report recommended a national negative formulary in place of a positive formulary program like MAC: "It is neither feasible nor desirable that studies of bioavailability be conducted for all drugs or drug products. Certain classes of drugs for which evidence of bioequivalence is critical should be identified. Selection of these classes should be based on clinical importance, ratio of therapeutic to toxic concentration in blood, and certain pharmaceutical characteristics."[52] By mid-1975, the FDA would extend this federal approach to a negative formulary by publishing a list of drugs thought to pose real or potential bioequivalence problems. This bioproblem list served at least three purposes. First, it identified for generic manufacturers those generic products that required bioequivalence testing to claim similarity. Second, it served as a model negative formulary for those seeking to create workable substitution programs. Third, it signified to the generic pharmaceutical industry at large that the FDA was consolidating its authority in the certification of therapeutic equivalence in the field of generic drugs.[53]

The FDA's entry into the world of published formularies generated several unintended consequences. Many physicians and consumers thought that, once a drug was published on the list of bioproblem drugs, *no* generic version could be trusted. Backpedalling, the FDA tried to clarify this by producing a second list, *Holders of Approved New Drug Applications for Drugs Presenting Actual or Potential Bioequivalence Problems*, beginning in January 1976 and revised regularly afterward.

Popularly known as the FDA *Blue Book*, the first version of this list included 117 problematic active ingredients in 173 dosage forms. Of the 173 dosage forms, 85 were available from only single firms (i.e., did not reflect brand/generic issues) and more than 50 were only hypothetical—leaving only about 30 actual products with demonstrated bioequivalence issues.[54] By 1978, the *Blue Book* divided drugs into a more confusing taxonomy of potential bioequivalence problems: group III drugs, for example, were those for whom "some but not all firms have demonstrated the bioequivalence of their product," as opposed to group II drugs, which had no demonstration of bioequivalence, because of "lack of methodology." A year later, FDA staffers proposed to further divide the *Blue Book* into nearly twenty categories of drugs, ranging from category A (product poses no therapeutic equivalence issues) to category X (radioisotopes), all of which demanded their own different bioequivalence considerations.[55]

Consumer advocacy groups such as the AARP worried aloud that such confusing taxonomies of equivalence would merely drive *all* consumers away from lower-priced generics, By the summer of 1978, a consortium of consumer groups petitioned the FDA to move away from the complicated negative formulary of the *Blue Book* and develop instead a single nationwide positive formulary for generic substitution. As the AARP's Fred Wegner wrote to FDA commissioner Donald Kennedy in April of 1978, "It seems to us that the logical end result of FDA's ultimate authority and responsibility over the U.S. drug supply is the publication of such a formulary of the most prescribed drugs. Because of its essential importance to all who prescribe, dispense, and use prescription drugs, it is difficult to understand why it does not already exist. Here then would be a list of drug products which the FDA has approved as meeting all requirements for quality, safety, and therapeutic effectiveness."[56] In his public appeal to the FDA to put forward a single positive formulary, a set of universal rails to speed the progress of generic substitution nationwide, Wegner had not revealed all of his cards. One of Wegner's informants, the investigative journalist turned generic policy advocate, Bill Haddad, had told Wegner that the FDA already had such a list close at hand.

Who had told Haddad about this hidden list? None other than Marvin Seife, the first head of the FDA's Office of Generic Drugs, who had maintained a close correspondence with the former journalist as Haddad became an influential figure in New York State politics. Hidden in the files of the FDA in the offices of the federal Parklawn Building, Seife told Haddad, was a secret document that generic advocates and consumer activists longed to see: not the *Blue Book* of problems but a universal formulary for generic exchangeability. Such a list could resolve all problems of generic substitution by identifying a definitive set of quality generic distributors whose drugs were deemed therapeutically equivalent. Officially, however, the FDA could not admit that such a list existed, for reasons Seife attributed to the powerful influence of industry within the FDA.

All of this would change in 1978 when the state of New York implemented its own law favoring generic substitution—built around an unearthed copy of the FDA list.

CHAPTER 9

UNIVERSAL EXCHANGE

To proceed state by state with laws that may or may not address themselves to interchangeability, with one state having a "negative" formulary, another a "positive" one, and a third none at all, with no consistency in which products are to be substituted if formularies exist, is to play a form of geographical roulette with too many people in too important a part of their lives.

—FELTON DAVIS, VICE PRESIDENT OF CIBA, 1977

At first glance, the passage of the bill encouraging the substitution of generic drugs in the state of New York might not stand out as an important event. The law appeared in 1977, almost halfway between the first (1972) and the last (1984) state repeals of antisubstitution laws. New York was not the first state to mandate (rather than merely permit) substitution, nor was it the first to use a formulary. But the passage of the New York bill would mobilize generic substitution on an unprecedented scale. Where other states had listed twenty to fifty exchangeable drugs, New York listed more than eight hundred.[1] Furthermore, the central advocate for the New York law, William Haddad, was not content to let the law apply to New Yorkers alone. Instead, Haddad claimed that his office in Albany had unearthed a universal solution to the problem of drug substitution: a solution that could be generalized to every state in the nation, if not in the entire world.

PORTRAIT OF A GENERICIST: WILLIAM F. HADDAD

Like the New York law itself, Haddad's life story moved quickly from local issues to global networks. To his supporters, he was a scrappy and passionate defender of the common consumer. To his detractors—most notably the leadership of the Pharmaceutical Manufacturers Association—he was a "vociferous journalist-politician with a bias against the brand-name industry that ran as wide as a legislative chamber." By the time he was appointed to head up the New York State Assembly's generic drug investigation, Haddad's career had already moved from a start in the merchant marine to working as Senator Estes Kefauver's administrative aide during the drug industry hearings to writing for the *New York Post* and *International Herald-Tribune* to leading the early Peace Corps as associate director under Sargent Shriver.[2]

In 1967, Haddad published what would become a widely circulated piece of investigational journalism exposing a tetracycline price-fixing ring among major American pharmaceutical manufacturers. In the aftermath of this publicity, he ran an unsuccessful campaign for Congress in a Brooklyn district. His campaign focused on the intersection of health and consumerism and included an exposé of price differences in prescription drugs in different parts of New York City, noting a fivefold difference in price for "the same drugs."[3] After losing his bid for Congress, Haddad served as president of the Citizens' Committee on Metropolitan Affairs and continued to use his investigative and publicity skills to further critique the price of drugs and their variability among cities. Comparing New York to Atlanta, Miami, Portland, and San Francisco, he found a fortyfold disparity in the purchase price different municipal governments paid for the same drug.

After Haddad was invited by Senator Gaylord Nelson to present this research to the Subcommittee on Monopoly and Antitrust in November of 1967, he first rated the attention of the executive director of the PMA, C. Joseph Stetler. The PMA chief dismissed Haddad and his study as a set of "warmed-over, worn-out charges": a faint echo of his political mentor, Estes Kefauver. In a series of initially promising but ultimately failed interactions with New York City mayor John Lindsay, Haddad urged the city to strengthen its generic-name purchasing system for welfare recipients. When the Lindsay administration backed away from the issue, Haddad shifted his political theater from city to state and petitioned Governor Nelson Rockefeller to support a bill for generic substitution in the 1968 New York State legislative session.[4]

This bill failed, but Haddad's efforts attracted the attention of the New York Democratic political machine, most notably two old New Deal Democrats: Stanley Steingut and Harvey Strelzin. The Albany veterans advised Haddad that the state would only make positive movement toward generic drug purchasing if he could find a universally acceptable table of interchangeable drugs. Six different bills to support generic substitution had been filed—and all six had failed—when Steingut and Strelzin appointed Haddad to an independent investigator's office in Albany, granted him subpoena power and a small staff, and tasked him with producing a scientific document that would provide undeniable support for generic substitution in the state of New York.[5]

THE NEW YORK STATE GENERIC DRUG INVESTIGATION

Haddad's first attempts to hold hearings in New York yielded little. His handwritten notes—preserved in the massive Generic Drug Investigation Files of the New York State Archives—document that these attempts were "soundly beaten by lobbyists in 1976 legislative session using FDA *Blue Book* and Warner Chilcott film with Frank Blair narration. Members/staff totally confused by industry statements re: FDA stand on bioequivalency." Memos written by Haddad to his staff vowed that they would be able, by the 1977 legislative session, to "lay to rest, once and for all, the bioequivalence myth; pass strong legislation."[6] A newsletter distributed through the State House concurred that "a formulary specifying medications which have equivalent therapeutic effects as generic drugs is imperative" before any generic substitution bill could be passed.[7]

As the head of the New York State Assembly's legislative investigation committee, Haddad's approach to consumer investigations was to triage by (a) relevance to public interest, with (b) the potential for substantial publicity, utilizing (c) the power of the subpoena. "My office is rather unusual, it has the derivative power of all the committees . . . The secret is the subpoena," Haddad described in a letter to a colleague in 1977. "We do it with mirrors and volunteers. Every time they cut my budget, I get another bored Wall Street lawyer to work for me for free."[8]

Haddad did have a gift for getting people to do work for him for free. His good fortune came in the form of a mole named Marvin Seife, a career FDA staffer with a long-term interest in generic drugs. It was Haddad who provided the description of Seife (used in the introduction to this book) as "a laconic Jewish New Englander who wears a Carter-type sweater with a hole in the elbow." Along with his immediate boss in the monographs division of the FDA, Gene Knapp—whom Haddad described as "a cantankerous doctor who quit as head of the new drug division of the FDA because he couldn't get along with Nixon"—Seife had organized the FDA's response to the Military Medical Supply Agency to determine which generic drug products could be considered interchangeable for military purchase.[9]

Seife informed Haddad in early 1977 that the FDA had collected what amounted to a vast positive formulary of exchangeable drugs evaluated for the Department of Defense. While Seife and Knapp were not technically permitted to publicize this list, they were willing to help. In furtive tones, over a series of handwritten letters, Seife guided Haddad and his team of interns and pharmacy students through a complex series of requests for materials under the Freedom of Information Act that—if properly disaggregated and reassembled—would reveal a hidden cache of FDA evaluations of producers of generically available drugs that had been cleared as therapeutically exchangeable for military purchase.[10]

By the middle of 1977, Haddad had extracted a large positive formulary from the FDA's own files. He had generated a narrative with a touch of cloak and dagger, which would in turn help to generate publicity. And he was ready to preach the value of this new formulary—which he called the *Green Book*—as a bible for generic substitution.

THE NEW YORK FORMULARY AND ITS PUBLICS

Haddad orchestrated the unveiling of the *Green Book* with Barnumesque attention to publicity. He staged an event at the World Trade Center for April 28, 1977—shortly before the New York State Assembly was set to discuss Strelzin's latest bill for a generic substitution law—with the provocative title *Are Generics Safe?* Executives from the PMA and major pharmaceutical companies were asked to come and present their best cases for the inferiority of generic drugs. And then, in what he confided to Harvey Strelzin would be the "most dramatic moment," he would reveal his key witness, Marvin Seife, who in turn would produce the New York State *Green Book* and officially declare the generic equivalence of the eight hundred drugs it described.[11] In anticipation of this event, Haddad seems to have written letters to personal contacts at nearly every major organ of the New York media. In an attempt to get *60 Minutes* to televise the hearing, Haddad described the story with expert journalistic pitch:

Key to the solution is a list of safe and effective drugs which everyone claims does not exist. It does. It is contained within two small cardboard boxes, sitting atop a file cabinet within the Food and Drug Administration. Inside are five-by-seven handwritten cards. Whenever there is a check alongside a drug and a company, it is deemed safe and effective as produced by that company. That little red check is the final step of one of the most sophisticated technologic processes anywhere in the world. Keeping that list from public view is worth billions to the drug companies.[12]

Similar letters were sent to personal contacts at *Newsweek*, the *Washington Post*, local and national television newscasters, and local and national political figures. If his ear for pitching an abstract policy issue as a compelling story was spot-on, Haddad's sense of fidelity for specifics was not always as well honed. In a similar letter to Representative Lester Wolff, Haddad described the list not as a set of file cards but as a set of electronic files "buried in their computers and hidden in the bureaucracy which lists every safe and effective drug in this country, tested not only as to product but as to company . . . The military uses this 'non-existent' list."[13]

But the theatrics of the April conference did not settle the issue quite as clearly as Haddad had hoped. In early May representatives from Eli Lilly called Gene Knapp to ask whether the FDA's *Blue Book* (of bioproblem drugs) had really been replaced by New York's *Green Book* (of therapeutic equivalents). Knapp himself seemed to have been nettled by Haddad's public statements and complained to Haddad's assistant in Albany, David Langdon, that it was a bit deceptive to claim that "the list that FDA was evaluating for the State was a list of therapeutic equivalent drug products."[14] When the vice president of Ciba received a copy of the list, he wrote back to Haddad to correct a series of errors in the list of several out-of-date marketed products and insisted "that you understand that by 'correcting' the list we do not indicate agreement with how you have characterized that list . . . Specifically we do not consider the list to be one of 'Safe, Effective and Interchangeable Prescription Drugs.' By interchangeability we mean that when a patient receives a copy instead of the original drug or copies made by two different manufacturers, there is no discernable difference in terms of the physician's expectations and the patient's response."[15] While Ciba was prepared to admit that therapeutic equivalents *could* exist, and that substitution *might* well be important, they pointed out that New York's list had yet to be certified by the FDA. Indeed, it would take about ten months for the FDA to come out with an official verdict on the accuracy of the *Green Book*.[16]

Haddad called a second public hearing at the World Trade Center in late May to showcase the first edition of the *Green Book*, also known as the *New York Formulary of Safe, Effective, and Interchangeable Drugs*:[17]

> The 20 year debate over the safety and effectiveness and interchange-ability of generic drugs will be definitively ended at a 15 minute hearing Tuesday, May 31 ... when the FDA will certify a list of prescription drugs prepared by the New York State Assembly. In that approved list will be 99.9 percent of all drugs now approved by the FDA. The drug industry maintains that such a list does not exist. As a result they have argued that doctors cannot prescribe generic drugs safely. The absence of this list has meant billions of dollars of profit for the drug companies each year.[18]

Haddad had a gift for translating pharmaceutical policy into human interest narratives that appealed to a broad readership—and to magazine and newspaper editors as well. For example, in a letter to contacts at *Time* and *Newsweek*, Haddad pitched a story of subterfuge and plucky local resourcefulness:

> New York State decided to go for the eye of a dragon and produce the list of safe, effective, and interchangeable drugs, and either through legislation or executive order establish that list as a formulary for the state and the nation. As we have done on other issues (notably natural gas and the Arab boycott) we have done for ourselves what the federal government failed to do. There is a story in how the list was produced, hidden from view of the political level of the agency, working with courageous bureaucrats who risked their jobs to help, using Freedom of Information requests to protect them (and disguising our interest because the drug companies monitor, on a daily basis, the Freedom of Information requests) and then compiling the list and coming in the front door with a *fait accompli* and asking the FDA to verify what we had done from their records.[19]

Generating fast and favorable publicity was of the essence. Two very different approaches to substitution were still floating around Albany:

Strelzin's Assembly bill, which would *mandate* generic substitution at the pharmacy, and an opposing Senate bill that would merely *permit* pharmacists to dispense generics in place of brands as long as they received permission from the physician. A compromise bill was signed by Governor Hugh Carey in the summer of 1977 with a "two signature" solution: if doctors signed on one side, substitution would be mandatory, if they signed on the other, pharmacists were required to dispense the brand as written or face possible penalties.[20]

The New York law in general, and the *Green Book* in particular, continued to attract controversy. Back at the FDA, Bureau of Drugs head Richard Crout voiced increasing discomfort with Haddad and Langdon's claims that the New York *Green Book* had been somehow "certified" by the FDA merely because it drew on FDA information. Crout particularly objected to the language of interchangeability. The FDA already had its hands full trying to describe bioequivalence, and Crout believed it had no business extending beyond this concept into a much murkier judgment as to what could make two drugs fully interchangeable. He suggested instead that New York settle for therapeutic equivalence: "We do not approve generic drugs as being 'interchangeable,' rather we approve such products as being therapeutically equivalent for their approved use. For example, two pediatric syrups which are therapeutically equivalent may differ as to taste and palatability. The agency imposes no test for taste equivalence or taste acceptability. We are prepared to address the issue of whether these two products are therapeutically equivalent, but we would not certify their interchangeability."[21] Mildly chastened by this encounter but still keen to use FDA endorsement as a promotional tool, Haddad's office published a second version of the *Green Book* in early October 1977 under the new title *The New York Formulary of Safe, Effective, and Therapeutically Equivalent* Prescription Drugs*. To mollify the FDA, the asterisk in the title clarified that "these drugs present no known issue as to interchangeability." Bundled into the front matter was a new "approval letter" signed by the FDA's Gene Knapp. Hospital pharmacy journals predicted that most pharmacists in the country would soon be using the New York *Green Book* as an authoritative reference for generic substitution.[22]

Opposition from the organized pharmaceutical industry was swift and fierce. After the *New York Times* published an editorial in favor of the New York formulary, PMA head Joseph C. Stetler shot back that the list was "being used by substitution advocates in a highly misleading way" and provided no "assurance of interchangeability."[23] The PMA questioned

Haddad's cloak-and-dagger narrative, asking the FDA how it was possible that such a "federally-certified secret list" of interchangeable drugs could have been prepared at "great risk" by courageous bureaucrats. "Lists of this sort," Stetler snorted, "can be assembled and exploited by anyone who wishes to read FDA press releases." How could there be a uniform list of bioequivalent drugs when there as yet was no agreed-upon definition of bioequivalence?[24]

Even Marvin Seife seems to have had some concerns that Haddad might have overreached in his claims for the *Green Book*. In a series of letters to Haddad, he warned that the head of the FDA's Division of Bio-equivalence, Bernard Cabana, would find some way to use the changing definition of bioequivalence to quash the New York formulary. After Haddad sent in a copy of the second edition of the *Green Book* for FDA review, Seife wrote, "Unfortunately, one copy was given . . . to our irrational friend Bernie Cabana. Cabana immediately went 'bananas' because the volume didn't follow his screwy notions." Confidentially, Seife also expressed fears that he was becoming isolated within the FDA hierarchy as a result of his public support for the New York formulary. During an opportunity when Seife was out of town, bureau chief Richard Crout, Gene Knapp, Bernard Cabana, and FDA counsel Paul Bryan met to discuss what to do with the revised *Green Book*. On his return, Seife described that Crout, who "at first . . . was all for the publication," had "after listening to Cabana . . . made a 180° turn in his thinking." Seife warned Haddad that "Cabana is now spending most of his time writing a critique of the N.Y. publication" and suggested that the head of bioequivalence testing was trying to enlarge his own fiefdom by expanding the list of "problem drugs."[25] In a subsequent letter, he noted, "Bernie's game plan, intentional or otherwise, is simple—with one hand, remove a few drugs on the 'bio' list and with the other add more than he removes . . . To hell with the consumers who pay Bernie's salary, as long as he and [his supervisors] keep the confusion going in the guise of science or keep their positions until they retire."[26] From Seife's admittedly jaundiced perspective, Cabana was focused only on the technical dimensions of bio- and had no interest in the clinical or public health value of -equivalence; his mistake was to think that all medical decisions "revolve around his all-consuming bioavailability biopharmaceutics-pharmacokinetic world." Left unchecked, Cabana's actions threatened to extend the universe of "bio problem drugs in the '*Blue Book*' (of unhappiness)" at the expense of the pill-consuming public.[27]

It is easy to imagine the conflict between the two men as one of so many conflicts between ambitious bureaucrats who—while rising at the same time through the ranks of a large institution—are only able to see each other as competitors and therefore depict the other as his nemesis. But the conflict between Seife and Cabana over the relative roles of the FDA *Blue Book* and the New York *Green Book* also reflects a broader set of conflicts written into the architecture of the FDA: between its de jure role as a science-based consumer regulatory agency and its de facto role as a key element in the structure of American public health and health policy.

FROM NEW YORK TO WASHINGTON: MOBILIZING THE INTERCHANGEABLE DRUG

This conflict between the FDA's commitments to developing new regulatory sciences of bioequivalence and the agency's responsibility to consumer and patient publics would continue to play out even after Seife's position won and the FDA publicly issued a statement in support of the New York *Green Book* on January 23, 1978. "It is our understanding," FDA commissioner Donald Kennedy explained, "that this list is intended to contain all products approved by the FDA as being safe, effective, and bioequivalent where bioequivalence is a real or potential issue." The resulting *Green Book*, he later concluded, was a syncretic product of FDA science and state prerogative over the practice of substitution in the pharmacy "determined by a combination of the New York law and FDA policies."[28]

A few days after formally endorsing the *Green Book*, Kennedy found himself seated next to PMA president C. Joseph Stetler on NBC's *Today Show* to discuss the impact of the New York formulary. Stetler had publicly warned that the FDA's endorsement of a list of equivalent drugs would "raise false hope among consumers about price savings that will not materialize in the real world." Such behavior, he warned, was well beyond the bounds of responsible activity for both the FDA and the state of New York; as Stetler would later chide Kennedy, "FDA action on this issue has given the New York list a stature well beyond what it deserves and a credibility well beyond what it merits."[29]

Kennedy and other FDA officials repeated that while the agency could not comment on the interchangeability of a drug (a concern for state lawmakers, not federal regulators), it was uniquely able to comment on the scientific basis supporting therapeutic equivalence. Occasionally, however, FDA officials would slip up, as Kennedy himself did while testifying later that year before congressional hearings regarding a bill proposed by Representative John Murphy (D-NY) to enact a nationwide generic drug substitution law. After Kennedy interchanged the word *interchangeable* for *equivalent* on a number of occasions, Representative Matthew J. Rinaldo (R-NJ) asked the FDA chief for a clarification of the distinction between therapeutic equivalence and interchangeability:

> KENNEDY: We don't use the word "interchangeability" except by mistake. Interchangeability is the . . .
> RINALDO: You've made a few mistakes!
> KENNEDY: I'm sorry. I hope very much that the record will correct that if I did so. Sometimes when people say "interchangeability" to me, I don't notice it. What we say we will provide . . . are lists of drugs that are therapeutically equivalent. The reason for that is that interchangeability is a political decision. It's the decision of the states whether to interchange drug A for drug B . . . To use an example that my colleagues are fond of, if you had two different pediatric formulations of an antibiotic, and one was banana flavored and the other was orange flavored, then we would call them therapeutically equivalent, but a state might not want to decide that [these two drugs] were interchangeable.[30]

As much as Kennedy might seek to divide the role of states (over substitution) and the role of the FDA (over equivalence), the states increasingly turned to the FDA for guidance on both. When Vermont passed its own mandatory substitution law in July 1978, it similarly limited substitution to drugs listed in the *Vermont Formulary*—a list of generically named drugs that "includes <u>only</u> those drugs which may be safely and effectively interchanged"—and asked the FDA to ratify its list just as it had ratified New York's.[31] The result was a formulary identical to the New York *Green Book*.

Kennedy soon found himself pleading for other states to stop asking his overworked staff for exhaustive reviews of their own formularies. As Kennedy wrote in a letter to state health officials, the FDA simply could not handle many more such requests. Instead, it would publish its own list to be adopted by all states—a universal lexicon of generic substitution—which he defined as "a list of therapeutically equivalent drug products for use by all states with laws relating to generic prescribing and drug product substitution."[32] As an unintended consequence of "signing off" on the *Green Book*, the FDA found itself pushed into publishing its own, federal standard for therapeutically equivalent drugs, a volume that would eventually become known as the *Orange Book*.

To Haddad, the *Orange Book* was the logical extension of the *Green Book* from the state to the national scale, and he worked behind the scenes to build a national network from contacts at the Federal Trade Commission and consumer advocacy organizations and Capitol Hill staffers. In May of 1978 Haddad announced the formation of a National Consumer Alliance on Prescription Drugs, which demanded that the FDA "end the generic-tradename controversy by publishing a formulary of interchangeable drugs." The alliance would eventually include the Consumers Union, the Consumer Federation of America, the National Consumer's League, the American Association of Retired Persons, the United Steel Workers, the Environmental Defense Fund, and the Health Research Group of Ralph Nader's Public Citizen. In June, Haddad convened the First National Conference on Generic Drugs. Hosted at the Mayflower Hotel in Washington, DC, the conference was a Who's Who of consumer activists, academic physicians, key lawmakers, and figures from the Department of Health, Education, and Welfare and the FDA. Senators Gaylord Nelson and Edward Kennedy (D-MA) served as the keynote speakers on the first day and explicitly linked the generics movement with a sweeping legislative proposal for FDA reform then in committee in the Senate.[33]

Paraphrasing Hunter S. Thompson, the physician's monthly *Private Practice* headlined its coverage of the event "Fear and Loathing and Generic Drugs." "Perhaps a somewhat similar atmosphere prevails in the meeting halls of the Ku Klux Klan or the Weather people," the article's author wagered, "but even such groups as these would be hard put to outdo the consumerists in terms of sheer suspicion and hostility." Worst of the consumerists were "the members of the genericist wing of the consumerist movement." To conservative physicians, genericism was a cult and Haddad was its high priest, gathering his acolytes in a place of power to expand the federal health bureaucracy and ensure further mediocritization of medical practice.[34]

Publicly, the PMA alternated between brushing off the generic conference as laughable and describing it as a threat to the practice of medicine in America. C. Joseph Stetler penned an essay in the trade journal *Medical Marketing and Media* that warned of "bureaucratic fiat" that increasingly threatened the practice of medicine and pharmacy: "Nowhere is this alarming trend more obvious than in the issues surrounding the

New York State list of so-called 'equivalent drugs.' On the one hand we find New York officials content to ignore reality and define the world the way they wish it to be. On the other hand, the Food and Drug Administration proclaims the universal equality of all prescription drugs, while its scientific and medical staffs struggle to justify their agency's gratuitous decrees."[35] As Haddad was working to telescope the New York list outward to broader and broader audiences—his correspondence in the spring of 1978 contains efforts to export the *Green Book* to the Ministries of Health in Argentina, the Bahamas, Bolivia, Brazil, Chile, Colombia, Costa Rica, Ecuador, El Salvador, Grenada, Honduras, Jamaica, Mexico, Nicaragua, Paraguay, Peru, Trinidad and Tobago, Uruguay, and Venezuela—Stetler sought to put the *Green Book* under a microscope of sustained and detailed criticism.[36]

Shortly after the National Conference on Generic Drugs, the PMA filed suit against the state of New York for unfair trade practices, claiming that the law unfairly handicapped research-intensive firms.[37] Beecham presented data that its version of amoxicillin suspension retained its potency if unrefrigerated for longer than generic versions produced by Biocraft, Robins, and Squibb. This form of difference would be undetectable when medications were refrigerated but might occasionally be significant when bottles were unrefrigerated for long periods.[38] Even a sympathetic journal like *Pharmaceutical Technology* had to labor to produce a scenario in which such a trivial difference might be understood to be significant: "Such a case might occur when, prior to a camping trip, parents discover that their child is ill and needs a course of amoxicillin or the trip must be postponed. 'Not to worry,' a doctor might advise. 'Here's a wonderful product you don't have to keep in the refrigerator. When it's empty, get a refill at any pharmacy.' The refill, however, isn't a Beecham or Roche product." On one level, this case delineates Kennedy's distinction between therapeutic equivalence and interchangeability: two drugs that might produce the same result in controlled clinical situations might actually perform differently in the lived experience of patients. Nonetheless, the editors of *Pharmaceutical Technology* were careful to note that such distinctions bordered on the ridiculous: "But, as the FDA officials point out, by the time the significant potency loss occurs, the course of amoxicillin therapy has extended to at least 15 days, and if junior isn't well by then, he or she shouldn't be camping anyway."[39] Other critiques of the list were more or less picayune.[40] Over the next eighteen months, all legal challenges to the New York law would be dismissed, the *Green Book* upheld as an instrument of generic substitution, and a nationwide *Orange Book* eagerly anticipated by consumer advocates and health system reformers.

FAILED STANDARDIZATION: NATIONAL POLITICS AND THE SPECIFICITY OF SUBSTITUTION

The year 1978 looked to be good for promoting a universal system of generic substitution. FDA staffers were scrambling to put finishing touches on the *Orange Book*—a universal guide to generic drugs that would replace the negative formulary of the *Blue Book* and the more provincial positive formulary of New York's *Green Book*. At the same time, HEW and the FTC were each drafting different model generic drug substitution bills to put before Congress by summer's end. This process came to a head when John Murphy, a Democratic representative from New York, sought to translate the New York proposal into national policy with the Substitute Prescription Drug Act of 1978.[41]

This was not Murphy's first attempt to pass a nationwide generic substitution law. Several bills had been proposed to this effect in the Ninety-fourth Congress (H.R. 882, 998, 999, 3546, 3987, and 6603) by a variety of representatives, but they had met opposition from industry, the medical and pharmacy professions, and the White House. But in 1978, with New York's *Green Book* in hand and the FDA's *Orange Book* visible on the horizon, Murphy thought that prospects were good for a single law to standardize generic substitution across the country. "Today's mobile society," he continued, "means that many travelers find the need to fill a prescription in another city on short notice," regardless of which state they were in.[42] Murphy's proposal drew immediate support from consumer advocacy groups; a spokesperson from the AARP testified that consumers deserved "a national minimal level of uniformity in substitution among all of the States."[43]

The bill met with strong opposition from organized medicine and organized pharmacy. The American Medical Association's John H. Budd complained that "the obvious result here is that Congress, if this bill were to be enacted, would enter directly into health care treatment and become the prescriber of the medication most desirable for treatment of the patient."[44] The PMA suggested that a universal formulary would cost the state more to produce and maintain than it would reap in savings. "We cannot think of a single reason," counsel William Patton testified in

1978, "for assigning this important responsibility to the legislature or to a State bureaucracy."[45] The APhA's William Apple also testified against the bill. Although organized pharmacy had led the drive for state substitution laws, it opposed nationwide legislation based on the New York example. Apple specifically judged that "the New York law is a bad law" because of its reliance on a bureaucratic *Green Book* rather than the clinical judgment of the pharmacist.

In the end, Murphy's universal substitution bill was isolated from almost all of the constituents who would be needed to support it: physicians, pharmacists, industry, the FDA, HEW, the White House, and even prominent progeneric senators such as Gaylord Nelson and Edward Kennedy. With no prospects of passing, the bill died in committee.[46] As a result, generic substitution in the United States has remained a patchwork landscape of positive and negative formularies, mandatory and permissive substitution, one- and two-sided prescription pads.

Generics are available everywhere, but the rules of substitution are never the same.

THE GEOGRAPHY OF SUBSTITUTION

In light of the special interests that presided over the political death of universal generic substitution, it is worth revisiting the testimony of the one expert who commented on Murphy's bill and claimed to have no interest whatsoever: health services researcher Theodore Goldberg. Goldberg had created a scholarly center at Wayne State University studying the effects of drug product substitution laws after Michigan passed its law in 1975. He repeatedly claimed that academics "have absolutely no ax to grind nor any vested interest to protect. We can and have viewed the data that have been produced with an unbiased perspective as we believe it is possible to achieve."[47]

Goldberg and other health services researchers were interested in developing a science for comparing the public health effects of prescription drug law changes in different places. They noted differences between positive and negative formularies and between the permissive and mandatory substitution arenas, and they worked to quantify "the gap between the potential for savings and the actual amount of savings" when such substitution regimes were enacted.[48]

For health services researchers, the paired success and failure of generating universal principles of generic substitution in 1978 would come to constitute a form of "natural experiment." The existence of a single standard of therapeutic equivalence (the FDA's *Orange Book*) coupled with the staggering heterogeneity of programs of therapeutic substitution (a different law in each of the fifty states, Washington, DC, Puerto Rico, and Guam) provided a robust tool kit for Goldberg and his intellectual progeny to develop a science of substitution. Goldberg described a robust field of research in 1986, in which "specific analytic techniques also must be developed to determine the influence of industry, physician, pharmacist, and consumer characteristics on the potential for rational prescribing and dispensing of drugs."[49]

Though the heterogeneity of state laws provided good material for the academic field of health services researchers in the late twentieth century, it continues to confound the delivery of health services and access to cheap generic drugs in the early twenty-first century. As latter-day health services researchers like William Shrank, Michael Fischer, Aaron Kesselheim, and Niteesh Choudhry have recently documented, the cost savings of generic substitution now appear to benefit populations in inverse proportion to economic need.[50] The current map of variable generic drug use is built in part on a legacy of earlier state-based heterogeneity: the persistence of comfort with federal structures of *equivalence*, but not of *interchange*.

In recent years, continued advocacy on the part of disease-based interest groups—most notably the American Epilepsy Foundation—has accentuated this heterogeneity by fighting rear-guard actions over the exchangeability of specific generic drugs. Ever since the antiseizure medication phenytoin (originally marketed by Parke-Davis as Dilantin) was first labeled in the *Blue Book* of bioequivalence problem drugs in the 1970s, neurologists and patients treated for epilepsy have been concerned over the putative inequivalence of brand-name and generic antiepileptic drugs. Because in the case of epilepsy the human costs of inequivalence are so high—a single breakthrough seizure can cost patients with epilepsy their driver's license, job, or even their life—advocacy groups have argued that this class of drugs should meet a higher standard of proof, even after several successive studies of the biopharmaceutics and pharmacoepidemiology of this class of drugs have found no difference between brand-name and generic versions.

While these arguments initially gained little traction with the FDA, they did find sympathetic audiences in several state legislatures. By 2007,

Hawaii and Tennessee passed laws restricting generic substitution in the case of epilepsy, and between 2007 and 2011 forty legislative proposals in twenty-four states sought to roll back substitution laws to prohibit the generic substitution of drugs for epilepsy. Another twenty-nine bills filed in 2007–2008 sought to create a similar rollback of generic substitution for immunosuppressant drugs used in posttransplant patients. Partly in response to these new state challenges to generic substitution, the FDA has since dedicated new funds generated from the Generic Drug User Fee Amendments of 2012 to try to determine what new forms of regulatory science might increase the confidence public officials could have in the equivalence and interchangeability of these products.[51]

In the legal battles over substitution, past and present, far more than the comparison of two or more drug products was at stake. Debates over the ability to substitute one drug for another were inextricably tied to larger questions of whether one could understand the health-care system as a series of interchangeable parts—to be systematically analyzed and tweaked as one would seek to better engineer an inefficient machine—or as a set incommensurable objects that could only be properly used by individual practitioners. Even as the passage of state laws set in motion broad engines of generic substitution, they did not immediately produce either the behavioral buy-in from physicians and consumers or the economic incentive for pharmacists to begin dispensing generic drugs in large amounts. If some patients and physicians from the 1980s onward embraced generic substitution, others fought it tooth and nail. Physicians learned to write "no substitution" or "dispense as written" on their prescription pads; patients learned to ask their physicians to document fraudulent "allergic to generics" letters to maintain reimbursements for brand-name pharmaceuticals.[52]

"Although the battle is not won," the founder of the newly prominent generics firm Barr Pharmaceuticals wrote to Haddad in the aftermath of the legislative failures of 1978, "I believe that the Generic Drug Concept has become an integral part of consumerism, and in time the public will prevail."[53] These lines would prove more aspirational than descriptive. Not all consumers in 1978 embraced the concept of the generic. To understand why generic substitution continued to be resisted long after it had become the law of the land, we must try to understand the paradoxical figure of the generic consumer.

PART V

PARADOXES OF GENERIC CONSUMPTION

CHAPTER 10

LIBERATING THE CAPTIVE CONSUMER

Your physician, therefore, may represent a hurdle to your getting a generic drug . . . So once again, you must take the initiative.
—CONSUMERS UNION, *THE NEW MEDICINE SHOW*, 1989

Somewhere in the middle of the twentieth century, the American patient was rediscovered as a consumer of health care. Nobody, however—not medical professionals, pharmaceutical executives, regulators or policy-makers, nor the newly classed consumers of health care themselves— seemed to be able to define exactly what this shift meant or how they should behave in this new regime.[1]

The problem of understanding the new medical consumer became particularly visible in the new market for generic drugs. Consider a curious series of letters sent to various government officials in 1967 by an aspiring generic consumer, E. F. Trapp. Trapp's first letter, addressed to the commissioner of the Food and Drug Administration, demanded that the agency "furnish me with information as to where I can secure the generic medicines, which are much reduced in price than medicines made by established firms, how do I go about securing prescription for same?" This letter was followed by a second, emphasizing the urgency of Trapp's need to locate "the cheaper in price, but equality of drugs, medicine." Receiving in reply a mimeographed form letter suggesting he speak with his doctor, an indignant Trapp sent another volley of correspondence to the secretary of Health, Education and Welfare, John Gardner, and Senator Gaylord Nelson, complaining that it was no use "to ask my Doctor about the generic medicines, he like myself has no knowledge about how to secure." The government, Trapp insisted, had a responsibility to tell him "where and how to go about securing the generic medicines which I believe are in pretty general use thru out the Federal Hospitals."[2]

These letters all found their way to the desk of Edna M. Lovering, whose Consumer Inquiries division within the FDA was struggling to handle an increasing volume of correspondence from people like Trapp, who were interested in learning more about generic drugs. Lovering tried to redirect Trapp's conviction that there was a secret government supply of generic medicines. She patiently set forth an account of complexity in the generics sector: all drugs have a generic name, many have a brand name; some but not all generically named drugs are available from multiple producers as "generic drugs"; a generic prescription can lead to a cheaper version being filled at the pharmacy, but only if such a version exists and the pharmacist stocks it and chooses to fill the prescription with it and the purchase takes place in a state that permits generic substitution. In closing, Lovering hoped she had clarified his confusion regarding "how to get 'generic medicines,' which you feel are in general use and are cheaper."[3]

She had not.

Stung by the suggestion that he was a confused, childlike consumer, Trapp lashed out at Lovering:

I think you are well aware of the intent of my inquiry, however, I will explain that what I wish to know is the price of the cheapest but equal quality medicines now and for some past being produced by firms other than the well known brands, such as Lilly, Parke Davis, Roche, and etc.

Sometimes back it was publicly announced that while in Hospital our president was treated with the cheaper but equal in quality medicine. I want the name of such medicines that the U.S. Govt has approved and I want to know where I can purchase such medicines.

It is indeed strange that when one writes to his Government for information, he gets the usual Pass the Buck dope. If you can not answer above send this inquiry to the proper person, I will appreciate prompt action, Rspt, EF Trapp.[4]

Trapp would not find relief. As it turned out, not a single medicine on his list was available in generic form, and most were still protected by patents. But every explanatory letter from Lovering seemed only to inflame Trapp's suspicions of a conspiratorial market where, hidden from view, a set of cheaper generic equivalents were being made available to federal employees yet denied to hardworking American consumers like him.

Trapp's quest for generic drugs that did not yet exist traces a paradox already latent in the reconceptualization of patients as consumers by the late 1960s. Even those inclined to see their own interests reflected in the consumption of "cheaper but equal in quality medicines" encountered a deep chasm of confusion and misunderstanding regarding what generic drugs were and what kinds of cost savings could be accomplished through generic consumption. Like Trapp, many petitioned federal and state government officials for information about generic drugs while simultaneously expressing deep suspicions of the government's role in the consumer marketplace. In turn, self-styled consumer advocates in the health arena were surprised to find resistance from the very consumers who stood to benefit most from policies promoting access to low-cost generic drugs.

Potential patients were not the only kind of consumers confused by the prospect of consuming generic drugs. Physicians, too, increasingly understood themselves as a sort of *expert consumer* who had the ability to designate the purchase of specific therapeutic products for the good of their patients. Though many within the medical profession denounced generic consumerism as a corrosive force that threatened the doctor-patient relationship, their arguments in this arena came to represent an alternate form of consumer activism—one in which physicians asserted their own role as expert consumers in the medical sphere.

Although the consumer movement of the 1960s and 1970s borrowed rhetoric and tactics from other movement cultures of the time—from civil rights to second-wave feminism—the historical development of the generic consumer presented a distinct series of paradoxes.[5] Did the choice of a generic drug empower individual patients to become savvy consumers and take more control of their own care? Or was generic consumption a gray and chilling extension of the state into the previously private realms of patient and healer? These questions, emergent at the time of Trapp's inquiry in the late 1960s, would become increasingly visible in the 1970s and 1980s, as general tensions over the role of consumerism in medicine took on concrete form in public debates over the safety of consuming generic drugs.

THE CAPTIVITY OF THE PHARMACEUTICAL CONSUMER

As she neared the conclusion of her testimony before Senator Estes Kefauver's investigational hearings in May of 1960, Mildred Brady paused to consider the agency of consumers in the field of health care. Although Brady and the organization she helped to found, the Consumers Union, believed strongly in empowering consumers with information, its flagship journal *Consumer Reports* had "not tested or reported by brand or prescription drugs for the very good reason that such reports to consumers would not be useful." Prescription drugs were unique in the world of goods in that they presented no toehold for consumer activism. "There is, in fact," she concluded, "no other product or service necessary to the maintenance of life that so completely escapes the exercise of consumer sovereignty as does the prescription drug in the circumstances under which it is sold today."[6]

Brady's presence on Capitol Hill that day was itself a product of the broader development in American political and popular culture in the 1960s and 1970s that consumer historian Lizbeth Cohen has named "third-wave consumerism." From the 1960 consumer message of John F. Kennedy until the downsizing of the protective state under Ronald Reagan two decades later, consumerism became a keyword of almost universal valence across all sectors of American political life. In Congress, few appreciated the political salience of consumerism better—or earlier—than the senior Democratic senator from Tennessee, Estes Kefauver.

As we have already seen, Kefauver's considerable publicity skills allowed him to leverage the hearings of his Subcommittee on Antitrust and Monopoly into a continuing stream of newspaper headlines and evening TV news programs. When Kefauver described the plight of the pharmaceutical consumer with the pithy axiom "he who buys does not order and he who orders does not buy," he was using the drug market to illustrate the broader problem of monopoly power in late twentieth-century consumer markets.[7] Pharmaceutical consumption was not *exceptional* but *exemplary*. Popular coverage of the hearings, such as *Life* magazine's 1960 feature "Big Pill Bill to Swallow" amplified and repeated Kefauver's depiction of pharmaceutical consumers as "captive consumers," paradoxically stripped of sovereignty precisely when they faced the most vital of consumer decisions.

In a nation of captive consumers, Kefauver realized, the consumer of prescription drugs was the most captive of all.[8] This line was echoed by

Mildred Brady in her testimony; Brady's characterization of pharmaceutical consumption as an act that "escapes the exercise of consumer sovereignty" was not an expression of nihilism but of activism. Like Kefauver, Brady intended to advocate for consumers in precisely the spheres where they currently held no degree of freedom.[9]

Although the figure of the "captive consumer" would continue to animate discussions of pharmaceutical marketing throughout the decade of the 1960s (figure 12), to Brady, the concept of the captive consumer did not do *enough* to describe the utter dependency of the patient when consuming prescription drugs. "Better informed consumers" of prescription drugs could not simply "insist on being less captive." Unlike an automobile consumer, who could "choose, within his captivity, among competing brands and prices of the goods and services essential to running a car," the pharmaceutical consumer had no choice but to accept the products already selected by his physician. Brady's argument was informed by the dominance of "sick role" theory in contemporary medical sociology, in which the agency of the patient was limited to complying with medical advice.[10]

FIGURE 12. Illustration from the proconsumer Credit Union National Association, Inc., criticizing the captivity of the pharmaceutical consumer.

"Drug Buying: The Captive Consumer," Everybody's Money, *Autumn 1968, p. 18. Copyright Credit Union National Association. Reprinted with permission; all rights reserved.*

As Kefauver's hearings were gaining momentum in 1960, CU members were working under the surface to subvert the passivity of the sick role. In 1961, CU published *The Medicine Show: Some Plain Truths about Popular Remedies for Common Ailments*, which claimed to be the first "guide for laymen to the half-revealed world that lies beyond the prescription pad." In this and many books to follow, the captive consumer would live on as a form of social problem that could be solved by a series of educational self-help books. The passive patient could become a more active consumer by learning how to shop for generic drugs. The new generic consumer would be a savvy consumer, piercing through the gauzy mystification of pharmaceutical marketing with a clear and steady gaze.[11] That gaze needed to be critical, but not too critical.

A USERS' GUIDE TO GENERIC DRUGS

To understand the CU's position on pharmaceutical consumerism in the 1960s, we must recall the role of drugs in the foundational consumer movement texts of the 1930s. Arthur Kallet and F. J. Schlink's *100,000,000 Guinea Pigs* revealed to consumers the dangers that lay behind the pharmacist's counter as well as the grocer's, butcher's, and baker's.[12] But in this text, the ethical marketing of drug firms like Squibb, Lilly, Upjohn, and Smith, Kline & French were highlighted to better contrast the darkly lit sellers of dangerous patent medicines. The CU's first explicit foray into health-care consumption, the 1940 *Good Health and Bad Medicine*, by and large channeled the expert recommendations of the American Medical Association's Council on Pharmacy and Chemistry. Strident in its declamation of fraud and deception in the marketing of patent medicines and personal hygiene products, the CU's critique of medical marketing in the 1930s and 1940s coincided almost entirely with the antinostrum and antiquackery propaganda of the AMA and the firms that would eventually comprise the Pharmaceutical Manufacturers Association. As the preface of *Good Health and Bad Medicine* concluded, the book's primary goal was to help the family "choose its home remedies wisely so that their use will supplement, and not interfere with, the physician's advice."[13]

"What Dr. Spock is to freshman parents," the *New York Times* reported in 1959, "the Consumers Union is to bewildered housewives." A substantial residue of this paternalistic, expert-driven sensibility was present when the CU's *Medicine Show* was published two years later. The critical gaze of the consumer advocate was still focused largely on the nostrum, or what editor Dexter Masters called "the popular products which have occupied the drugstores, which stare at us from advertisements in the newspapers and magazines, and which give us our instructions from the television screens they now command." As with the CU's works from the 1930s and 1940s, the bulk of *The Medicine Show* was devoted to over-the-counter products for conditions that were not clearly diseases: gargles for sore throats, elixirs for coughs and colds, antacids and laxatives, vitamin supplements, reducing drugs and devices, antiaging pomades.[14]

Yet the careful reader could find a tonal shift in *The Medicine Show*. Emboldened by the widespread public critique of the prescription drug industry (on the one hand) and the medical profession (on the other) during the Kefauver hearings, Mildred Brady and others insisted that the time had come that such guidebooks do *more* than just supplement physicians' advice. If the CU's ambition to provide a "guide for laymen of the half-revealed world that lies beyond the prescription pad" was unfulfilled, it was not insincere.[15]

In the final section of the book, Masters hoped to teach consumers how to critically view the semijournalistic cheerleading for new "miracle drugs" in articles in the mainstream press, which Kefauver's investigations had linked to industry ghostwriters and public relations efforts.[16] CU was particularly concerned that this form of public relations helped to support high prices for brand-name versions of drugs that could be purchased in more inexpensive form by generic name. The importance of generic consumption was emphasized in the chapter on how to buy drugs:

> When you get a prescription from your physician, request him to direct the druggist to label it with the names of the drugs, their concentration, and the expiration date . . . It makes no sense to pay more for them than you need to. These drugs are available usually in two forms, as trademarked, brand named products, or packaged simply under a generic name. Both types conform to U.S.P. [*United States Pharmacopoeia*] or N.F. [*National Formulary*] standards. In nearly all cases the non-branded drugs sell for less than their highly advertised counterparts. Thus, if you request drug products by asking for them by generic rather than by brand name, and if you insist on the lowest priced package available, you can frequently save some money.[17]

The Medicine Show represented an early attempt to advise consumers how to gain a modicum of agency as patients through principles of generic consumption, but its tone was often tentative and its aims more illuminating than activating. Subsequent revisions over the course of the 1960s and 1970s did little to alter this message.[18] As we saw in chapter 3, however, the CU's hesitant tone would soon be countered by a more radical entry into the new genre of medical consumer self-help books: Richard Burack's *Handbook of Prescription Drugs*.

A HANDBOOK FOR GENERIC CONSUMPTION

Even before its release in the spring of 1967 Burack's *Handbook* became the subject of heated controversy. Senator Robert Byrd (D-WV) introduced a prepublication copy of Burack's book to the Senate in April 1967 with the injunction that "if this book is fairly considered by the American press and the public, it will shake the drug industry and the American Medical Association just as Ralph Nader's book [*Unsafe at Any Speed*] shook the automobile industry."[19] Acting as a form of congressional book club, other senators compared Burack's paperback with Rachel Carson's *Silent Spring*. Praised by those closely identified with the consumer movement—such as Edward Kennedy and Gaylord Nelson—Burack's book would be angrily waved by the conservative physician-representative Durward Hall (R-MO) on the floor of the House as an omen of the alarming spread of consumerism in medicine, which Hall considered "a bad prescription" for the health system. With tongue only partway in cheek, editors of *BusinessWeek* asked their readers, "Has drug industry met its Nader?"[20]

The *Harvard Crimson*, which interviewed Burack twice in the spring of 1967, flatly concluded he was "no Ralph Nader." Rather, as the college weekly duly reported, Burack was a genial general internist at Cambridge Hospital who had studied and taught pharmacology with Harvard Medical School's Otto Krayer. Hardly the demagogue, by the time the book was published, Burack had already laid out rather bucolic plans to move from Cambridge to become a general practitioner in rural New Hampshire after a sabbatical in Oxford. At the same time, the *Crimson* thought that it needed to qualify Burack as "not like your family doctor at all" but somehow, to the extent it was possible for a Harvard professor to be, a "rebel."

Burack's most rebellious quality was his willingness to disclose the secret laws of therapeutic equivalence known to doctors but not patients and to encourage readers of his book to "talk back" to their doctors.

The book that has the drug industry worried—it's selling at the rate of 8,000 copies a week.

Has drug industry met its Nader?

Boston doctor turns out a book that shows wide price spread
between brand name and generic drugs. It hits
industry at time when shots are coming from all directions

FIGURE 13. "Has drug industry met its Nader?"

BusinessWeek, *June 10, 1967, pp. 104-10. Used with permission of Bloomberg L.P.*
Copyright © 2014. All rights reserved.

"A patient should shop around for drugs," Burack claimed in an interview
with the *New York Times* in May of 1967, shortly after his *Handbook* was
launched, "*because* he is a captive consumer and has the right to know
what he is buying and how much it costs."[21]

Why would this simple statement make Burack such a controversial
figure? It is important to recall that, as late as 1967, consumers of pharmaceuticals were not only limited by their physicians' prescriptions but
often had no idea what compound they were purchasing in the first place.
Prescription bottles were not required—and in many states were formally
forbidden—to divulge the name of their contents until the late 1960s.[22] If
patients could not even know what category of thing they were consuming, Burack asked, how could they possibly make reasonable decisions as
consumers? He urged his readers to demand to know at least which medicine they were taking, so that they could then look it up in his *Handbook*:

> *For the reader who is a doctor*: a reminder to instruct the pharmacist
> to label each prescription with the generic name of the drug . . . *For
> the reader who is a patient*: when a doctor hands you a prescription,

see whether it instructs the pharmacist to label, and if it does not, ask
the doctor please to do so. He will almost always be glad to, if only
because it is a good safety practice to include it. The cost of drugs
will begin to tumble only after it becomes common practice to label
... Should the druggist not like the idea, take your business to one
who will cooperate; in a free enterprise system, competition works
miracles.[23]

Patients could be savvier consumers of doctors as well, and Burack urged
patients to warn physicians who refused to prescribe generically that they
might lose their business. Likewise, Burack argued, conscientious doctors should acknowledge their "responsibility to all aspects of a patient's
immediate welfare—including his pocketbook."[24]

Like the *Medicine Show*, Burack's *Handbook* relied on a form of substituted paternalism: for his readers to be empowered to take a more
"egalitarian" position vis-à-vis their own doctors, they had to trust *him*.
His arguments for the equivalence of generic drugs were reassuring but
largely anecdotal.[25] Curiously, readers responded to the supportive affect
in Burack's accessible prose to petition him for medical advice beyond
that contained in his book. Burack's personal papers reveal an animated
correspondence with hundreds of readers who sought individual medical
advice on how to manage their own prescriptions, ailments, or physicians. If, in his replies, he insisted that his book did not substitute for the
judgment of a personal physician, he also suggested that the physical
presence of his book might serve to empower patients in their encounters
with their own physicians. As one patient wrote Burack the morning of a
doctor's appointment, "Right now I got the book out, to have my wife take
to our physician this afternoon, when my wife goes to an appointment.
I have learned that our physician has not seen the book [and] I consider
the book very valuable."[26]

Burack's guide went far beyond *The Medicine Show* in its efforts to
make patients into active consumers of generic drugs. In turn, the CU
was chary in its own published review of Burack's volume in *Consumer
Reports*, warning that although the book "may help consumers save
money on prescription drugs; [there] are pitfalls in a book that seeks to
address both patients and doctors at the same time."[27] In particular, CU
criticized his "wholesale endorsement of all generic drugs, per se," in the
face of the active debates over the therapeutic equivalence of generic
drugs—as were already detailed in chapters 6 and 7 of this volume.[28] "The

consumer, unfortunately, is in no position to take sides in disagreements among physicians about the efficacy of this or that course of treatment for this or that condition. CU's medical consultants feel you're better off going along with your own physician than, armed with a book for laymen, disputing his judgment or shopping around for a doctor who agrees with the book."[29]

Within academic medicine, the *Handbook*'s critics included the co-author of the iconic textbook *Principles and Practices of Pharmacology*, pharmacologist Alfred Gilman. In an open letter to Gaylord Nelson, Gilman objected to the widespread uptake of Burack's *Handbook* among clinical and policy circles so soon after its release. Gilman echoed the CU's concern that the *Handbook* model of consumer empowerment went too far and suggested that generic consumerism would "be a source of infinite trouble with respect to [the] doctor-patient relationship." It was, quite simply, "not the sort of thing to be placed in the hands of the layman." Burack, wincing at these accusations, shot back that Gilman himself was being a sore loser, since Burack's *Handbook* was clearly outselling Gilman's much drier *Principles and Practices of Pharmacology.* Gilman's further conflicts of interest were revealed a few months later when an aide to Senator Gaylord Nelson discovered that Gilman's letter had been "suggested" (and partly drafted) by C. Joseph Stetler, the executive director of the PMA.[30]

Peevish and interested as it may have been, Gilman's complaint was not entirely without merit. Burack's book had indeed become a publishing phenomenon well in advance of its publication. Book-of-the-month clubs were promoting the *Handbook* months before its release, and once launched the book sold out run after run. Paperback copies were positioned by the checkout lines of pharmacies and grocery stores alongside popular potboilers. This marketing mix reached a strange blend of audiences: medical professionals and medical students, concerned consumers, and a skeptical public increasingly captivated by the new pharmaceutical muckraking literature of the 1960s.[31]

By the time the fourth (and final) version of the *Handbook* was published in 1976, Burack could point with pride to the widespread acceptance of generic drug consumption.[32] Between 1966 and 1974 the number of generically written prescriptions had increased from 6.4 percent to 10.7 percent—which he read as both good news in relative terms (almost doubling in eight years) but also bad news in absolute terms (still only one out of ten prescriptions were filled generically). "A critical mass of influential people," he claimed, "has begun to realize that the brand-name drug industry is, as has been charged, the last of the robber-barons and has begun to catalyze action to bring some order out of the disorder and chaos arising from the Organized Medicine—Drug Industry 'connection.'" Indeed, he added, "the 'great generic drug controversy' which raised its head first when Senator Kefauver was probing in the late 1950s and which heated up white-hot during the earlier years of the Nelson hearings, has now been settled to the satisfaction of nearly all unbiased experts."[33]

Why, then, was a handbook of generic drugs still necessary? What would it take to encourage the remaining nine out of ten prescriptions to be generically filled? The problem, Burack conceded, was that physicians still believed in the power of brands, kept alive by the pharmaceutical industry as a "diversionary smoke screen because it does not wish to allow time for thought and discussion of a far more basic issue, the matter of *rational prescribing*."[34]

THE IRRATIONAL PRESCRIBER

Burack relied on transcripts of the Kefauver investigations for evidence of a pervasive irrationality which brand names inculcated in physician prescribing behavior. Speaker after speaker hauled before Kefauver's Senate subcommittee in the early 1960s testified to the status of the physician as an *inexpert* consumer, whose delusions of hyperrationality rendered him all the more susceptible to the blandishments of pharmaceutical marketing.[35] Mildred Brady had noted as much in her own testimony: the role of the physician as learned intermediary was confused by "a second assumption that is also unrealistic: namely, that even if so pressured by 'better informed consumers' the doctors themselves are well enough informed about prices to exercise that 'finer discrimination.'" Brady continued:

> To those of us who have watched the development of product promotion closely, and who have followed the spirit of the promoter from the medicine man to the TV program of today, one very unhappy development has been the subjection of the physician since World War II to one of the most offensively lavish promotional outpourings yet witnessed. And we who listen hours a day to the conflicting claims made by various brands of deodorants, cold remedies, toothpastes,

hair dyes, headache remedies and laxatives—claims made frequently by the same companies which also promote and sell brand name prescription drugs to doctors—we television viewers and radio listeners cannot help but feel a sympathetic uneasiness for our physicians, subject as they are to the equally intensive and extravagant hard sell directed at them.[36]

Called upon to defend the position of physicians as expert consumers, Chauncey Leake (professor of pharmacology and medical jurisprudence at University of California, San Francisco) insisted physicians could not be "persuaded to use new drugs in the same way that people generally are persuaded to buy new clothes and new automobiles."[37] Indeed, such comparisons were often—in the words of Kefauver aide Paul Dixon, "predicated upon the proposition that you can't fool a doctor or an expert."[38]

Yet by 1960, study after study had demonstrated just the opposite— that physicians *were* persuaded in their consumer decisions in exactly the same way that the average consumer was persuaded to make decisions in the grocery store. Kefauver's committee critically examined the AMA-funded Fond du Lac Study, a major investigation funded by the professional organization in the late 1950s to examine physician prescribing habits in relation to marketing efforts. As the study concluded, the physician was not just an expert but a human being: "As a human being, he is comparatively quick or comparatively slow; he is comparatively hard-working or he is comparatively lazy; he is friendly or crabby; social or solitary; happy or unhappy. His morning contacts with his wife and children affect in a greater or lesser degree his attitude toward patients, toward co-workers, and toward detail men. His basic temperament, modified by his daily interpersonal relations, influence all his actions and attitudes to some extent. His human-beingness is modified by his being a physician."[39]

A contemporaneous study conducted by Ernst Dichter's Institute for Motivational Research for the Pharmaceutical Advertising Club of New York mapped the mind of the physician as a split brain with a wide gulf between conscious and unconscious selves. "A basic conflict arises from the physician's self image as a rational scientist," Dichter noted. "He expects himself to make up his own mind on the basis of objective evidence ... and yet he finds himself confronted, like a housewife in a supermarket aisle, with a misery of choice which he tends generally to resolve by irrational and emotional factors."[40]

To equate the physician to a "housewife in a supermarket aisle" in the year 1955 was to make a powerful statement conflating the gendered stereotypes of the (male) professional and the (female) consumer. If physicians were just as susceptible to the brainwashing of modern advertising as any other boob, the role of the physician in pharmaceutical consumption threatened to dissolve into an infinite regress of consumers suspended in a medium of marketing.[41] Popular trust in the unbiased expertise of the physician consumer had not improved in the intervening years, certainly not after the thalidomide tragedy, the feminist critique of the safety of the birth control pill, and the revival of a literature of pharmaceutical muckraking, including journalist Morton Mintz's *Therapeutic Nightmare.*[42]

By the time Gaylord Nelson began his hearings into competitive problems in the pharmaceutical industry in 1967, consumer advocates like Richard Burack, the CU, and the health wing of the more radical Naderite group, Public Citizen, were conflicted. The authority of all of these groups in advising patient-consumers was predicated on there being such a thing as expert consumer of medical knowledge, even if everyday physicians were no longer fulfilling that role. To confront this dilemma, consumer advocates began to insist that patients had a responsibility to take on additional agency as consumers *precisely because the ability of their physicians to act as expert consumers had become compromised.* It was like being on a bus with a sleepy driver—the concerned passengers might not know how to drive the bus themselves, but they could do their best to keep the driver awake.

ESSENTIAL GUIDES TO INESSENTIAL DRUGS

As the genre of consumer drug handbooks grew over the 1970s, subsequent entries would repeat this refrain but shift its tenor from the polemics of political economy toward a more pragmatic and technical modality. James Long's *Essential Guide to Prescription Drugs,* first published in 1977, grounded its more sober, pragmatic argument in concern for consumer safety. Long held it to be self-evident, by the late 1970s, that new drugs brought unknown risks to consumers and that physicians were imperfect vessels for communicating such risks to consumers. This new order of things required a reconceptualization of the patient as a consumer concerned with health risks, not cost savings. "It is now obvious," Long explained, "that the traditionally paternalistic and uninformative attitudes of some physicians have no place in the context of today's powerful and complex drugs. At the same time, the patient has become increasingly aware of how dangerous it can be to continue in the passive role of unquestioning drug consumer.[43]

Long's book assumed that the responsibility for drug consumption belonged to both doctors and consumers. "If the goals of drug treatment are to be achieved," he added, "adequate information and its proper use by all concerned are indispensable . . . We have now reached the point where the use of prescription drugs must become a *shared* responsibility, a collaboration between physician and patient."[44]

Long's *Essential Guide* assumed that an era of more egalitarian medical decision making had already arrived and brought with it what sociologist Renee Fox would later call the "tyranny of choice": a double bind in which patients now bore new responsibility for the outcomes of their own health decisions. To perform effectively in this new era, consumers needed a crash course in basic pharmacology, with access to a cross-index of brand and generic names of drugs and a set of "drug profiles" that helped explain mechanisms of action, side effects, and contraindications of most common drugs. Long's book provided all of the above, with disclaimers that the book was *not* a complete "do-it-yourself manual" that could substitute for professional medical care. If these books were not intended to substitute for the physician's judgment, they were explicitly intended to draw attention to the physician's fallibility as an expert intermediary and to the importance of viewing the prescription as an exercise in negotiated therapeutic authority. Long's *Essential Guide* would be followed by Bantam's 1979 *The Pill Book: The Illustrated Guide to the Most Prescribed Drugs in the United States*. The first edition alone would run through seventeen printings and more than one million copies.[45]

Following the passage of the Hatch-Waxman Act of 1984, the genre of guidebooks to prescription drug consumption continued to expand. When M. Laurence Lieberman published his own *Essential Guide to Generic Drugs* for a wide audience of consumers and physicians, the contrast even with Long's effort of the same name from a decade earlier was striking. Lieberman was a pharmacist and a writer who penned a number of columns and op-eds on the popular discussions of generic drugs surrounding the Hatch-Waxman bill. Differentiating himself from Burack's sweeping endorsement of generic drugs, Lieberman asked consumers to make decisions on a drug-by-drug basis. As his book was going to press, he wrote in an op-ed to the *New York Times* that the "brand vs. generic" debate had painted a far more Manichean picture than was warranted. Some generic drugs were useful, some were not. The ability to differentiate between a bargain (the same but cheaper) and a lemon (cheap in all senses) was a teachable skill.[46] The *Essential Guide* proposed to teach consumers to separate the wheat from the chaff.

With a pharmacist's prerogative, Lieberman suggested that even most doctors did not really understand the pharmacology of generic drugs enough to be able to meaningfully advise their patients. Lieberman listed, for each individual drug, a digest of FDA rulings on bioavailability testing along within an estimation of the reliability of individual firms' Good Manufacturing Practices and an assessment of the significance of the price differential between branded and generic versions. This book allowed a layperson to perform a decision analysis of generic purchasing of particular drugs, weighing the possible risks of generic consumption against the possible benefits of cost savings.

A typical entry, for the anticholinergic agent dicyclomine, noted that although generic tablets were relatively easy to make and bioequivalence issues were minimal, the fact that the drug was used for only a short period obviated the opportunity for any real cost savings from generic consumption. In contrast, Lieberman's entry for the antihypertensive drug glutethiamide recommended generic consumption even in the face of known biopharmaceutical absorption problems in tablet form. "*Brand or Generic?* The generic can yield savings of up to 60 percent and, with many generics considered bioequivalent to the brand, is the best choice. Consult your doctor. The tablet form has the potential for erratic absorption in the body, with wide variation in dose. (This drug also comes in capsule form by brand name only and is not available as a generic.) Generic savings are worthwhile provided the same manufacturer's product chosen from the list below is used with each prescription refill."[47] Lieberman demanded a good deal of effort and a priori health literacy from his readers. The book was a tome of substantial girth, too big to slip into a purse or a jacket pocket and bring along to the pharmacy. His dense prose contained contradictory recommendations, especially when standards of bioequivalence differed between federal and state regulatory authorities, as with this asthma/allergy medication:

> The generic tablet is not considered bioequivalent to the brand by the
> FDA. There is almost a 50% savings possible with generic use, but with

three active ingredients, the generic may not give the consistent and predictable performance needed when this medication is called for. Unless used on a long-term basis where savings really add up, stick to the brand. Several generic syrups are considered bioequivalent to the brand by some states, although not by the FDA. There was once a dye in the syrup that was implicated in allergic reactions, but this has been removed from the brand product. A clear, colorless syrup would indicate removal of this dye from generics. Generic use is a good alternative, offering savings approaching 40%. Consult your doctor."[48]

Lieberman's text cast the pharmaceutical consumer as an actor with far more dimensions of possible discernment when shopping for generic drugs. This required a high degree of health literacy and tolerance for ambiguity—and was still, nonetheless, tightly scripted.

BECOMING A GENERIC CONSUMER

By the late 1980s Lieberman's book was only one entry on a crowded bookshelf of paperbacks promising to help patients become more astute consumers of prescription medicines. By that time, a genre unthinkable at the dawn of the 1960s—the consumers' guide to prescription drugs— had become commonplace, uncontroversial, and even banal. Alongside revised editions of Burack's, Long's, and Lieberman's volumes, new entries to the field were popping up nearly every year. The first edition of Max and Betty Ferm's popular *How to Save Money with Generic Drugs* came out in 1988, and 1989 saw a complete overhaul of the CU's original volume, emphatically retitled *The New Medicine Show*. Emboldened, perhaps, by its competitors, *The New Medicine Show* expressly identified the importance of questioning one's prescriber about generic drugs in being a savvy medical consumer. "Your physician, therefore," the book concluded, "may represent a hurdle to your getting a generic drug . . . So once again, you must take the initiative."[49]

This chapter has traced a shift from paternalistic to more egalitarian models of the flow of medical information by tracing the linked discourses of consumerism and genericism in a series of manuals for patients learning to become savvier health care consumers. We must be cautious not to read into this genre history more than it can actually tell us. These texts are not transparent windows into actual consumer practices, but a scripted form of "how-to" literature that retained key paternalistic qualities at the same time that it helped to bring into focus a new image of the patient as an empowered consumer of generic drugs. Because these books ran though many, many editions and because this genre grew steadily from the 1960s through the 1980s and ramified further in the 1990s and 2000s, we assume there was and still is a market for this kind of advice. But it is difficult to read from the books alone *who* exactly sought out this advice and *how* this advice was actually used in practice.

The genre of the consumers' guide to prescription drugs does make clear, however, that the project of expanding consumerism in health care in the 1960s, 1970s, and 1980s was tightly linked to an expanding regime of genericism at the pharmacy. An initially timid set of arguments intended to supplement physicians' advice became increasingly bold in arming consumers with tools to question that advice. As this self-help literature became more confident, the goal of supplementing physician advice shifted to displacing physician advice, or at least providing consumers with a set of tools so that they might better assess the kind of medical advice they were being given.

CHAPTER 11

GENERIC CONSUMPTION IN THE CLINIC, PHARMACY, AND SUPERMARKET

Generics aren't really new. Products of similar quality have probably been in your supermarket for a long time. What's new is the way they're being marketed.

—*CHANGING TIMES*, 1978

By the close of the twentieth century, consumers' guides to prescription drugs, much like consumers' rights to their medical information, had become a rather uncontroversial issue. When, in 2004, *Consumer Reports* began to publish *Best Buy Drugs*, a new serial whose first installment featured a comparison of cholesterol-lowering drugs alongside its ratings of dishwashers, refrigerators, and toaster ovens, few eyebrows were raised. Nor would anyone protest, two years later, the publication of the *Complete Idiot's Guide to Prescription Drugs*.[1]

Consumers of generic drugs were not experts, perhaps, but neither were they complete idiots. Advice to generic consumers crossed two important boundaries that had seemed clear and concrete in the middle of the twentieth century. As they worked to provide expert advice to the patient as consumer, these self-help guides increasingly questioned the expertise of the physician as consumer. By placing prescription drugs on par with home appliances and breakfast cereals, the expanding sphere of generic consumption hacked away at the barriers separating pharmaceuticals from the broader world of goods.

THE PHYSICIAN AS CONSUMER

Too little attention has been paid to the role of the physician as a consumer, and we are increasingly aware of the dangers of this omission. In recent years evidence of the extent to which physicians' cognitive structures and perceptual environments have been shaped by powerful marketing organs of the pharmaceutical, food, and tobacco industries has become undeniable. This reach extends far beyond journal advertisements and sales representatives to the processes of ghostwriting, infiltration of popular science writing, and the structure and content of continuing medical education.[2] Only in the early twenty-first century has the majority of the American public learned to see physicians as the pharmaceutical industry saw them for at least a half a century: as the most influential consumers in the therapeutic marketplace.

If physicians were conscious of their status as consumers, they seldom wrote about the topic explicitly. The few mentions of physicians as consumers in the medical literature of the 1970s and early 1980s refer to physicians who themselves needed medical treatment and were thus physician-patients, assumed to be "the ultimately informed 'consumer' of medical and surgical services." Most authors assumed that physician-patients made for model medical consumers because they were so well informed. Occasionally researchers turned up surprising data suggesting that most physicians did *not* make for model patients, nor were they always well informed. In a whimsical article entitled "The Physician as Consumer of Medical Literature," medical librarian Estelle Brodman complained that the average physician knew very little about how to obtain information on new therapeutics. "How can the physician make informed decisions about health and disease," Brodman complained, "if he is not informed correctly and completely and cannot even check up on so called statements of facts?"[3]

In contrast, for pharmaceutical marketers the role of the physician as consumer was a well-described entity already by the 1950s, as the discussion of the Dichter Study and the Fond du Lac Study in the last chapter noted. In the eyes of market researchers, the prescription did not divide doctors (as producers of prescriptions) from patients (as consumers of prescription medicines); it linked them. As historian of medical consumerism Nancy Tomes notes, with the codification of prescription law, "physicians became the funnel through which the most expensive of drug products flowed to patients." Over the course of the 1950s and 1960s, the pharmaceutical industry learned to concentrate its marketing efforts on this easily targeted and well-bounded group of elite consumers whose decisions influenced the entire market for pharmaceutical consumption.[4]

Paradoxically, a key means for the pharmaceutical industry to mobilize the physician as an expert consumer was to express concern over the increasing role of "consumerism" in medicine. In the eyes of skeptical physicians, consumerism connoted "pushy patients" who demanded a more active role in decision making. As consumers, physicians believed patients to be penny wise and pound foolish, interested in short-sighted savings that might ultimately be far costlier in terms of health outcomes.

Patients as consumers often asked for generic drugs. Physicians as consumers, however, could be counted on to know the value of pharmaceutical brands. Here lies one of the principal paradoxes of consumerism in late twentieth-century medicine: as doctors came to exert more and more of their agency as consumers, they became increasingly hostile to what they perceived as medical consumerism on the part of patients. Having internalized their own role as learned intermediary in the consumption of therapeutic goods, physicians came to jealously guard that position as a natural right. They were supported in this effort by the lobbying and public relations of the organized pharmaceutical industry, who crafted anticonsumerist and antigenericist campaigns designed to cast consumerism as a pernicious social process that threatened the rightful expertise and agency of the physician.

As the pharmaceutical industry realized that consumerism wasn't merely the watchword of the Kennedy and Johnson administrations but would also carry over into the Nixon years, the Pharmaceutical Manufacturers Association initiated a series of campaigns seeking to unite physicians and pharmaceutical manufacturers against the perils of consumerism in medicine. "Consumerism isn't anti-business," Richard Nixon's appointee to the Federal Trade Commission, Caspar Weinberger, had remarked on assuming office in 1969; it is a "necessary part of living in these days." Nixon's new special assistant for Consumer Affairs, Virginia Knauer, made early ripples that fall with a speech on consumerism to the National Association of Retail Druggists. The *F-D-C Pink Sheet* excoriated Knauer for her "intense, 'nothing-sacred' consumerist inclination," largely because she endorsed generic drugs.[5]

Knauer was no Ralph Nader, and she frequently expressed concern over the limits of consumerism in medicine. She did not, for example, believe that patients as consumers had a clear right to know what was in their prescription bottles. "I believe a person should know what he is getting, *if possible*," she waffled, but she was concerned that consumers might "be literally scared out of their minds" once they looked up drug names and saw the possible side effects associated with them. "As a non-expert," Knauer demurred in an interview, "I wonder if pharmacists could not make a prudent judgment and list the generic name when their discretion indicates such action would not be detrimental."[6] But she did work effectively to insert a progeneric message into Nixon's speeches on consumerism. When Nixon delivered his first consumer message to Congress that year, he announced that he wanted all pharmacy bottles to contain the generic names of their products. Pharmaceutical trade journals warned that the Food and Drug Administration under Nixon was prey to this "step toward genericism" and that "no one will become commissioner if he isn't a dedicated 'consumerist.'" In the pharmaceutical trade press, *consumerism* and *genericism* became dirty words, largely exchangeable one for the other.

As some pharmaceutical executives warned that "rampant consumerism" was taking over the health field, others worked to excise the consumer from *consumerism*. Complaining that drug manufacturers had unfairly been made into the "doormat of activists," the marketing director of Lederle Laboratories presented data from focus groups suggesting that *real* consumers were not really interested in consumerism, at least not in the kind of consumerism propounded by groups like the Consumers Union and Public Citizen's Health Research Group. Unlike consumerists, actual flesh-and-blood consumers "have little experience of doctors or pharmacists who recommended generic drugs and only in the case of aspirin or vitamins have a few seemed willing to buy generically."[7] Yes, one advertising executive told pharmaceutical executives in an attempt to navigate the turbulence that the consumerist and feminist movements were causing marketing executives, "the consumer of the 70's is going to be more sophisticated in medical matters." But that did not mean she was a frothing "consumerist." "The consumer is not a moron," he concluded. "She is your wife. Don't insult her intelligence."[8]

The medical profession was generally sympathetic to this normative and highly gendered reading, often preferring to distinguish consumerism as the work of agents provocateurs moving in from outside the clinic over admitting that their patients legitimately preferred alternate forms of medical authority. As the editor of *Archives of Internal Medicine* noted in 1971, the medical consumer was an anomaly, "because for generations it has been an accepted truism that the complex professional elements that determine the decisions of the physician can only be completely handled by individuals who have been trained as physicians."[9] Like other professional warnings against consumerism in medicine, his argument hinged on the naturalization of the physician as an expert consumer.

The anticonsumerist stance of organized medicine and the pharmaceutical industry only intensified in the early 1970s. In 1971, the trade journal *F-D-C Pink Sheet* warned its readership that consumerists were using "increasingly sophisticated approaches" in their efforts, including "freshly-researched and documented legal briefs and meetings with FDAers to which the consumerists come well-prepared for initiating new ideas."[10] By 1974, the head of public relations for the PMA, William C. Cray, on *Meet the Press* cautioned that consumer advocacy groups advancing consumer education on generic drugs were metastasizing across the country, with organizations in forty-eight states, including twenty-two Naderite Public Interest Research Groups. These "consumer groups, labor unions, senior citizens groups are really hot about [generic substitution]," Cray continued. "They got to the point of picketing some firms and things of that sort. It's a very serious problem right now."[11]

By late 1975, with the repeal of antisubstitution laws in full swing, the American Pharmaceutical Association's Edward Feldmann wrote to the organization's president, William Apple, confident that the movement was now sure to "snowball among the remaining states." But Feldmann was concerned about a new strategy among physicians to focus opposition to generic substitution. As he noted, "It now appears that a shift in the strategy of opposition is developing in which medical organizations have essentially concluded that if they cannot prevent amendment of antisubstitution laws in the legislative process, they can still thwart the intent of the amendment by inducing physicians generally to regularly and consistently prohibit drug product selection by entering the designation 'DAW' on each prescription order written." DAW stood for *dispense as written*: the American Medical Association had passed a resolution in the summer of 1975 to urge physicians to write this acronym on each prescription they wrote, and state medical societies had taken out full-page advertisements in medical journals urging the same. This strategy, Feldmann worried, posed a grave threat to generic substitution efforts, unless "some forceful and positive effort" was made "to head off this latest maneuver before it spreads further or becomes an entrenched practice."[12]

DAW and "no substitution" campaigns were indeed part of an explicit strategy to blunt the rapid uptake of generic consumption. As Florida considered passage of its new substitution law in 1974, the AMA and PMA pressed hard for a "two-line" prescription pad with spaces for physicians to sign on one side if they approved of substitution, on the other if they did not. Hugh Hussey, former head of the AMA and current editor of *JAMA*, argued that only if a physician was able to designate certain pre- scriptions as "dispense as written" or "do not substitute" would such a law be an "acceptable solution to the physician."[13] Years later, the PMA's William C. Cray would claim that the two-line prescription form had been a PMA concept, as part of an offensive maneuver in its fights against substitution legislation:

> PMA was anxious to swing away from the "firefighting" and defensive battles to positive proposals. In that quest, it came up with an extraordi- narily simple yet superb piece of legislative draftsmanship—the two-line prescription form. It was to become a landmark item in PMA history. Brennan, Patton, and other PMA company lawyers forged the concept. The Board first regarded it skeptically, then strongly endorsed it in 1975. The form preserved physician prescribing and patent rights, and its fair- ness appealed to legislators. Around it PMA built its approach to leg- islators. The PMA's model form attenuated the impact of substitution, with the market studies showing that 70 percent of doctors specifically prohibited it. Without the two-line form, up to 95 percent of those eligi- ble in many states could or would have obtained substitution products. Eventually 29 states accepted the two-line provision.[14]

As medical historian Dominique Tobbell has demonstrated, the vaunted autonomy of the prescribing physician became a rallying point for orga- nized medicine and the pharmaceutical lobby in several major health policy debates in the 1960s and 1970s.[15] Yet the development of the no substitution movement illustrates the extent to which the articulation of physician autonomy was predicated on the physician's role as expert consumer. Only in this role did the physician have the autonomy to di- rect or designate the consumption of a particular drug for the patient in the face of the generic substitution laws advocated by an anonymous pharmacist, anonymous state bureaucrat, or anonymous manufacturer of generic drugs. In resisting the control of their actions by the bureau- cratic structures of the state, physicians did not seem to realize how much they in turn allowed their own agency to be redefined in terms of the bureaucratic structures of an industry promoting a particular set of consumer goods.

The physician's autonomy as a practitioner was bound up with the physician's consumption of drugs and accompanying drug-related information in a set of key legal challenges in the 1980s. By the end of the 1970s, courts sought a language and logic for mediating physician liability for adverse outcomes suffered by patients whose physicians did not know enough about the risks of the drugs they were prescribing. The 1982 decision in *Oksenholt v. Lederle* appears to be the first to define the rights of the physician as a consumer of medical information. Erling Oksenholt, an Oregon physician, was sued by his patient after she went blind following a course of treatment for tuberculosis with Lederle's Myambutol. After settling charges of negligence with his patients for $100,000, Oksenholt turned around and sued Lederle for failing to notify him of the known risks of Myambutol (ethambutol); adding to the bill of $100,000 he had paid an additional $150,000 for damages to his professional reputation. The case made its way to the Supreme Court of Oregon, which sided with Oksenholt, arguing that the physician's rights as a consumer of medical information were also protected by the FDA's regulatory requirements, and these rights had been violated through misrepresentation of product safety in Lederle's marketing materials. As the *Northwestern University Law Review* commented on the case, physician-consumers had a "right to rely" on pharmaceutical marketing materials "just as consumers are entitled to rely on a producer's advertisements that guarantee product safety."[16]

Certainly the rhetoric used by physicians in their opposition to generic substitution was high-handed: autonomy, ethics, intrusion of the state into the sanctity of the doctor-patient relationship. But we should be careful not to merely accept physicians' rhetoric as an accurate description of their political agency. Passionate articulations of the autonomy of the physician to select the brands they believed in served in part to disguise the extent to which the physician's own role was now increasingly understood as that of a consumer.

SPACES OF GENERIC CONSUMPTION: THE PHARMACY AND THE SUPERMARKET

Although much ink would be spilled in medical and pharmaceutical periodicals over who best understood the needs of the medical consumer—the state, the industry, the profession, or consumer advocacy groups—these debates took on their most material consequences in the market spaces where pharmaceuticals were bought and sold. In early 1971, pharmaceutical industry journals reported with some alarm that the city of Boston now required all stores with pharmacy counters, including several chain supermarkets, to publicly post the prices of one hundred key drugs. Other cities and states began to float policies for prescription price posting and mandatory reporting of generic names on prescription bottles.[17] New York governor Nelson Rockefeller included his support for generic drug consumption—through generic labeling and pharmacy price posting— as a key part of his moderate Republican consumer revolution message in 1971. As more states passed laws mandating drug price posting and generic labeling and other states passed laws explicitly banning the practices, the landscape of the pharmacy itself quickly became a battleground for proconsumerist and anticonsumerist lobbies.[18]

Chain drug stores and supermarkets saw an opportunity to build bigger markets by explicitly identifying their own brands with the politics of consumerism. The pharmacy chain Revco began to advertise its comparative prices to consumers in 1973; the next year Walgreens announced its own *Consumer Rx Price Books* on 1500 pharmaceuticals products, both branded and generic (though explicitly excluding any scheduled narcotics), through a series of advertisements and "Dear Consumer" letters stressing the quality of Walgreens generics. Charles Walgreen III himself held a press conference at which he distributed a pamphlet called *How to Save Money on Rxs* and announced that "both brand name and generic prices will be listed in a way which makes it easy for consumers to compare one with the other."[19]

The generic consumer here was not a cheapskate or a second-rate purchaser but a savvy consumer, a model consumer, an educated consumer: the kind of consumer that national chains wanted to attract. As the Revco and Walgreens campaigns suggest, the subject of generic consumerism was not defined by public, professional, and consumer advocates alone. Increasingly it would be shaped by the marketing interests and strategies of pharmacy chains and supermarkets as well.

The career of Esther Peterson traces these different terrains of generic consumerism in the 1970s. The former special assistant for Consumer Affairs in the Johnson administration, Peterson had left the position of "consumer czar" to Virginia Knauer when Nixon took office in 1969, but she did not leave Washington, DC. Instead, Peterson joined the staff of the Washington-based Giant Food supermarket and pharmacy chain as their permanent vice president for Consumer Programs. In her years as a local consumer czar at Giant, Peterson became a strong advocate for the expansion of generic drug promotion by chain drug stores. Peterson and Giant perceived a win-win situation for the American medical consumer and the chain store: consumers would gain greater ability to get good value for their medical dollars, while the firm would gain added sales volume and consumer trust. Giant rolled out its first generic advertisement on February 2, 1977, with a voiceover by Peterson: "What about quality? In our quality control laboratory, Giant tests generic prescription drugs to be sure they are consistent in quality batch-by-batch. Not all drugs are available as generics. If you have questions, call any Giant pharmacist."[20]

Broadcast media were followed by full-page advertisements (figure 14) in the *Washington Post* featuring a picture of Esther Peterson under the banner "Let's Talk about Generic Drugs." In the accompanying text, Peterson assured consumers that "batches of generic drugs purchased by our Drug Distribution Center are analyzed in our own Quality Control Laboratory with modern, sophisticated drug testing equipment."[21]

"We do this," Peterson attested, "even though the law does not require it."[22] In a joint statement the FDA and Giant Food announced that these advertisements "may explain, for example, the generic drug concept" to consumers.[23] Just a few days after the first Giant ad appeared, the rival Washington-based pharmacy chain Peoples Drug launched a generic drug campaign in the DC metropolitan area. Since a "recent survey of MDs in the area," according to Peoples' president Bud Fantle, "showed a high percentage are currently prescribing generically, when possible, we now feel it is appropriate to assist in the education of consumers, reassuring them of quality and potential saving afforded by using generic drugs." Citing recent legislation in support of generics in DC with substitution laws already in effect in Virginia and Maryland, Peoples began to send cards to doctor's offices with the names of all commonly prescribed generic drugs and the Peoples logo upon it. Though Peoples did not run quality assurance tests on its generics, as Giant did, Fantle pointed to the reputation of their generic supplier, Purepac, as "the largest full-line generic manufacturer in the country, with an outstanding record for quality during the past 45 years."[24]

Esther Peterson and Bud Fantle were immediately targeted by the PMA's Joseph Stetler, who insisted that the reputation of Giant and Peoples was not sufficient to guarantee the quality of their generic prod-

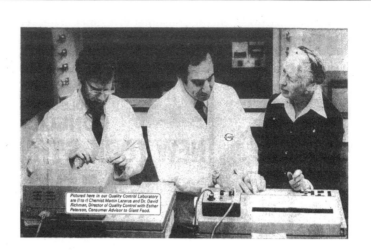

Pictured here in our Quality Control Laboratory are (l to r) Chemist Martin Lazarus and Dr. David Richman, Director of Quality Control with Esther Peterson, Consumer Advisor to Giant Food.

Let's talk about Generic Drugs

what is meant by the term "Generic"?

ge·ner·ic (jĭ-nĕr'ĭk) ~~~~~~~~~~ ~~~~ting to or ~~~~~ Of or relati~~~~ of an ~~~~ ~~~~~~ ~~y trade~~ a ~~~~~~ ~~~~~~ from **3. Commonly available; not protected by trademark; nonproprietary.** Said of drugs. ~~~~~~ The qual~~ gen·er·o~~

A drug's generic name is its common chemical name, not protected by a trademark. For example, HydroDIURIL® is a brand name; hydrochlorothiazide is its generic name.

are Generic Drugs as effective as brand name drugs?

Yes, provided they meet certain criteria. All drug manufacturers must meet certain standards which are enforced by the Food and Drug Administration (FDA). In addition, most drugs (generic and brand name) require approval by the FDA prior to marketing.

With the exception of antibiotics, which are certified batch to batch by the Food and Drug Administration, batches of generic drugs purchased by our Drug Distribution Center are analyzed in our own Quality Control Laboratory with modern, sophisticated drug testing equipment. We do this even though the law does not require it.

what tests does Giant perform?

Our professional Quality Control staff, consisting of a Ph.D. in Pharmacy and two chemists conduct a number of tests to insure the quality of generic drugs.

SOME OF THE TESTS WE PERFORM INCLUDE:

- **CHEMICAL ANALYSIS**-- quantitative and qualitative
- **TABLET DISINTEGRATION**--the ability to disintegrate when ingested.
- **TABLET OR CAPSULE DISSOLUTION**--an indication of adequate absorption of the drug into the bloodstream.

We require a certificate of analysis for each batch of generic drugs. This certificate is the manufacturers' evidence that their products meet government standards. The data in this document are compared to our test results. We then accept or reject the batch, based upon our own findings.

do Generic Drugs cost less?

YES. Brand names cost the pharmacy more than generics, and so the consumer pays more--around twice as much. At Giant the savings with generic drugs average 47.6%. But remember, not all drugs are available as generics.

HERE ARE SOME EXAMPLES

QUANTITY	BRAND NAME	OUR SELLING PRICE	GENERIC	OUR SELLING PRICE
100	Pavabid,* 150 mg	$10.59	Papaverine, 150 mg	$5.35
100	Serpasil* 0.25 mg	$6.78	Reserpine 0.25 mg	$1.47
100	HydroDIURIL* 50 mg	$6.05	Hydrochloro- thiazide, 50 mg	$3.11

always remember...

When it comes to taking drugs, your doctor is the one to make the decision on when generics are to be dispensed. Check with your doctor and see if your prescription can be filled with a generic drug.

Esther Peterson
Vice President,
Consumer Programs

DEDICATED TO BETTER HEALTH

GIANT GR PHARMACIES
THE PRESCRIPTION PEOPLE
A DIVISION OF GIANT FOOD

Our commitment is simply to make available to physicians and their patients quality generic and brand name medicines, at the lowest possible cost.

FIGURE 14. Giant Food advertisement for high-quality generic drugs.

Washington Post, *February 4, 1977, A14. Reproduced with permission of the copyright owner.*

ucts. "I do not mean to minimize Giant's efforts to check on the quality of its generic products," Stetler explained in a letter to Peterson, "but consumers need more assurance of the quality of those drugs than can be offered through the chemical analyses your lab conducts." Yet many consumers seem to have found assurance in this campaign, and both firms found their generic lines to be enormously successful. Within a few months, industry trade journals reported with concern that Peterson had left her position at Giant to resume her role as consumer czar in the new Carter administration. "We believe it is likely that Mrs. Peterson," the *Wood-Gundy Progress Report* noted, "who had been the spearhead for the Giant generics program, will advocate a similar program nationwide to save consumers' health dollars."[25]

CELEBRATION OF AUSTERITY

By the end of the 1970s, Wall Street had marked the generic drug sector as a clear opportunity for growth. But such growth depended on an expanding market of generic consumers, a population that was poorly understood. Who were these generic consumers, and what factors led them to generic consumption? Investors, manufacturers, and managers of pharmacy and supermarket chains commissioned study after study to explore the psychology and demographics that explained the growth patterns of generic consumption.

For most of the 1960s and 1970s, marketing theory had been dominated by Pierre Martineau's doctrine of segmentation. Martineau and others argued that the market was not a single stable pool but a series of niches, each of which could be sold a subtly different brand-name version of automobile, shoe, or cigarette. By 1960, the one-size-fits-all approach to mass-produced goods, like the Model T, was replaced with a diversity of brands intended to appeal to different subsections of the consumer market. General Motors' multiple brands (Cadillac, Pontiac, Chevrolet, Buick, Oldsmobile) were all made in the same factories, yet they inspired differential brand loyalty among different social and demographic strata. But in the phenomenon of generic drug consumption, market analysts of the late 1970s saw an alternative strategy emerging, which could be called "countersegmentation." Consumers in the age of Jimmy Carter, living in an era of rampant inflation and decreased purchasing power, were increasingly critical of brands and marketing. Barry James, author of the first textbook on the marketing of generic drugs, suggested that marketers needed to respond to these discerning consumers with more generic marketing, promoting "low-priced, austerely packaged goods of standard quality, identified only by their content and not supported by branded advertising."[26]

Generic drugs were joined by other generic goods. As a later market analysis noted, these generic products were chiefly "food and household staples characterized by little or no advertising, absence of brand names, and plain, stark, black-and-white labels." Swedish chain stores had begun to market goods with intentionally plain packaging in the early 1970s. In 1976, the French supermarket chain Carrefour introduced fifty generic items in thirty-eight stores; within a year these generic products made up nearly half of all sales volume. After Leo Shapiro, a marketing consultant for the Chicago-based Jewel Food Stores, adopted the concept in 1977, generics soon made up more than 40 percent of volume in certain commodity classes in Jewel's stores. By 1985, one out of three of the nation's thirty-three thousand supermarkets carried a generic line of foods and one out of every ten products sold in an American supermarket was a generic product.[27]

Most market analysts initially dismissed the generic marketing concept as a passing fad, but generic goods expanded in volume and presence nonetheless. Generic products soon included canned fruits and vegetables, paper products, and other household staples. Even the cigarette—perhaps the most iconic example of the role of branding in the American marketplace—could be made generic. Liggett & Myers produced a line of 14 mg tar cigarettes, wrapped in black and white packaging, priced roughly a dollar less per carton than other national brands. By 1982, 90 percent of all US supermarkets were selling generic items, accounting for some $2.2 billion in retail sales (2 percent of all supermarket sales); this was predicted to grow to 10 percent of all sales by 1985. Prominent professors of marketing soon began to refer to the generic marketing concept as "one of the industry's most successful innovations."[28]

Who was this new generic consumer? Certainly not all consumers trusted generics to the same extent. Supermarkets and pharmacy chains conducted surveys to target which consumers were more or less likely to buy generic items for their refrigerator or medicine cabinet. These surveys used available markers of difference—race, sex, age, income—to predict how likely a given consumer was to consume generically. "Blacks and minorities feel they are being put upon" when they receive potentially inferior generic drugs, one survey reported, as did many elderly and "low income" or "low educational level" consumers studied. Middle-aged, middle-class, white consumers with more education were far more likely to consume generically. Although these researchers strove for an objective tone in their published reports, the sum total of their findings was nonetheless inflected with a thinly veiled moral register: well-educated, middle-class, white consumers were often described as more sophisticated generic consumers, while populations marked as marginal—African Americans, the elderly, the poor, the less educated—were not.[29]

Research into generic consumption addressed a new paradox as more and more states passed progeneric substitution laws. Consumers had been expected to immediately rally behind the cost savings such laws promised, but consumer advocacy groups soon realized that only a subset of pharmaceutical consumers were generic consumers. Pamphlets provided by state public interest groups entitled "How Do You Best Use the New Generic Drugs Law?" tried to exhort patients to take advantage of these new frameworks as active consumers. Consumers were urged to ask doctors *not* to write "DAW" or "no substitution" on their prescriptions. Consumers were urged to shop around for pharmacists who stocked lower-price generic drugs and to report noncompliant physicians and pharmacists to relevant state boards. But even though more than thirty states had passed permissive or mandatory substitution laws by 1980, large parts of the population were relatively apathetic about generic drugs. As a survey that year indicated, "except for the chronically ill and a small price conscious minority of the population, consumers infrequently request the lower priced drugs and physicians and pharmacists are still reluctant to prescribe and dispense generically."[30]

Market researchers were vexed by the subject of the generic consumer. Although many studies in the late 1970s and early 1980s tried to carefully characterize the demographic and personality profile of the generic consumer, a single definition proved elusive. Teams of social scientists approached consumers of generic drugs and grocery products to interview them in middle-American urban and suburban environments held to be representative of US society, like Tulsa, Milwaukee, and Akron. But generic consumers in Milwaukee did not necessarily look like the "typical Tulsa user of generic products." Some analysts thought generic consumers were not one group at all but an amalgam of "professional college-educated consumers" and "lower-income, blue-collar shoppers" who collectively "exhibit a willingness to substitute some branded items in exchange for reduced prices."[31] James Henson, president of Jewel Food Stores, challenged market researchers to do more to unlock the "generic mentality," or a generic way of thinking about the world of goods, that was increasingly compelling to certain groups of consumers. "To what extent," marketing professors Martha McEnally and Jon M. Hawes asked in 1984, "is generic mentality a generalizable concept that can be applied in the marketing of other products?"[32]

It is important to remember that these generic goods were actually heavily promoted. "In introducing their own no-frills lines," *FDA Consumer* reported, "none of the stores have been shy about promoting them, no matter what other claims they have made to the effect that the lack of frills means lower prices to the consumer."[33] Generic "popped" as a marketing strategy. "Their stark labels," *Changing Times* reported in 1978, "jump out at you from the supermarket shelves."[34] As *Fortune* told its readers, generic marketing invoked a paradox, in that "generics are not so much an anti-brand brand as a new kind of brand. Astute retailers clearly identify themselves on the labels."[35] Likewise, one could still distinguish oneself as a generic consumer by choosing between Pathmark's No Frills line over Ralph's Plain Wrap products or Safeway's Brandless line. In a particularly delightful demonstration of the continuing valence of the oxymoron, the phrase *Generic Grocery Products* was itself trademarked by Jewel Food Stores.

What new language could accommodate the admittedly oxymoronic concept of the generic brand? One team of market researchers suggested the term *neogeneric* to describe trademarks like Kroger's Cost Cutters, and Safeway's Scotch Buy, which suggested the thrifty qualities valued in generic consumption. Even the FDA complained that the concept of generic consumer products was applied in different ways in different places, each chain store presenting "its own idea of what no frills means, its own list of kinds of no-frills foods (and often nonfood products) offered, and usually its own name for the practice."[36]

Generic brands shared what one journalist called a "celebration of austerity": plain wrapping, muted color schemes, fewer inks, and prices 30–40 percent lower than national brands and 20 percent lower than prior in-house brands. Yet this reduction in packaging and advertising costs accounted for only a small part of the price reduction of generic grocery products. In an interview in 1978, a spokesperson for A&P admitted that the black and white labels were only a few cents per thousand cheaper than brand-name or store-brand labels: "the real savings are strictly in quality." Standard Grade fruits and vegetables were cheaper to buy and process than Extra-Standard or Fancy Grade. No-frills products were explicitly *not* the same as their brand-name counterparts: they were made of inferior materials. Jewel's no-name ketchup had fewer tomato solids and was therefore runnier than national brands, and it contained fewer herbs and spices; the generic brand also used less sugar in its fruit cocktail syrup and used broken peanuts to make its peanut butter. A&P's No-Frills Trash Bags were 35 percent thinner than name-brand bags, while No-Frills Mac & Cheese had less cheese and more macaroni than competing brand-name products.[37]

Generic products were not the same as brand-name products, but consumers did not seem to mind the differences—or at least they found these differences trivial compared to significant cost savings. Paradoxically, the generic marketing concept was at times so successful in supermarkets of the late 1970s as to be occasionally self-defeating. As each store scrambled to stock low-quality consumer goods, they soon found that there were not enough of them. The national stock of standard-grade fruits and vegetables was rapidly exhausted. As one journal reported:

> Since there's so much scrambling now to dress food in black and white wrappers, competition for lower-grade products is on the rise and shortages sometimes occur. When they do, stores have to choose between dropping a generic item or paying more to put a higher-grade product inside the can. A&P ran into that problem last summer. "The only corn we could find was Extra-Standard," a spokesman reported, "and that's not what we wanted to put in the generic can." But they did. So for a while, at least, shoppers buying A&P's generic corn were getting a higher-quality product than the no-name's stark label implied.[38]

Generic consumerism required that consumers understood the strengths and limitations of the generic concept. Michel Hatt, a spokesperson for Jewel, told *Changing Times* that generic peas work fine for a casserole, but a consumer might want to purchase name-brand peas for individual servings. The FDA, however, was becoming increasingly uncomfortable with the proximity of generic drugs and the more general generic marketing concept. The FDA entered the debate over generic foods late in 1978, objecting to calling such products "generic" since they were not, in fact, the same. "Some news accounts about this new marketing concept have called these products 'generic' foods, but FDA officials have discouraged the use of that name for this type of food because it may confuse consumers, who will think the term has a meaning similar or equivalent to generic drugs, which is not the case."[39]

"The concepts of generic foods and generic drugs are totally different," another FDA official stressed. The difference here was that in the eyes of the FDA, generic *drugs* were of the same level of quality as brand-name drugs, where generic *groceries* clearly differed in quality compared to brand-name groceries. If the two kinds of generic goods were equated in the eyes of consumers, much of the FDA's efforts in promoting generic drug equivalence could be lost.[40] As *Food and Drug Letter* reported in 1978:

> As the result of FDA's highly publicized campaign for generic drugs, consumers have come to interpret the term "generic" as meaning equivalent . . . It is the difference in quality of no-brand foods which has FDA worried, because even if all grades of the foods are nutritionally equivalent, the concept that generic foods are lower quality than brand-name products could affect FDA's efforts in the generic drug area.
>
> Bureau of Drugs staffers say they want the agency to advise consumers of the differences between generic drugs and foods, and they want the term "generic food" replaced by expressions like "no-frills," "plain-wrap," or "plain label" foods to distinguish generic food and drug issues. "We don't care what they call them (no-brand foods), just as long as they're not called generics."[41]

Conversely, the FDA worried that experience with the different quality of generic groceries might undermine consumer belief in the equivalent quality of generic drugs. A survey by the American Association of Retired Persons in the mid-1980s found that many American consumers were still "fairly skeptical" about generic drug quality. While a majority of consumers were now familiar with the term *generic*, more than one in five respondents believed (erroneously, AARP pointed out) that brand-name drugs were FDA approved to meet higher standards than generic drugs, much the same way that brand-name frozen peas, ketchup, or peanut butter tended to be made of higher-quality materials. The AARP study likewise characterized the kind of consumer who was more likely to be an "aggressive generic user" as "the better educated, generic-knowledgeable respondents." Yet awareness did not mean action. More than 75 percent of respondents who answered that they *were* informed about generic drugs "admitted they *never* ask their doctors or pharmacists to write or fill prescriptions using generic drugs."[42]

GENERIC DIVERGENCE

In retrospect, the relationship between generic drugs and generic groceries turned out to represent a brief intersection of two distinct trajectories, rather than part of a pervasive cultural trend, as marketing analysts of the late 1970s and early 1980s had predicted. Generic drugs would increase in relevance and utilization over the 1980s (from 10 percent to 40 percent of prescription sales by volume by the end of the decade) and continue to increase in following decades (to represent more than 80 percent of the prescription drug market by the year 2010). In contrast, during this time generic consumer goods largely disappeared from supermarket shelves. As the overall economy expanded once again in the 1980s and then in the 1990s, the value of branding (and segmentation) proved to be quite durable indeed, and the generic logo faded from middlebrow sensibility to ironic hipsterism. As general interest in generic marketing dwindled, pharmaceutical marketers grew more and more interested in characterizing generic consumers.[43]

Why did the generic form continue to take on value in the sphere of prescription drugs, while it lost value in the more general world of consumer goods? I have argued in this chapter that the rise of generic drugs was closely linked to the emerging definition of the patient as consumer in a newly egalitarian model of medical care taking shape in the 1970s and 1980s. As discussed in detail in the previous chapter, as patients learned to think of themselves as consumers, they thought of themselves increasingly as generic consumers. As the role of the patient as consumer gained ground in the 1980s and 1990s, the role of generic drugs in the pharmaceutical marketplace likewise grew and grew. Exploring generic drugs from a consumer history perspective also provides a perspective on the increasing uneasiness of physicians forced in the late twentieth century to come to terms with the fact that their own political and economic power was increasingly being framed in consumer terms as well.

Studying generic drugs and their consumers also provides an opportunity to re-examine the exceptional rules by which we assume that medical marketplaces must function differently from other economic sectors. Following Kenneth Arrow's work in the 1960s, health economists have claimed that medical decision making requires a different set of rules than other "rational consumer" models of economic activity, because people fundamentally value health differently than other goods and services. Following Peter Temin's work in the 1970s, historians have repeated as a catechism that the pharmaceutical consumer does not have the agency of the typical consumer because physicians, not patients, make the consumer decisions for the pharmaceutical marketplace. And since Paul Starr's work in the early 1980s, sociologists have taken for granted that the medical profession had been able to "escape the corporation," at least until recently, because the scientific and moral authority of the medical profession allowed it to be regulated by something more than just the marketplace.[44]

Consuming health care, we tend to assume, is not like consuming other goods and services. Yet studying the consumption of generic drugs reveals just how porous the relationship between medical markets and other markets of household goods has been and will continue to be. In the earlier sections of this book, it has become evident that one cannot speak of *generic drugs* without speaking of *the consumer*, even if one also cannot define precisely who or what such a term denotes. As we have explored in this section, the converse is also true: in a time in which patients and health policy were increasingly defined in consumerist terms, it was difficult to speak of the consumer in health-care politics of the 1960s, 1970s, or 1980s without speaking of the generic.

PART VI

THE GENERIC ALTERNATIVE

CHAPTER 12

SCIENCE AND POLITICS
OF THE "ME-TOO" DRUG

This unlimited modern pharmacopoeia is like an oriental bazaar. If in a supermarket 15 kinds of bologna are displayed, it gives the dazzled consumer a delightful feeling of having a wide choice. It's still bologna to be sure but it does not harm. However, in the drug supermarket we should not be permitted to treat drugs like bologna or we may do real harm.

—WALTER MODELL, DRUG INDUSTRY ANTITRUST ACT HEARINGS, 1961

As a cheaper alternative, the generic called the value of more expensive brand-name pharmaceuticals into question. As an older alternative, the generic drug also called into question the value of newer drugs and the patents that protected them. If newer, patent-protected pharmaceuticals were truly superior to generically available drugs in safety, efficacy, or palatability, they represented true innovations. If, however, these newer drugs were *not* significantly different—if they were merely similar drugs sold at higher prices—they represented false innovations, or "me-too drugs." Over the past half-century, as the "me-too drug" became a symbol of the false promises of biomedical progress, the generic alternative presented a litmus test for skeptics of the true value of therapeutic innovations.

Take, for example, the cholesterol-lowering agent Livalo (pitivastatin), newly approved by the Food and Drug Administration in 2009. At the time of its launch, pitivastatin was the eighth arrival in the class of cholesterol-lowering drugs known as statins, which already held three generically available forms (lovastatin, simvastatin, and pravastatin) and another whose patent was set to expire shortly (atorvastatin). The potential value of the new drug was complicated by the ready availability of several inexpensive generic counterparts. Why, critics asked, did the world need another me-too statin, if they all basically did the same thing?[1]

The critique of the me-too drug is historically linked to the generic drug on a number of registers.[2] When the US Task Force on Prescription Drugs offered a public definition of the *me-too drug* in 1968, it used the term to cast suspicion on the value of newer, patent-protected brand-name drugs relative to older, generically available drugs in the same therapeutic class. "Whether such additions make the armamentarium *better* or merely *bigger*," the report concluded, "has not been objectively determined."[3] Like generic drugs, me-too drugs challenged the relationship between innovation and imitation. Is the imitative process of inventing "me-too" drugs merely wasteful, or is it a necessary and productive part of therapeutic innovation?

Me-too drugs also posed new problems for advocates of generic substitution. The suggestion that different drugs within a therapeutic class might be more or less interchangeable with one another—that one statin might be as good as any other statin, one allergy pill as good as any other allergy pill—represented a powerful extension of the principle of generic substitution discussed in earlier chapters. The relative success and failure of this broader project of *therapeutic substitution* has varied substantially by institutional and geographic location and by therapeutic class. The interchangeability of antiallergy pills and agents that reduce heartburn, for example, has been relatively uncomplicated, while the interchangeability of cardiovascular and anticancer drugs has been far more contested.[4]

In the closing decades of the twentieth century, as several stakeholders pushed for the expansion of principles of generic substitution from chemically identical drugs to chemically similar drugs, the fate of generic drugs and me-too drugs became linked in new and more challenging ways. Therapeutic substitution of me-too drugs provoked powerful concerns about how tightly or loosely biomedical knowledge could be connected to biomedical objects. What proofs of equivalence, what rules of substitution, could govern the swapping of a generic drug for a me-too drug? When was a medicine, if not the same, nonetheless good enough?

MOLECULAR MANIPULATIONS

In her bestselling 2004 account *The Truth about the Drug Companies: How They Deceive Us and What We Can Do about It*, Marcia Angell sharply criticized the American drug industry for devoting most of its research and development effort toward making more me-too drugs. The following year, *Consumer Reports* revealed that three out of every four drugs recently approved by the FDA were me-too products instead of breakthrough drugs. Many contemporary observers—even some with industry funding—agreed that something needed be done to separate true from false innovation. Even the typically proindustry Tufts Center for the Study of Drug Development has suggested this should occur at the level of FDA approval—that a drug be approved for sale in the US market only if it shows superiority to existing therapies or at least compared to similar drugs in the same therapeutic class.[5]

The same arguments had been floated in almost identical terms nearly a half century earlier. The politics of the me-too drug became headline news in 1961 when Estes Kefauver introduced S. 1552, his star-crossed Drug Industry Antitrust Act. At the conclusion of fourteen months of hearings, Kefauver proposed to rewrite pharmaceutical intellectual property law so that patents on new drugs would be granted only if the FDA or another arm of the Department of Health, Education, and Welfare could ratify that the drug represented a "true innovation" and not mere "molecular modification." "In our drug hearings," Kefauver explained, "it has been shown that many U.S. patents have been issued on slight molecular modifications of drugs about which medical expertise testified there was no difference in the therapeutic effect as compared with their predecessors already on the market."[6] The proliferation of false innovation muddied the waters so much that it became hard to perceive the true innovations on the much rarer occasions that they actually occurred. As Kefauver continued, "New drugs, if truly more efficacious to the patient, are of great interest to doctors, patients, and all of us. Significant advances have been made in recent decades, and we are fortunate in having these available. At the same time, it is unfair to the public and to the great many doctors who are in no position to determine the truth, to permit the grant of patents unless their added therapeutic value has been determined by experts ... We are not dealing with gadgets, but products of health, life, and death."[7]

Several academic pharmacologists testified before Kefauver's sub-committee that many new so-called wonder drugs were only trivially different in form and function from existing therapeutics. Louis Lasagna, professor of pharmacology at Johns Hopkins, agreed that it would "be terribly helpful to the medical profession when molecular modifications come along, to have the ways in which these molecular modifications do or do not differ from the original product clearly spelled out." Without some kind of filter separating true innovation from trivial innovation, he continued, "it is a source of great confusion and disturbance to the physicians faced with a dozen different thiazide diuretics, most just about the same, to be caught in this cross-fire of conflicting claims as to the merits of one over the other." Lasagna's counterpart at Cornell, Walter Modell, likewise complained that the average physician had almost no reference point in evaluating the "numerous slight molecular modifications which pharmacologically are virtually the same," comparing the befuddled physician to a consumer lost in the marketplace. In the memorable quote that served as epigraph to this chapter, Modell warned that though this redundancy gave the "dazzled consumer a delightful feeling of having a wide choice," the overall effect of me-too drugs was to produce redundancy and confusion.[8]

Yet Kefauver found that academics had difficulty translating their negative critique into a positive one. Critical as Modell might be of the bazaar of "phony bologna," he also demurred that "slight molecular modifications" could on occasion produce highly significant therapeutic effects. Modell's colleague Louis Goodman, coauthor of *The Pharmacological Basis of Therapeutics*, warned that any attempt to implement a law "which provides that a close chemical congener of an existing drug must be proven to have superior therapeutic efficacy in order that patent rights be granted" would ultimately be both impracticable and unenforceable. While it was entirely possible to insist on proof of efficacy for new drugs, it would be unwieldy and impractical to require *comparative* efficacy among slight molecular modifications.[9]

"First," Goodman asked, "what is 'molecular modification,' and how slight is slight?" Efficacy alone, he suggested, was not enough to evaluate a drug.

> What if the close chemical derivative is not superior therapeutically, but is definitely more acceptable to the patient, from the standpoint of odor, taste, aesthetic appearance, etc? What if a slight chemical

modification is not clearly superior therapeutically but does have physical-chemical properties which make it more stable, more easily placed in infants' formulas, more—or less—rapidly absorbed by the intestine, longer in its sojourn in the blood so that not so many doses need to be given per day, etc.? How does the panel of experts handle these questions, all of them related in one way or another to usefulness and flexibility, even if not clearly related to therapeutic efficacy, of a drug?[10]

This point was amplified by Vannevar Bush, a prominent architect of postwar American science policy then serving as chair of the board at Merck, who pointed out to Kefauver's committee that, "in the first place, I don't know what molecular modification means, and I don't think anyone else does."[11]

Other experts called to testify by Kefauver's committee challenged the opposition of me-too drugs and meaningful innovation, suggesting instead that imitation in some form was essential to innovation. "The molecular manipulation of molecules," University of Maryland pharmacologist J. C. Krantz argued, "has been the inexhaustible spring from which a constant supply of new drugs has flowed." Imitation of molecular forms was necessary for drug development and ensured that "the physician may use his therapeutic skill and judgment to select the best drug for his patient from an increasingly deep and broad reservoir." To Krantz, there was no shame in molecular mimicry: "The speaker is a molecule manipulator. Ten years ago, we put fluorine into the molecule of ethyl ether. This removed the explosion hazard, thereby saving many lives and introduced a new era in the field of general anesthesia. But I manipulated the molecule further and produced hexafluorodiethyl ether (Indoklon). An anaesthetic—no, a convulsive drug that has now taken its place as a substitute for electroshock therapy in the treatment of the mentally ill. And because of this, scores of people have been released from mental hospitals."[12] In 1931, Krantz had published a volume titled *Fighting Disease with Drugs* that showed the limited state of clinical pharmacology. Only seven diseases, at the time, were curable with pharmaceutical agents. Since that time, Krantz argued, "molecular manipulation" had produced the most dramatic therapeutic revolution known to humanity.[13]

Krantz's claims for the clinical value of molecular manipulation were echoed by organized pharmacy. When questioned by Kefauver about the redundancy of molecular modifications on the drug market, American

Pharmaceutical Association executive director William Apple shot back, "Senator, I would say that the pharmacist would prefer to work with a smaller inventory, but there are people alive in the United States today because of molecular modifications." Nor did any federal agency—from the Patent Office, the FDA or HEW seem to want the responsibility of defining and policing molecular mimicry or distinguishing between significant and insignificant innovation. Kennedy's new appointee to HEW, Abraham Ribicoff, agreed with Kefauver that me-too drugs created some waste in the drug market but protested that distinguishing between true and false innovation was not really in HEW's bailiwick. By the time that the president of the Pharmaceutical Manufacturers Association, Eli Lilly's Eugene Beesley, appeared in front of Kefauver's committee, the proposal was all but dead in the water. Defining a federal standard for drug efficacy was difficult enough, Beesley argued, but setting any standards for comparative efficacy would be "completely impracticable."[14]

Kefauver wearily declaimed that he was "somewhat perplexed" that innovative pharmaceutical firms did not join him in his efforts to stamp out me-tooism. "I should think," he added in an overture to Beesley, Eli Lilly, and the PMA, "that the drug company which made an important innovation would want to be protected from those coat-tail riders whose method of riding the coat-tails is to make a slight molecular modification of their own, show the Patent Office that the chemical structure is slightly different, submit a few animal tests, and thereby secure a patent of their own."[15] But what concerned the PMA at the time was *how*, and *by whom* the distinction of true versus false innovation would be determined. Members of the PMA did not trust the FDA with this decision, nor did they trust panels of "ivory tower" academic physicians. They trusted the market—in combination with an intellectual property regime that protected individual compounds regardless of their relative value. In early 1962, as Kefauver attempted to respond to critiques from the patent commissioner, his bill was effectively gutted in a meeting of his own subcommittee to which he had not been invited. The version of S.1552 that emerged from that meeting, which eventually became the 1962 Kefauver-Harris Amendments to the Food, Drug, and Cosmetics Act, was stripped of all language of molecular modification.

The critical language of the me-too drug disappeared from health policy discourse almost entirely until James Goddard, Lyndon Johnson's iconoclastic appointee as FDA commissioner, took office in 1966. Goddard was the first FDA commissioner to be appointed from outside the agency, and he vowed to shake up the FDA's somewhat cozy relationship with the pharmaceutical industry. Goddard was a highly visible figure in the late 1960s, featured in magazines such as *Esquire* and on television news programs, where he had ample opportunity to publicize his critiques that most new drugs were pale "isomer shadows" of existing remedies.

In a characteristic interview in late 1967 on NBC's *Today Show*, Goddard declared that the industry needed to "get away from 'me-too' research" and focus instead on "really new products." Goddard elaborated that a superstructure of deceptive pharmaceutical marketing was built on a substructure of me-too research that "pushes drug companies into false advertising because they are forced to create non-existent differences between the new product and others already on the market." Unlike his predecessors, Goddard declared that this *was* the business of the FDA. In fiscal year 1967, he declared, sixty-two of eighty-three New Drug Applications approved by his agency had been for me-toos.[16]

Goddard's positioning of the me-too as a link between false innovation and false advertising helped to add momentum to Senator Gaylord Nelson's initial hearings into the competitive problems in the drug industry, as did Goddard's endorsement of the critical HEW report on public sector overspending by the Task Force on Prescription Drugs (discussed in more detail in chapters 6 and 7 and mentioned at the beginning of this chapter). As Task Force chair and former assistant secretary of HEW Philip Lee announced on a 1969 PBS TV program called *Rx Drugs: Prices and Perils*, most R&D funding by the pharmaceutical industry was dedicated "not to finding new solutions to complex problems, new drugs for coronary disease or diabetes or arthritis, but rather to the development of ... 'me-too' drugs, or minor molecular modifications of existing drugs."[17]

Goddard and Lee, speaking respectively for the FDA and HEW, heralded the emergence of a broader critique of me-tooism in US health policy. In the time he was FDA commissioner, "on the whole, the actual improvement in the drug supply [was] quite small," Goddard complained, while "enormous outlays of drug advertising give the illusion of extensive improvement, convincing, if possible, all prescribing physicians that 'me-too' products are new and different." And yet Norway, by contrast, had only eight hundred drugs in commerce at the time, and a new drug was only allowed to be marketed if "it [could] be proved to be better—safer, more effective" than an existing drug on the market. "If this approach of a relative criterion were adopted here," Goddard concluded, "it would conserve research manpower and facilities and yield only those drugs that advance the practice of medicine."[18]

Compared to their predecessors in the early 1960s—who would not touch the concept of me-too drugs with a ten-foot pole—Lyndon Johnson's FDA and HEW appointees were far more willing to take on the issue of me-too drugs within the new infrastructure of the Great Society. But in the meantime their counterparts in academic medicine had become much more cautious. In the first days of the hearings in spring of 1967, Gaylord Nelson invited therapeutic reformers like Walter Modell to talk again about me-too problems under the new terminology of *molecular shuffling*. Yet Modell's reply reminds us how academic language is often unhelpful to policy makers, even though Modell supported Nelson's overall desire to reduce unnecessary drug use in the name of more "rational" prescribing.

> SENATOR NELSON: Is there any value to this molecular shuffling?
> DR. MODELL: Oh, yes, as a scientific procedure, it is of value because from time to time, quite unexpectedly shifting the molecule produces either a completely different action or a superior action. It is not always quite the same, so that from the point of view of research, the study of what we call structure action must be pursued. But it doesn't necessarily follow that every new chemical that is made must then be marketed. Only the best of the litter ought to be used in medicine.
> SENATOR NELSON: If the purpose is, and the result is, to produce the same therapeutic effect, there wouldn't be, appear to be, any value?
> DR. MODELL: Well, it is not precisely the same.[19]

Modell's careful assessment of the similarity of me-too drugs—as not precisely the same—illustrates the kind of problems that resurfaced in trying to harness abstract critiques of the me-too into concrete policy reforms. Things that are not precisely the same are by definition different, perhaps even significantly different. Any proposal to eliminate me-too drugs as false innovations might do away with some real innovations as well.

The PMA would use a similar "baby and bathwater" argument in its own testimony, provided by Leonard Scheele, the former surgeon general of the United States, now president of the Warner-Lambert Research Institute. During Scheele's tenure as surgeon general, a wave of me-too research effectively established the major new classes of postwar pharmaceutical innovation:

Studies in which attempts were being made to simplify the chemical structure of quinine led to synthesis of Atabrine and other antimalarials. Later studies of pharmacologic properties of the synthetic antimalarials led to observation of an unexpected, new effect which was interpreted as an antagonistic action to histamine. These observations culminated in synthesis of an important new class of drugs, antihistamines. Later observation of the sedative effect of one of these on mental patients in France led to discovery of a very useful tranquilizer, chlorpromazine. This drug has played a major role in decreasing the number of patient beds in use in mental hospitals for 11 consecutive years. Thus, antihistamines and tranquilizers can trace their history to an ancient remedy, quinine, and its molecular modifications.[20]

Like a series of stepping-stones across a wide river, Scheele argued, a chain of me-too drugs formed a bridge between the nineteenth-century drug quinine and the therapeutic revolution of mid-twentieth-century pharmaceuticals from antihistamines to antipsychotics. How could one think of outlawing molecular modification when this very practice produced the wonder drugs that potentiated US military victory in World War II and enacted the "chemical revolution" that liberated the chronically mentally ill from the confines of the asylum?[21]

Even Richard Burack, whose critical *Handbook of Prescription Drugs* (discussed in chapters 3 and 10) was a key reference point of the early Nelson hearings, echoed the PMA in cautioning that eliminating all me-too drugs might dispose of more baby than bathwater. In a letter to an admiring colleague who was casually dismissive of molecular modification, Burack wrote, "Don't denigrate 'molecule manipulating' as a form of research because an occasional important drug has come out of this kind of activity, but it's perfectly fair for doctors to understand that much of the 'research' is stimulated by a need to 'box in' patents lest some other company grab off a part of the market with a manipulated congener."[22] Burack illustrates the paradox of the me-too critique: valid in the abstract, it was difficult to concretize in policy form without collateral damage.

As his hearings into the drug industry rolled on through the late 1960s and early 1970s, Nelson was not able to put forward any legislative solutions to combat me-too drugs. Nonetheless, his continuing investigation of the marketing practices of the pharmaceutical industry helped to assure that me-too drugs would maintain a steady position in critical discourse about pharmaceutical marketing practices. When Senator Edward

Kennedy introduced a new series of investigations into the drug industry in 1973 (under the Health Subcommittee, which he chaired), he began by asking "why so much industry research and development focuses on the development of me-too products, rather than on new areas?" Did we really need "five or more companies manufacturing identical products?"[23] Yet Kennedy's call would be answered with familiar arguments that "most of the major pharmaceutical advances since 1940 have come about not through targeted basic research, but through molecular modification."[24] Imitation, in other words, was not inimical to innovation—it was a crucial part of the innovative process.

Another defense physicians mobilized of the me-too drug hinged on the logic of personalized medicine. It was a long-held precept in clinical practice that one size should not be expected to fit all. As one psychiatrist testifying at the Nelson hearings explained, me-too drugs produced a needed variation within the field of therapeutics. General critiques of me-too drugs, he claimed, needed to be weighed "against the flexibility their availability offers to the M.D. searching for the best drug for his individual patient."[25] Without a diversity of drugs that did similar things, the physician would have no freedom to tailor therapies to individual patients. As a California practitioner wrote in *Private Practice* in 1976 in a piece entitled "Is There a Place for Me Too Drugs?," the alternative to me-too medicine was totalitarian medicine: "In some totalitarian countries, only one car is manufactured. The price is prohibitive, the construction slip-shod, and the waiting period interminable. And no one, other than the consumer, cares."[26] These later defenses of the me-too drug preserve key points from Scheele's arguments a decade earlier—or Krantz's arguments a decade earlier still. But to read this homology as pure continuity would be a mistake. By the late 1970s, the debate over me-too drugs was no longer a debate over a proposed overhaul to the patent system. Critiques of me-too drugs had become linked to a new, more threatening logic of similarity in American medicine: the prospect of therapeutic substitution.

WHEN IS A MEDICINE GOOD ENOUGH?

Therapeutic substitution entails a set of rules for exchanging one drug for another *even when they are different molecules*. To its supporters, therapeutic substitution clamps down on waste and offers a streamlined, more cost-effective approach to care. To its detractors, it is a dangerous precedent that heralds the end of therapeutic innovation and physician autonomy.

As with generic substitution, protocols for therapeutic substitution first took shape in the hospital and then spread outward into the community. As we saw in chapter 8, the first rules of generic substitution were governed by hospital formularies. By the late 1970s, a few ambitious hospital pharmacy and therapeutics committees suggested that by the same logic pharmacists should be able to swap therapeutically similar drugs—even if their chemical structures were not exactly the same.[27]

This opened a set of legal, professional, and epistemological conflicts. Legally, hospital formularies were allowed to substitute generic for brand-name drugs on the basis of prior consent. According to this rule, in joining the medical staff of a hospital, a physician agreed contractually to allow hospital pharmacists to substitute generic versions of the therapeutics they prescribed. Could hospital pharmacists use the same institutional logic of prior consent to substitute similar but chemically inequivalent products?

The issue first came to a head in Oregon in 1979 after a hospital pharmacist asked whether he could fill a prescription for an antibiotic with a functionally similar but chemically distinct drug from the same class of antibiotics. The state legislature took up the issue, and in 1981 Oregon amended its generic substitution law to allow substitution of "a drug product that does not have the same generic name but is considered to be chemically related and therapeutically equivalent by the Medical Staff of the hospital."[28]

The Oregon case was fought over a specific class of antibiotics—the cephalosporins—that had witnessed a veritable explosion of me-too variation over the 1960s and 1970s. The first cephalosporin antibiotic was grown from mold found in a sample of seawater off the coast of Sardinia in 1945 by Giuseppe Brotzu, the director of the Instituto d'Igiene in Cagliari. An extract from this mold, *Cephalosporinium acremonium* was shown to be effective against a broader spectrum of bacteria than penicillin. When the active molecular entity, cephalosporin C, was first isolated in Oxford in the early 1960s, researchers recognized that the structure could be produced semisynthetically and was therefore an ideal target for molecular manipulation. The core of the cephalosporin structure, in the words of its describer, Sir Edward P. Abraham, "could be the source of a great variety of cephalosporins with potentially useful properties."[29]

The first cephalosporin to be marketed in the United States was Eli

Lilly's Keflin (cephalothin) in 1962, available only by injection. Glaxo's Loridine (cephaloridine) was launched in 1964; and Lilly followed with the oral agents Kafocin (cephaloglycin) and Keflex (cephalexin) in 1965 and 1967. The first "second-generation" cephalosporins (more active against gram negatives, less active against gram positives) followed in the early 1970s, along with more "first-generation" compounds: cefadroxil, cefaclor, cephadrine, cephazolin, cephamandole, cephapirin, cephanone, cefoxitin, cefuroxime. By the end of the 1970s, extended spectrum "third-generation" cephalosporins had been added to the mix, from cefotaxime, cefodizime, cefitizoxime, ceftazidime, ceftriaxone, cefpodoxime, and cefsulodin. If the list of similar *words* in this paragraph is at all challenging for the twenty-first-century reader, one can only imagine the condition of hospital pharmacists in the late 1970s trying to accurately stock and fill prescriptions for the things themselves.[30]

By the end of the 1970s, the first-generation cephalosporins were routinely prescribed before and after major surgeries as prophylaxes against infection. These drugs soon accounted for the single largest line-item outlay for pharmaceutical purchasing at most academic hospitals. In particular, hospitals with busy surgical services needed to stock large volumes of cefazolin, cephalothin, and cephapirin, which many physicians and pharmacists believed to be clinically undistinguishable in their efficacy and safety. Without some program of rationalization, many hospital administrators feared that continuing to stock all of the available cephalosporins would "break the back of hospitals."[31]

Oregon's experiment with automatic therapeutic substitution among first-generation cephalosporins was watched with great interest and quickly imitated. By the early 1980s, nearly half of all hospitals had instituted therapeutic substitution protocols for this workhorse class of prophylactic antibiotics. By the end of the decade, first-generation cephalosporin exchanges had become nearly omnipresent in academic hospitals, seemingly without complaint from physicians.[32]

The Policy Committee on Professional Affairs of the APhA reviewed the success of the cephalosporin exchange in 1982 with an eye toward expanding therapeutic substitution beyond the hospital. Like generic substitution a decade earlier, therapeutic substitution offered pharmacists an expanded role in clinical and health policy arenas. The APhA should "support providing pharmacists with independent authority to select and dispense alternate drug entities (therapeutic alternates) from within the same general pharmacologic and therapeutic classification."[33]

But phrases like *automatic substitution* and *independent authority for pharmacists* soon rekindled the same political and professional conflicts seen a decade earlier in fights over generic substitution. The American Medical Association in turn resolved "to vigorously oppose any concept of pharmaceutical or therapeutic substitution of drugs by pharmacists" and proceeded to take the fight against therapeutic substitution to the state legislatures.[34]

If therapeutic substitution was the second stage in a broader turf war of substitution, many physicians feared that the third and ultimate stage would be for pharmacists to gain the right to prescribe themselves—a right that many nurse practitioners and physicians assistants had recently realized. "History," one physician concluded, "shows yesterday's extremes often become tomorrow's commonplace. Certainly, there is no question that those who dream of pharmacists replacing physician prescribers in total or part are among the most ardent advocates of generic or other drug substitution. Physicians, be warned!"[35] As a contributor to the conservative physicians' journal *Private Practice* warned, "TS" was only the first step toward the total loss of physician control over therapeutics.

> TS is the interchange of chemically *non*-equivalent drugs which have therapeutically similar properties . . . Here's how it works: A hospital requests bulk prices for cephalosporin antibiotics. Bids come in. Let's say cefazolin is the best buy of the month. The hospital formulary is revised and all other cephalosporins are removed. When any of them is prescribed, the hospital pharmacist "substitutes" cefazolin automatically without contacting the physician for permission. The legality of the procedure is doubtful; it could never invade the private sector; physicians would never accept it—right? Wrong. TS is already established in 40 percent of all the short-term hospitals in America. The figure is expected to rise to 50 percent by years' end. How long before the corner drug store dispenses penicillin VK when the doctor orders ampicillin?[36]

In 1986, the PMA and the AMA declared in a joint statement that therapeutic substitution was "not negotiable." Speaking at the PMA annual meeting that year, the AMA's James Sammons described the spread of therapeutic substitution as "the most dangerous problem to quality health care delivery in this country that I have ever seen."[37]

Pharmacists attempted to define *therapeutic interchange* as a collaborative process in which "filling pharmacists provide information and advice to willing prescribers, resulting in drug therapy decisions more appropriate to patient needs than decisions either could have achieved alone, it is therapeutic *synergism*; it flows from the voluntary combination of the knowledge and skills of two complementary professionals."[38] But physicians and pharmaceutical manufacturers saw in therapeutic substitution something far more threatening to medical practice and medical industry. Could all drugs be reduced, like the first-generation cephalosporins, to class alone? Would the universe of therapeutics be reducible to a set of interchangeable types, in which the prescribing habits of a given physician or the marketing strategies of a given brand would be rendered meaningless?

The AMA and PMA fiercely resisted the concept of therapeutic substitution and worked through various means to deny the possibility of therapeutic class being a marker of therapeutic equivalence. Surely not *all* categories of drugs were as freely swappable as first-generation cephalosporins. Retasked with a new battle against substitution, the National Pharmaceutical Council found its poster child in another class of wonder drugs: the beta-blockers.

WHEN IS A MEDICINE NOT GOOD ENOUGH?

Beta-blockers were the cardiovascular wonder drugs of the late 1960s. By the end of the 1970s, they had been credited with the reversal of the modern epidemic of coronary heart disease—which the Centers for Disease Control would later call one of the top ten public health accomplishments of the twentieth century. As early beta-blockers went off patent in the 1980s, public health advocates looked forward to making even broader use of these drugs to enact even more meaningful public health gains.[39]

If the early development of cephalosporins could be characterized as finding several superficially different ways to make basically the same thing, the early development of the beta-blockers was predicated on noticing significant differences among drugs that superficially seemed to be the same. Following the early twentieth-century success of adrenaline—one of the earliest innovations from the R&D labs of an American pharmaceutical firm—the pharmaceutical industry in the interwar decades produced a number of adrenaline-mimicking drugs known as sympathomimetic amines. In the postwar period it became clear that different drugs in this class produced different effects in different parts of the body depending on which kinds of adrenaline-specific (or adrenergic) cell-surface receptors they fit into. By 1948 R. P. Ahlquist had differentiated two different kinds of adrenaline receptors—alpha and beta—through which otherwise similar drugs produced very different effects on the heart. When an adrenergic drug bound to an alpha receptor in the heart, it caused the heart muscle to relax; when it bound to a beta receptor, it caused the pulse to quicken and the heart to contract more forcefully.[40]

A drug that selectively blocked a beta receptor could be expected to help slow the heart rate, decrease overall cardiac effort, and help improve survival after a heart attack. Though dichloroisoproterenol, the first beta-blocker to be marketed, was of limited clinical utility, the me-too drug propranolol was met with great enthusiasm by cardiologists.[41] By the early 1980s, six different beta-blockers had been approved by the FDA, the first four (propranolol, metoprolol, timolol, and nadolol) already in wide use, with five more (pindolol, atenolol, labetalol, oxprenolol, and esmolol) soon to be launched.[42]

These drugs were similar, but not exactly the same. The variations in their molecular structures—some minor, some more radical—were accompanied by variations in their effects on heart rate, heart muscle contractility, and blood pressure. Clinicians were at odds over how much these differences really mattered in clinical use. On the one hand, each beta-blocker had been officially approved by the FDA for a different set of purposes (propranolol for angina pectoris, arrhythmias, hypertension, migraines; nadolol for angina pectoris and hypertension; timolol for glaucoma). But they could all still be considered to be beta-blockers and therefore in some ways interchangeable. In a pharmacological review of the drug class for the *New England Journal of Medicine* in 1981, the cardiologist William Frishman concluded that "there are no studies suggesting that one of these drugs has major advantages or disadvantages" in treating cardiovascular disease, as long as the drug was used in the right dosage. "When any beta-blocker developed since the synthesis of propranolol is titrated to the proper dose," he continued, "it can be effective in patients with arrhythmia, hypertension, or angina pectoris."[43]

Yet as the fields of cardiovascular physiology and pharmacology developed further, it became apparent that alpha and beta were only the first-order cuts in a rapidly branching tree of receptor subtypes and sub-subtypes. Laboratory work in the late 1960s suggested differences between a set of beta receptors involved in cardiostimulation and another set of beta receptors involved in vasodilation and bronchodilation. Research units at Smith, Kline & French and the British pharmaceutical firm ICI began to consider the differential utility of selectively blocking beta-1 versus beta-2 receptors. In the late 1970s, ICI gained approval to market Tenormin (atenolol), the first "cardioselective" beta-1 blocking drug, followed soon by Lopressor (metoprolol), a slightly different beta-1 blocker. Moreover, even though metoprolol and atenolol had similar *pharmacodynamic* profiles in terms of their function at the beta-1 receptor, they had important *pharmacokinetic* differences in how they were metabolized by the body. Metoprolol was a shorter-acting drug that needed to be dosed three to four times per day; atenolol was a longer-acting drug that could be taken one to two times per day. Metoprolol was cleared from the body by the liver; atenolol was cleared by the kidney. To some patients these differences were trivial—but to others they could be a matter of life or death.[44]

By the end of the 1980s beta-blockers had emerged as one of the most promising and broadly utilized new therapeutic classes in modern medicine—but a class that challenged the coherence of therapeutic class as a concept. In a repeat review on beta-blockers in 1987, William Frishman reversed his earlier stance on the interchangeability of beta-blockers and offered the many differences within this therapeutic class as an object lesson in pharmaceutical policy. Beta-blockers were "a representative class of agents having similar chemical structures and pharmacological properties, *which are nevertheless not exchangeable.*"[45] Over the course of the decade, clinical experience with these apparent me-too drugs had revealed their differences to be nontrivial.

Ironically, as Frishman noted, it was the very proliferation of me-too drugs within the class of beta-blockers that illustrated how little had been known previously about receptor subtypes. Only after the introduction of multiple beta-blockers did the relevance of differences in qualities like lipid solubility (and with it, mobility across the blood-brain barrier) become apparent in the different side-effect profiles of lethargy, depression, impotence, and psoriasis these drugs could cause. Likewise, only after the increasing array of uses of beta-blockers by 1987—now used for more than twenty medical conditions—did enough knowledge emerge to demonstrate that any given beta-blocker could *not* be considered substitutable for any other beta-blocker for a particular condition. Where previously he had asked for proof of difference among beta-blockers, Frishman now insisted the burden of proof lay with those who would claim similarity. "Substitution of one beta blocker for another," he concluded, "should be based on published evidence of efficacy in treating the condition for which it is being used."[46]

Frishman's research was now funded by the NPC—a proindustry professional body first described in chapter 2 that had waged a highly successful campaign against generic substitution in the 1940s and 1950s. By the mid-1980s, the NPC had conceded generic substitution to be an inevitable part of the consumer landscape, and it refocused its efforts at opposing therapeutic substitution. For the NPC, the story of beta-blockers illustrated the dangers of assuming similarity in a world of clinically important differences. How many patients with heart attacks would have died had Medicaid simply demanded that propranolol be used for all cardiac patients instead of atenolol or metoprolol? Two years after Frishman's essay on the nonsubstitutability of beta-blockers, NPC staffer Richard A. Levy published a critical essay on the nonsubstitutability of nonsteroidal anti-inflammatory drugs (NSAIDs). This was followed a few years later by a general tract on the clinical dangers of therapeutic substitution. Beta-blockers, it seemed, were only the tip of the spear in a countering thrust against therapeutic substitution.[47]

POLITICS OF SUBSTITUTION, REDUX

By the end of the 1980s, medical and industrial lobbies had closed ranks against therapeutic substitution. John Ballin, the head of the AMA's Department of Drugs, denounced therapeutic substitution as an unethical "usurpation of the physician's prerogative"; his counterpart at the PMA, Gerald Mossinghoff, voiced agreement.[48] To many observers, the recapitulation of professional turf battles seemed inevitable; as one pharmacy journal reported:

> Therapeutic interchange! The battle-lines have been reconnoitered and the major protagonists and antagonists are headed for the front. The old nemesis "substitution" has been dressed up as "interchange." They are in fact equivalent terms in this therapeutic field of battle. As health professionals we are so used to making war on cancer, battling AIDS, holding the line on arthritis, and bombarding malaria-infested jungles, it is no surprise that we struggle tenaciously for a limited

health care turf and love to do a little interprofessional skull-bashing on Saturday nights.[49]

As with genetic substitution a decade earlier, the AMA, APhA, and PMA, could not find an easy consensus that would satisfy pharmacists, manufacturers, physicians, and consumers. Yet the stakes had changed since the 1960s and 1970s. At stake in the battles over therapeutic substitution in the 1980s was the broader interchangeability of biomedical therapeutics like cephalosporins or beta-blockers and the professional turf wars over who had a right to determine which specific therapy a given patient would get: the physician, the pharmacy, or the health-care system itself.

Unlike generic substitution, which promoted logics of cost containment predicated on equivalent care, therapeutic substitution suggested a far broader role for private and public payors to establish rules or rationales of what kind of therapies were, if not identical, "good enough" for daily clinical practice. The very slipperiness of the concept of therapeutic substitution made it hard for opponents to even define what it was they were arguing about. As one commentator noted in 1988, "One great difficulty in this debate is a working definition of therapeutic interchange. Generic interchange is much simpler: the active chemical entity, strength, dosage form, and bioavailability should be equivalent. The major argument relates to the methodology and frequency of bioavailability assessment. What does therapeutic equivalency mean and which data will the pharmacist need to make an informed decision?"[50] Therapeutic equivalency was not a unitary concept but a spectrum of concepts arranged on an axis from "the same" to "not the same." First-generation cephalosporins appeared to be similar enough to be interchangeable, more or less. Beta-blockers clearly were not. It was not at all clear where other drugs fell along this spectrum or which methods, or whose judgment, could distinguish between medicines that were "good enough" to be interchangeable and those that were not.

The me-too drug played an important part in debates over late twentieth-century health policy, debates that carry over to the twenty-first. As a form of critique, claims of me-tooism tend to be made from outside the industry. Rare indeed is the pharmaceutical advertisement that claims a new branded entry into the drug market is the most important me-too drug of the year. And yet me-too drugs do have value. Me-too drugs have had their defenders from Scheele and Krantz onward. Ironically, a world in which no me-too drugs existed—that is to say, a world in which every therapeutic class contained only a single drug—would afford no opportunities to try to achieve cost savings through therapeutic substitution.

Even the most cynical observer of the pharmaceutical industrial R&D process has to admit that me-too drugs often offer something. In many therapeutic classes, from beta-blockers to statins to H2 antagonists to cephalosporins, the first drug in a class is not always the most effective or safest. It is a common theme in the history of technology more generally that second or third entries into a new technological field often produce a product that is far more broadly useful than the innovator product itself. Historians of technology have referred to these second-wave innovators as "tinkerers"—the individuals who take a truly new idea and make it more broadly accessible. Yet the technological field is also rife with stories of planned obsolescence, of serial logics of next generations whose relative value becomes harder to calculate as minimal improvements come at greater costs. Imitation may be necessary for innovation, but imitation can also masquerade as innovation, and there are few yardsticks adequate to judging the difference.

Likewise, even the most committed therapeutic rationalist must acknowledge that a deep supply of me-too drugs has value when individual bodies do not perform according to textbook definitions of drug and disease. Some patients will find better relief from their allergies with cetirizine than loratadine; others will find better relief from their allergies with loratadine than cetirizine, even though all population-based data suggest that these drugs—indeed the whole class of nonsedating antihistamines—are functionally equivalent in terms of efficacy and safety. Some patients with arthritis will develop dyspepsia with one form of NSAID and not another. Some patients will find their depression is better managed on citalopram than fluoxetine. Although the field of pharmacogenetics (now pharmacogenomics) has promised to explain all such individual variation on the levels of hitherto misunderstood subpopulations of drug responders, most practicing physicians will suspect that some individual variation in response to drugs is as axiomatic as death and taxes.[51]

Quietly, as far away from the nihilistic rhetoric of their professional associations as they could huddle, small groups of physicians, pharmacists, and payors suggested that credible governance by a P&T committee operating with transparency and representation from all groups might be able to make meaningful decisions about where drugs are interchangeable and where they are not. After all, some health systems had functional P&T committees in which member physicians actually found the input of their pharmacist colleagues to be helpful. Key to the success of these functional P&T committees was the notion that interchange is not a monolithic or uniform practice. Darryl Rich, the director of pharmacy at the Boston Medical Center, began to argue in 1989 for a "two-tier" approach to therapeutic interchange, in which pharmacists might automatically interchange some classes of drugs (like first-generation cephalosporins) but contact physicians before interchanging other classes (like cardioselective beta-blockers). This approach—which would later be rebranded as a "tiered pricing" or "stepped therapy" approach, could differentially accommodate blanket substitution and intense personalization or tailoring of therapy, depending on the situation.[52] As long as physicians felt that the P&T process was collaborative, they were happy for automatic substitution to take place with basic agents like first-generation cephalosporins. As one physician commented on a particularly successful therapeutic substitution program at Virginia Mason Hospital, "Historically, when costs were retrospectively reimbursed, there was no incentive to control costs. Now when we're working within a system where up to 50 percent of the patients are enrolled in an HMO, PPO, or Medicare, it's apparent that we can't afford to function in the same way."[53]

As these comments suggest, unlike generic substitution in the 1970s, therapeutic substitution took shape in a new political economy of managed care emergent in the 1980s, as an ethos of cost containment previously associated with public payors like Medicare and Medicaid became articulated in new private formations of managed care, health maintenance organizations (HMOs), for-profit hospital chains, and pharmacy benefit managers (PBMs).[54] Managed care was heralded as an organizing focus that could finally trim the fat from me-too medicine. HMOs and PBMs were supposed to transform an unruly realm of therapeutics into a more rational system of managed therapeutic practice. In this new, rationalized form of care, the world of therapeutics could be parsed as a world of substitutable agents, whose use could be optimized across a healthcare system to bring down costs while bringing up therapeutic value.

That is not, however, what would come to pass.

CHAPTER 13

PREFERRED DRUGS, PUBLIC AND PRIVATE

The granularity of a formulary is not a scientific absolute, but a discretionary act with material consequences.

—LAWRENCE ABRAMS, *JOURNAL OF MANAGED CARE PHARMACY*, 2004

Like so many other elements of American health policy, the managed care revolution of the late twentieth century was a private sector solution to a public health problem. The initial terms of the Health Maintenance Organization Act of 1973—which introduced the term *health maintenance organization* to many Americans—was to enroll over 90 percent of the population into these new structures by the end of the decade. Early HMOs were modeled on the progressive outcomes–oriented examples of Kaiser Permanente in California and the Group Health Cooperative of Puget Sound in Washington; these HMOs were not permitted to be publicly traded, for-profit entities. Yet, as the number of Americans enrolled in managed care plans mushroomed in the late 1980s, the strategies of HMOs also shifted away from outcomes improvement toward cost containment. As the number of HMO enrollees rose from six million in 1976 to fifteen million in 1984 to nearly forty million by the end of the 1980s, therapeutic substitution of "preferred drugs" became a clinical reality on a previously unprecedented scale.[1]

The Pharmaceutical Manufacturers Association aired its concerns over managed care in 1986 after a report commissioned to the DC-based policy research firm Lewin and Associates showed that 90 percent of the nearly fifteen million HMO enrollees studied received outpatient drug benefits with "stricter enforcement of formularies, growing physician acceptance of generics, greater use of therapeutic substitution, more sophisticated drug utilization reviews, and aggressively discounted contracts with retail pharmacies."[2] In 1987, 35 percent of HMOs employed some measure of therapeutic substitution; by 1993, 70 percent had restrictive formularies in place.[3]

Restrictive formularies were risky endeavors for health insurers. Already by the mid-1990s a managed care backlash had become visible across many segments of American society. As more consumers came to view HMOs as for-profit entities that were more concerned with cutting costs than improving the health of individuals or populations, the HMO became a new symbol for unfeeling bureaucratic rationality in medical care. The theoretical fears of bureaucracy intruding between doctor and patient that had helped block passage of the Clinton health-care reform program in 1992 had nonetheless materialized in the private sector. By the end of the 1990s, some political analysts were calling this the "Jack Nicholson effect," for the actor's role in the 1997 romantic comedy *As Good As It Gets*.[4]

As Good As It Gets would become an important popular culture touchstone for the prospects of unequal substitution in the context of a managed care backlash. As the title suggests, acts of substitution permeate the film. Managed care serves as both a narrative villain and a thematic metaphor for the act of settling for something that is survivable but not optimal. One settles for something that is not as good as one might like, perhaps, but meets some basic standard of care. Most concretely, *As Good As It Gets* denotes the limitations of restrictive coverage decisions under HMOs. For many viewers, the most memorable line in the movie belonged not to Nicholson but to Helen Hunt, cast as the overworked single mother of an asthmatic son who was denied adequate care for years because of her managed care plan. When Hunt's character learns from a doctor acquaintance that her son's ill health could have been prevented years earlier with appropriate treatment, she lets off a tirade of explosive epithets: "Fucking! Bastard! HMO! Pieces of shit!" After she regains her composure and apologizes to the physician, he shrugs off the apology with an easy wave. "It's OK," he replies, "I think that's their technical name."

"Audiences in the Washington area have been erupting in whoops, whistles, and applause," the *Washington Post* reported, "when actress Helen Hunt, playing the single mother of a chronically ill child, denounces HMOs with a string of unprintable epithets." *As Good As It Gets* clearly hit home. The movie would receive multiple Academy Award nominations, and would land for Hunt and Nicholson matching Oscars for Best

Actor and Best Actress. As the *Post* declared in its Oscar-season headline, "Art Imitates Life When It Comes to Frustration with HMOs." Indeed the movie would resurface repeatedly as an icon of the managed care backlash over the rest of the 1990s, because of its articulation of popular frustration with therapeutic substitution and limitations. As a spokesperson for the Kaiser Foundation noted, "HMOs have become sort of the villain in our society—whether it's fair or not, they have become the easy target."[5]

As Good As It Gets reflects a backlash that had begun years earlier. Already by 1996, consumer advocates in eighteen states had succeeded in passing legislation against some of the more unpopular HMO measures, such as physician gag rules and "drive-through" maternity protocols. But no small part of the fear of "good enough" medicine under managed care lay in concerns over therapeutic substitution. If managed care substitution programs were devising a formula for determining when a medicine was, indeed, good enough to be substituted, they were not doing enough to convince the American public of that exchangeability.

How did the structures of "good enough" medicine take shape in managed care, particularly in the private sector of pharmacy benefit management firms through which pharmaceutical costs were rationalized? How did public programs respond to managed care formularies with their own, more transparent means of preferring certain drugs over others? As this chapter describes, the development of the field of therapeutic substitution in the 1990s and early 2000s took on very different forms in private and public arenas.

PHARMACEUTICAL LEGERDEMAIN

As health-care costs continued to rise under managed care in the 1990s—with pharmaceutical costs rising more quickly than other sectors—a new industry developed in the space between the HMO and the pharmacy: the pharmacy benefit manager, or PBM. By 1996, more than 50 percent of the US population filled prescriptions through a PBM, with nearly eight out of ten HMO enrollees doing so. The business of managing preferred drug lists for managed care organizations concentrated as it grew, just three companies filling one out of every three prescriptions in the U.S by 1998.[6]

The early history of Express Scripts, the market leader in PBMs by the turn of the twenty-first century, illustrates the origin and growth of this industry. The company was originally founded as a general managed care organization, but by the end of 1989 it decided to specialize in prescription drug programs, managing a list of roughly ten thousand drugstores. By the time the company went public in 1992, its prescription plans covered two million lives; by 1998, ten million, by 1999, twenty-five million, with further acquisitions in 2005 bringing the number up to fifty million lives. As Express Scripts CEO Barrett Toan later summed up, "PBMs focus on making drugs more affordable. They also make drug usage just a little bit safer through their drug utilization review systems." Toan was proud of Express Scripts' "generic fill rate" of 51 percent and hoped to bring that number up over 55 percent in the next year. Other early PBMs included Medco, which developed incentive discounts to pharmacists based on their rates of generic and therapeutic substitution. Pharmacists, physicians, and consumers could be steered toward a smaller list of preferred drugs that PBMs bought from manufacturers at discounted prices.[7]

PBMs like Express Scripts essentially sold prepackaged formularies to their clients, but that simple phrase belies a more complex phenomenon: To the PBM, the formulary was not just a cost-containing strategy but represented a market-shaping tool in its own right. A company like Express Scripts functioned by profiting from articulating both supply and demand in the pharmaceutical marketplace. Crudely speaking, health insurance companies want to pay as little as possible for favorable outcomes, while pharmaceutical manufacturers want to sell as many drugs as possible at the highest possible prices. Into this oil-and-water field, PBMs acted like the soap molecule, interacting with both manufacturers and insurers. Payor clients like Medicaid, Medicare, corporate health plans, and private insurers hired Express Scripts to steer providers and patients toward more cost-effective medicines. Pharmaceutical firms and large pharmacy chains, in turn, set up arrangements effectively paying PBMs through rebates and discounts to make sure that their products or stores would be featured in the preferred list of the formularies that PBMs produced.

The discounts, rebates, and relationships through which PBMs helped shape lists of preferred drugs were not public information. The true cost of any drug on any list could never be known in the absolute, and so the amount of price discount that was passed on to the payor (versus the amount retained by the PBM itself) remained a gray area. PBMs showed their payor clients price lists in which some drugs looked cheaper than others but had no need to disclose how much they were effectively being paid by the manufacturer of the "cheapest" drug to make that drug *look* cheapest. The business model of the PBM, therefore, worked both sides of the formulary as a form of pharmaceutical legerdemain.

Nor were the relative price of drugs the only item manipulated by PBMs. By shaping and reshaping the taxonomy of therapeutic classes themselves, PBMs were effectively able to control the field of perceived choice within which payors, physicians, and consumers experienced therapeutic substitution. PBM managers spent a lot of their time thinking about what kind of taxonomies of drug classes were more likely to achieve better arbitrage for themselves. As one observer characterized the therapeutic gerrymandering of "the formulary game," "Designers must choose the number of therapeutic classes in a formulary. We call that characteristic of a formulary its 'granularity'. . . In order to provide comprehensive care, they have to make sure that all nonformulary drugs have at least 1 formulary drug that is therapeutically equivalent. A therapeutic class can be viewed as a market—a place where choices are made from a set of substitutable goods. A formulary is a group of separate markets. One can reduce the competition for inclusion in the formulary by creating very narrowly defined therapeutic classes."[8] For example, a formulary manager could make a formulary with one broad class for all calcium-channel blocking drugs represented by one preferred, cost-effective, generic drug. Or that same manager could make a formulary with three therapeutic classes for each of the three subtypes of calcium-channel blockers. In the second example, a "highly refined formulary," the brand name drug in one subclass would *not* be exchangeable with a generic drug in another subclass—therefore steering purchasing toward a particular brand-name drug and driving up drug costs. The architecture of the formulary itself could steer toward a payor's interest (lower cost, more generic substitution) or toward a manufacturer's interest (higher cost, more brand-name drug sales) depending on which side of the PBM's Janus-like client interface was being prioritized.

As one HMO pharmacy director complained in 1995, PBMs shaped markets by defining the categories within which substitution and price competition take place—while keeping both clients in the dark about how this process took place: "The loss of control over pricing and rebates [is worrisome]. It's the accuracy of the rebate checks and the money you get. It's their say-so; you have no way to check if it's accurate. They say the pharmaceutical companies can't let you know the amount of the rebate. You have to take it on faith. It's based on the honor system. No other business in America involving millions of dollars is done on the honor system that I know of."[9] Likewise, as pharmaceutical companies became more aware of the role of PBMs in shaping which drugs became preferred in private payor formularies around the country, industry strategy switched from purchasing preferred status from PBMs toward purchasing PBMs themselves. Merck, one of the largest drug firms in the world in 1993, purchased Medco, one of the largest PBMs. SmithKline Beecham and Lilly likewise each purchased PBMs. Other companies tried to build up PBM capacity internally. "You don't need the PBM per se," one pharmaceutical executive noted in 1995. "You need the functions of a PBM. And if you can get that without spending the money on the PBM, you're ahead of the game."[10]

The Federal Trade Commission began an investigation into the proposed Merck-Medco merger but announced that the vertical integration of pharmaceutical manufacturers and formulary managers did not violate existing antitrust law. After the merger, the Food and Drug Administration issued a set of proposed guidelines in 1998 that, to the extent that PBMs conducted promotional activities for the pharmaceutical firms that "owned or influenced" them, they should be regulated as an extension of pharmaceutical marketing.[11] By the end of the decade, the path to preferred drug status had become another manageable element of pharmaceutical marketing. Concluding its investigation of Merck-Medco, the FTC's William J. Baer announced, "We have found that Medco has given favorable treatment to Merck drugs. As a result, in some cases, consumers have been denied access to the drugs of competing manufacturers. In addition, the merger has made it possible for Medco to share with Merck sensitive pricing information it gets from Merck's competitors, which could foster collusion among drug manufacturers."[12] Merck and Medco would eventually part ways in 2003. Just before the separation went into effect, a set of court documents leaked to the *New York Times* showed that some $3 billion in rebates received by Medco in the late 1990s to promote brand-name drugs in their formularies were not passed on to plans or consumers but merely retained as profit. Medco's formularies heavily favored Merck products: preferring Merck's Vioxx over Pharmacia's Celebrex; Merck's Prilosec over Abbott's Prevacid; and Merck's Zocor over Pfizer's Lipitor, even though Merck products were often more costly. Lawsuits against other PBMs suggested that such brand-brand substitution was indeed commonplace. Yet these documents—under the terms of a series of legal settlements—were sealed by the courts, not visible to payors, prescribers, or patients. After two decades of inquiry, the private logics by which PBMs turn profits through orchestrating institutional drug preferences remain largely shrouded in secret.[13]

PUBLIC PREFERENCE: THE DRUG EFFECTIVENESS REVIEW PROGRAM

By the turn of the twenty-first century, public spending on prescription drugs was also increasing at a fast clip. After a decade of increased generic substitution, managed care, and PBMs, state Medicaid spending on prescription drugs had not decreased at all but had instead increased fivefold from $4.8 billion in 1990 to $21 billion in 2000. In 2002, a group of key "reforming states" alarmed by the future impact of these increases on their solvency published a manifesto in *Health Affairs* on the importance of developing transparent, evidence-based means of comparing drug products to produce cost savings where equivalent care could be delivered. Existing placebo-controlled trial data did not do enough to provide "timely, accurate, and independent information for prescribers and purchasers on the relative effectiveness of different pharmaceutical agents."[14]

While a few prominent comparative trials had been federally funded— such as the University Group Diabetes Project and the Coronary Drug Project in the 1960s and 1970s, or the ALLHAT and CATIE comparative trials for antihypertensive and antipsychotic agents in the 1990s and 2000s, these represented only a few drops in the ocean of clinical trials, most of which were privately funded and noncomparative. Pharmaceutical manufacturers had little incentive to promote head-to-head research that risked exposing their products to unfavorable comparison with competitors. While some insurers created comparative effectiveness programs—such as the Blue Cross / Blue Shield Association's Technology Evaluation Center (TEC), founded in 1985—in general, insurers received little competitive advantage from producing comparative effectiveness knowledge once it entered the public domain.[15]

The TEC nonetheless became an early laboratory for using systematic reviews of the medical literature as a means of comparing evidence gathered separately about different drugs to compare them as a class. It also became a springboard for translating early comparative effectiveness methods into public policy. John Santa, who would later become an influential figure in the development of Oregon's public sector drug effectiveness program, recalled that he became familiar with comparative effectiveness research in the 1990s while a staff member of the TEC, using international research on drug effectiveness to help private health insurers make evidence-based coverage decisions. By 2002 Santa was working with the office of Oregon governor (and emergency medicine physician) John Kitzhaber on applying systematic reviews of data to evidence-based policy making. Oregon became the first state to publish its own evidence-based preferred drug list and was soon collaborating regionally to help Washington State and Idaho develop lists of their own. These lists formed a public, transparent, evidence-based analogue of the private formulary-shaping activities of PBMs; their goal likewise was to produce meaningful therapeutic substitution.[16]

As historian and Milbank Memorial Fund president Daniel Fox has noted, these lists allowed states to "invite manufacturers of drugs that were chemically equivalent to but more expensive than preferred drugs to acquire preferred status for their products" by paying rebates to states to effectively reduce the cost of their drug to that of the otherwise preferred item. As other states began to use these public lists of preferred drugs, the Drug Effectiveness Review Project (DERP) expanded quickly. By 2008, thirty-three states had operating preferred drug lists, and the project was fully supported by Alaska, Arkansas, California, Colorado, Kansas, Maryland, Michigan, Minnesota, Missouri, Montana, New York, North Carolina, Wisconsin, and Wyoming, along with the Canadian Agency for Drugs and Technology in Health. The project commissioned systematic reviews based on member interest and disseminated them widely. DERP officials spent a great deal of time working out a governance structure focused on transparency and the scrupulous management of potential conflicts of interest.[17]

As DERP researchers set out to define a methodology for building clinically meaningful comparisons of drugs in a therapeutic class, they focused on three core questions. First, how could one measure the comparative effectiveness of drugs for a given therapeutic class? Second, how could one evaluate the comparative safety of these drugs? Third, when two or more drugs were found to have similar safety and efficacy in a general population, were there any subgroups for whom that similarity did *not* apply?[18]

This last question—whether drugs might be exchangeable for some people but not others—opened up the DERP to withering critique from a constellation of industry actors and minority health groups. As DERP researchers tried to address subpopulation concerns by looking at demographic categories (gender, race, ethnicity, age), potential drug interactions, or comorbid disease conditions (e.g., liver failure, renal failure), that might be relevant to class comparisons, they often found themselves on the defensive. The PMA and the National Pharmaceutical Council, in turn, used the idea of ethnic and racial difference in pharmaceutical metabolism as a wedge issue to split minority health advocacy groups against therapeutic substitution programs.

In 1990 PMA executives had posed for photographs with leaders of the National Black Caucus of State Legislators to protest therapeutic substitution in state Medicaid programs. As the president of the National Black Nurses Association argued, "The patients who are sickest and most vulnerable will not get the medication that the doctor ordered and very likely will be given something that won't work as well." A few years later, Richard A. Levy, the VP of Scientific Affairs for the NPC, published a review of the implications of early research in the field of pharmacogenetics for the management of therapeutic exchange programs. Levy and the NPC shrewdly tied emerging identity politics (of racial and ethnic difference) to the critique of policies of exchange based on similarity. Although therapeutic substitution might pose no problem to generic white populations, Levy argued, "ethnic and racial minorities may be subject to greater risks if they are prescribed, or switched to, an 'equivalent' remedy." As an outgrowth of what sociologist Steven Epstein has documented as an emergent "inclusion/difference paradigm" in the last decade of the twentieth century, the assumed similarity of therapeutic products could be fractured anew on lines of race and ethnicity.[19]

The NPC, PMA, and AMA recruited minority caucuses and patient advocacy groups in the mental health fields to protest the DERP protocols for therapeutic substitution as monolithic and based on an antiquated notion of homogeneous mainstream society. By 2002, Richard Levy of the NPC had coauthored a series of articles with Valentine Burroughs and Randall Maxey of the National Medical Association, the leading professional association of African American physicians, claiming that "there is good evidence to show that therapeutic substitution of drugs within the same class places minority patients at greater risk." As they continued:

Physicians and managed care plans must be more alert under these new "formulary laws" for atypical drug responses or unexpected untoward side effects when treating patients from racial and ethnic minority groups. Dosage adjustments might be necessary in using generic drugs as therapeutic substitutions in untested racial and ethnic groups. There is a distinct possibility of a toxic accumulation of such drugs from slower metabolism, or the need for medical service substitutions to supplement the ineffective generic drugs in some racial and ethnic and minority groups. Either outcome will demand a greater use of health system resources and thus obviate the original purpose of cost containment.[20]

As Pfizer CEO Henry McKinnell told the *Medical Herald* in 2005, DERP-informed policy would specifically harm African American and Hispanic patients, whose racially or ethnically marked bodies apparently metabolized similar drugs differently. DERP staff noted strong industry lobbying of minority and patient interest group caucuses, alongside other forms of intervention, to forestall preferred drug lists, including direct intervention by companies in governors' offices, organized letter writing by physicians to try to get their favorite drugs preferred, testimony by company representatives and surrogates at state advisory meetings on preferred drug lists, and lobbying to exclude key categories of disease such as mental illness from policies of therapeutic substitution.[21]

Industry stakeholders took their objections to evidence-based preferred drug lists to the courts as well, most visibly in the case of *PhRMA v. Concannon*, which challenged Maine's DERP-inspired Maine Rx program. Maine Rx extended Maine's Medicaid preferred drug pricing list to all citizens of Maine, effectively making the state into a sort of public PBM: obtaining manufacturer rebates and pharmacy discounts and allowing all citizens to access that benefit. While the Supreme Court upheld the constitutionality of Maine Rx, they deferred to the Department of Health and Human Services (HHS) regarding the extension of these lists from the public to the private sector. With the encouragement of HHS, the state of Maine backed away from the full stance of Maine Rx and established the Maine Rx Plus program, which would be limited to a Medicaid-based income-eligibility scale.[22]

Nonetheless, many states saw the 2004 *PhRMA v. Concannon* ruling as a vindication for public preferred drug lists. As DERP programs expanded, estimates of cost savings ranged from $1 million to $80 million per state. In just under a decade, participating states had changed prescribing patterns to cheaper, effective preferred drugs with up to 80–90 percent penetration. Some states could even present evidence that linked cost containment with improving measures of quality care.[23] And as the Bush administration transitioned to the Obama administration, more than $1 billion in federal stimulus funding helped to expand the movement to achieve higher-quality, lower-cost health care in the United States through the conducting and dissemination of comparative effectiveness research (CER).

As with the DERP project, broader investment in CER was meant to assist clinicians, patients, and health systems in comparing the relative real-world value of therapeutic alternatives. Early plans for a national CER center (now called the Patient Centered Outcome Research Institute, or PCORI) looked from the DERP at home to other precedents abroad—particularly in Britain (National Institute for Health and Clinical Excellence), France (Haute Autorité de santé), Germany (Institut für Qualität und Wirtschaftlichkeit im Gesundheitswesen), and Australia (Pharmaceutical Benefits Scheme).

Nonetheless, even as the new federal program was still defining its initial priorities, a strong domestic opposition to comparative effectiveness quickly assembled. As conservative groups such as the American Enterprise Institute and the Heritage Foundation portrayed federal investment in CER as an entry point to state control of health care, articles in the mainstream press described CER as a program of rationing that might threaten the autonomy of the American physician, restrict access to life-saving medicines, and intrude on the doctor-patient relationship. Even among those committed to the success of therapeutic comparison and therapeutic substitution, there was considerable controversy over where a center for comparative effectiveness research would be housed, whether its findings should be tied to approval and reimbursement decisions, and whether there would be any role for cost in evaluations of comparative effectiveness.[24]

The argument that knowledge of therapeutic comparisons is a public good and therefore unlikely to receive sufficient funding from the private sector remains the strongest rationale for the establishment of the PCORI and sustained federal investment in CER. But it also underscores the vulnerability of programs like DERP or PCORI. As the recent recession combined with increasing pension crises has tightened state budgets still further, more and more states have dropped out of the DERP consortium. Why not? They can use the data DERP produces whether or not they help to pay for their production. Once spread over thirty states, DERP at the time this book went to press had only nine paying members, and its continued budgetary support—and ability to continue comparing drugs in useful ways for physicians and policy makers—are at risk.

PUBLIC AND PRIVATE RATIONALITIES

Studying me-too drugs and therapeutic substitution highlights another set of contests over the politics of similarity in modern medicine. The reduction of the world of therapeutics to a single representative per therapeutic class could be performed privately (as we saw with the financially robust yet secretive PBMs) or publicly (as with the financially unstable yet transparent DERP project). Each approach is incomplete; each raises thorny questions about the conflict between the rational design of health-care systems and the autonomy left to the individual prescribing physician. On a systemic level, prescribing decisions can be *rationalized*, through policies that favor the exchange of cheaper for more expensive drugs where safety and efficacy are the same. But it is equally important for such systems to not place individual physicians and patients in *unreasonable* situations, where drugs that are necessary are unavailable or unaffordable because they do not fit into a volume discount made between large institutional players.[25]

When, then, is a medicine good enough? The successes and failures of prior attempts to compare me-too therapeutics—from Kefauver to Kennedy, from DERP to PBMs—live on as a set of challenges in the current health policy landscape. This chapter has narrated a series of unintended consequences that has arisen as the role of the generic drug as a therapeutic alternative has seeped outward from the realm of the chemically identical into the realm of the chemically similar in the private (PBM) and public (DERP) policies of US health-care delivery. Yet in the closing decades of the twentieth century, as the role of the generic alternative was expanding into new therapeutic classes in North America, it was also expanding into new therapeutic geographies. By the end of the twentieth century, markets for generic drugs had shifted from regional to global, and the science and politics of the generic alternative had shifted to accommodate these new geographies of pharmaceutical production, circulation, and consumption.

CHAPTER 14

THE GLOBAL GENERIC

Spurred by incentives for prescribing, and demands for cost-effectiveness at the payer level, generics players are now flexing their considerable financial, legal, and marketing muscles across all geographies.

—IMS HEALTH, *BRAND RENEWAL*, 2007

Generic drugs emerged locally as a solution to problems of cost and access in health care earlier in the United States than in many other places in the world. The history of generic drugs narrated so far in this book illustrates a series of ironies specific to the American health-care marketplace: last among large industrialized democracies to mobilize a plan for universal health care, Americans have never and likely never will empower a single payor to negotiate lower prices of brand-name pharmaceuticals, as most European countries did from the mid-twentieth century onward. Americans consume more pharmaceuticals than people of any other nation, we spend more on pharmaceuticals than any other place in the world, and we have a greater tendency to pursue pharmaceutical solutions to existential and social problems.[1] It is precisely because the public sector has so little leverage in negotiating the prices of brand-name drugs in the United States that generic drugs—a private sector solution—became an operative mechanism for health-care cost containment.

By the early twenty-first century, however, the generic pharmaceutical had become visible as a solution to problems of cost and access in health care across a far wider range of geographies. In a 2007 report on the global pharmaceutical market, the market research firm IMS Health stratified countries by market share of generic drugs and amount of price savings generic drugs offer compared to brand-name drugs. The United States led the pack in both axes; laggards Spain and France demonstrate limited utilization of generic drugs, with other benchmark nations somewhere in between. In the descriptive language of the IMS report, the United States effectively became a new telos of pharmacoeconomic development: a "mature generics market."[2]

This phrase, and many others like it, repeats a geographical narrative of diffusion and development common in the history of biomedicine. Like other biomedical technologies in the late twentieth century, generic drugs and their accompanying scientific, legal, regulatory, and economic frameworks are understood to have developed in the United States in the early postwar era and then spread to other, more peripheral parts of the world over the remainder of the twentieth century. Yet we should be careful not to conflate history and geography too easily. Recent transnational perspectives on American history have shown that "domestic" events—such as the Hatch-Waxman Act of 1984, the Nelson hearings of the late 1960s and 1970s, and the Kefauver-Harris Amendments of 1962—were part of policy conversations that stretched far beyond the border of the United States. Likewise, "international" events—as we saw in chapter 1 with the production of the *Pharmacopoeia Internationalis*—were read closely by American actors in political, professional, industrial, and activist circles. The history of the generic drug in India, Israel, or Brazil is not merely a diffusion of an American concept into other geographies. But it is not entirely disconnected from that concept, either.

Generic drugs are now globally available, but the ways in which they are understood to be the same or not the same vary by local context. As this book is going to press, generic substitution is widely accepted in the United Kingdom, is somewhat less accepted in Germany, and remains a political third rail in France. These variable histories of generic drug sectors across the global North should alert us to an even more pronounced divergence in the generic histories of the global South as well. A recent review of generic pharmaceuticals across Latin American countries, for example, concluded that even economic and technical definitions of *generic* shifted markedly from country to country. This variability cannot be understood without attention to the specific historical context in which pharmaceutical markets have differentially developed across different geographies. By the 1980s, the assembled drug industries of former European colonies across Asia, Africa, and Latin America entailed a complex enterprise of variable quality and unclear provenance, with relative deserts of drug availability on the one hand and a surfeit of counterfeit goods on the other.[3]

The multicolored map of intellectual property structures in the late twentieth century likewise produced a complicated geography for the availability and unavailability of specific pharmaceutical products for imitation and replication. In some countries, like India or Brazil, local pharmaceutical firms could manufacture any medicine they were able to reverse engineer. In other geographies local industries were limited to drugs already off patent in the United States and Europe. The patchiness of the global map of local generic drug production helps explain why, by the 1990s, the US-led movement to "harmonize" intellectual property accords through the Agreement on Trade-Related Aspects of Intellectual Property Rights (TRIPS) had a differential impact on local pharmaceutical industries in different locales. Well before TRIPS, generic drugs had a different trajectory in Brazil than they did in India, or Israel for that matter, and those differences continue to matter in a post-TRIPS world.[4]

ESSENTIAL MEDICINES AND ACTIVIST STATES, FROM GENEVA TO RIO DE JANEIRO

The crowd greeting WHO director general Halfdan Mahler as he approached the dais to address the Twenty-eighth World Health Assembly in Geneva in May of 1975 had a strikingly different composition than the group of people who had attended the first World Health Assembly twenty-seven years earlier. In 1948, the World Health Organization had been largely a club of Northern empires. By 1975, however, the ranks of voting delegates were now swollen with representatives from newly independent former colonies, from Bangladesh to Zambia. It was with these delegates in mind that Mahler announced a shift in direction for the multilateral agency: the WHO needed to address the "urgent need to ensure that most essential drugs are available at a reasonable price" in generic form everywhere in the world.[5]

The essential drugs concept sought to carve out a core subset of pharmaceutical products and appropriate them from the world of private goods into a public health commons.[6] Mahler tasked the pharmaceutical division of the WHO to form an expert committee that would define an "essential drugs philosophy," create a list of essential drugs, and provide technical advice on how such a list could be translated into a global strategy for increased access to essential health commodities. The team visited twenty-five countries to interview Ministry of Health officials, doctors, pharmacists, and other health providers, and it convened a series of expert panels in Geneva in 1976 and 1977.

The formal definition of essential drugs that appeared in the 1977 *WHO Technical Report 615: The Selection of Essential Drugs* was closely informed by the experience of national drug programs in Sri Lanka and Pakistan, but it was also informed by North American debates on the equivalence of generic and me-too drugs. The American delegate to the expert committee, American Medical Association Council on Drugs member Daniel Azarnoff, drafted early guidelines for defining essential drugs and choosing a short list of necessary therapeutic compounds. Although the therapeutic revolution of the postwar pharmaceutical industry had produced a number of truly effective therapeutic agents, Azarnoff argued, their global public health impact had been muted because of the pernicious forces of the private market: "Pharmaceutical companies have undertaken extensive advertising activities in an attempt to coerce physicians to prescribe their generally more expensive brand name products . . . In order to provide the best health care feasible at the lowest cost it is necessary to restrict the availability of all drug products to those available for the health of the Public . . . A select list may not provide for the needs of every person, but certainly will for the vast majority although the choices in most therapeutic categories will be minimal."[7] The essential drug philosophy that Azarnoff helped to frame favored generic drugs over brand-name drugs and older products of proven value over newer products of uncertain value. The equivalence of generic drugs was assumed, as was the importance of reducing waste by eliminating me too medicines in favor of a simplified system of generic alternatives.

These developments were highly concerning to the pharmaceutical industry delegates present (but not voting) at the meetings of the expert committee. The interests of multinational pharmaceutical firms were represented in Geneva by the International Federation of Pharmaceutical Manufacturers & Associations (IFPMA), which wasted no time in declaring the essential drugs concept to be "completely unacceptable to the pharmaceutical industry."[8] Michael Peretz, the IFPMA's permanent vice president, later explained that "if the WHO was recommending a list of essential drugs it would follow that the WHO was implicitly arguing that all other drugs not included in the list were non-essential."[9] In a formal letter to Mahler in April 1978, the IFPMA outlined its "serious reservations" about the essential drugs policy, considering that, by promoting generic drugs over brand-name drugs, the WHO was exceeding its own technical capacity to monitor drug standards. "The pharmaceutical industry is not aware of any *developed* nation where regulatory authorities can provide assurance of the bioequivalence or interchangeability of the drug supply within their jurisdiction," they huffed, "and the state of regulatory effectiveness in most developing countries is substantially less advanced."[10] "Because of this reality, governments, the medical profession, and the patient must rely upon the reputations of companies with consistent histories of producing high quality products as the best assurance of safety and therapeutic effectiveness . . . To discourage [the use of brand names] as the WHO Report does would have grave repercussions for the quality of pharmaceutical supplies and health care, whether in developed or developing nations."[11]

Mahler's WHO nonetheless depended on cooperation from individual drug manufacturers even as it articulated a broad critique of the excesses of the multinational pharmaceutical industry. These partnerships were often strained. As companies pledged support for technology transfer for basic drug production through training programs, target countries complained that such programs created a new fistula in the global "brain drain" that sucked trained health professionals from countries of the global South to the global North. Utopian plans for affordable, scalable, easily transferrable pharmaceutical production plants were drawn up, such as the blueprints for the WHO Low Cost Pharmaceutical Formulation Plant (LCPFP) found preserved in the WHO archives, but the plants were not necessarily built. The LCPFP provided a generic model for a generic plant that might produce generic pharmaceuticals in developing countries. Each plant would be 2,800 square meters, maintain a staff of fifty to seventy persons, and provide, for an investment cost of $3 million, an annual output of two to three hundred million tablets, twenty-five to fifty million capsules, and fifty tons of necessary liquids, ointments, and powders as directed by the WHO List of Essential Drugs—enough to service a population of three to five million people (figure 15). While it is true that few countries, if any, achieved the goals of local production spelled out optimistically in the LCPFP, WHO policies nonetheless managed to support the development of local generic drug production in many parts of the world in the late 1970s and early 1980s.[12]

The essential drugs program also helped increase the visibility of the developing world as a vast potential market for lower-priced branded generics produced by multinational pharmaceutical houses. Even while the IFPMA was formally opposing the essential drugs concept, indi-

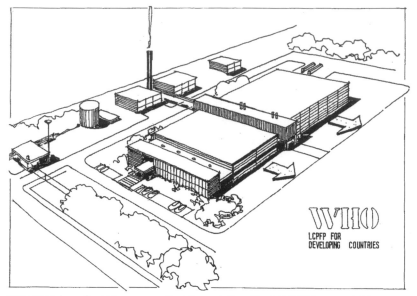

FIGURE 15. Low Cost Pharmaceutical Formulation Plant (LCPFP), a model for a generic plant that might produce generic pharmaceuticals in developing countries.

M. Stork, W. B. Wanandi, A. S. Arambulo, Guidelines and Recommendations for the Establishment of a Low Cost Pharmaceutical Formulation Plant (LCPFP) in Developing Countries *(Geneva: World Health Organization, 1980), cover image.*

vidual firms reached out to the WHO to probe possible access to these markets under the WHO aegis. Several firms created their own branded generic subsidiaries expressly for the purpose of distributing essential drugs to the developing world. In 1979, for example, Ciba-Geigy launched a new branded generic production unit called Servipharm AG "to open up the Third World markets and provide some of the basic drug requirement."[13]

Yet it remained a problem to multinational pharmaceutical firms that one country's patent-protected brand-name monopoly was another country's patent-free zone for generic competition. As global interest in generic drugs increased in the early 1980s, the IFPMA became increasingly concerned about the heterogeneity of global intellectual property law, as did its American component, the Pharmaceutical Manufacturers of America. After the Drug Price Competition and Patent Term Restoration Act of 1984 had legitimated the domestic generics industry and established a friendly patent environment for research-based drug firms within the borders of the United States, members of the PMA and IFPMA began to look toward international law to extend the US-style patent protections for their products in global pharmaceutical markets.

Brazil became an important early battleground in this fight. "As a newly industrialized nation," PMA president and former US patent commissioner Gerald Mossinghoff complained to the US trade representative in 1985, "it is time for Brazil, the eighth largest economy in the West, to start playing by the rules of the international trading system."[14] Mossinghoff had become president of the PMA in January of 1985, taking over from the previous president, C. Joseph Stetler, shortly after the Hatch-Waxman Act took effect. Where Stetler had cut his teeth as general counsel for the AMA (and therefore helped consolidate professional-industrial relations against generics in the 1960s and 1970s), Mossinghoff had trained as a patent lawyer in St. Louis, acted as congressional liaison for NASA, and served as the head of the US Patent and Trademark Office during the early Reagan years, where he came into close contact with growing issues of pharmaceutical intellectual property.

As head of the US Patent and Trademark Office, Mossinghoff called for the revision of the hundred-year-old Paris Convention for the Protection of Intellectual Property so that patent regimes around the world might better reflect the interests of American industries—a process that would lead to the formation of the World Trade Organization (WTO) and the TRIPS accords.[15] The choice of Mossinghoff as leader of the PMA in 1985 helped position the PMA as a vanguard actor in the extension of American-style intellectual property laws in increasingly globalized arenas. Shortly after moving his office from the Patent Office to the PMA, Mossinghoff made his first bid for pharmaceutical sanctions at a US Trade Representative hearing on Brazil, naming Brazil's nascent generic drug industry a set of dangerous "patent pirates." As Richard Furlaud, the CEO for Squibb, told the trade representative, Brazil's refusal to grant patent protection had cost PMA firms more than $160 million between 1979 and 1986. Brazil was not just a bad actor qua Brazil: the PMA hinted that Brazilian generics would quickly metastasize into other Southern markets through South-South distribution networks. As Mossinghoff spelled out, the PMA was "concerned over the potential for Brazil to become a market for and an exporter of infringing products" throughout Latin America and possibly the entire developing world.[16]

Brazil was not alone. The American pharmaceutical industry, Mossinghoff announced, was losing hundreds of millions of dollars to patent infringement in "patent pirate" countries with permissive intellectual property regimes from Mexico to Argentina to Taiwan to South Korea.[17] A few months later, at a conference on trade at the State Department, Mossinghoff hammered home the importance of international intellectual property protection to US business interests: "It's a crime to rob a bank, but it is not unlawful everywhere to pirate a patent. The theft of patents has been with us a long time . . . The difference in recent years is not the fact of the crime, but its magnitude and potential impact. For years, PMA members have competed in countries where patent protection has been weak, unenforceable, or simply nonexistent. I urgently hope that the failure of a nation to provide adequate intellectual property protection will be a trade-actionable item under the GATT."[18] The body Mossinghoff referred to as GATT—the General Agreement on Tariffs and Trade—had not yet grown any regulatory teeth to harmonize global intellectual property law by 1987. Patent, trademark, and copyright law still varied widely from country to country, especially in the pharmaceutical sector. In many countries pharmaceuticals had been excluded from patenting altogether because of their public health significance. In France, for example, pharmaceutical products had been unpatentable from 1833 to 1959 and in Germany from 1877 to 1969, while product patents on drugs did not appear in Japan until 1976, in Switzerland until 1977, and in Spain and Argentina well into the 1990s.[19]

In Brazil—the PMA's first target in their attempt to use trade pressure to secure harmonization of pharmaceutical intellectual property—pharmaceutical products had been explicitly excluded from patenting under President Getulio Vargas in 1945—with a goal of maximizing technology transfer of drugs developed abroad. A Drug Production Institute was created in the Oswaldo Cruz Foundation (often glossed as the Brazilian analogue to the US National Institutes of Health) by presidential decree in 1970, and the Health Ministry operated its own chemical synthesis laboratory through the 1980s with the explicit goal of copying the formula of drugs developed abroad for transfer to the Brazilian industry—a field that had expanded to nearly five hundred firms by 1985.[20]

Mossinghoff's complaint against Brazil was heard. In July of 1988, the US trade representative to Brazil initiated an investigation and by October invoked Section 301 to impose a 100 percent tariff increase on Brazilian goods. Speaking at a hearing of the House Subcommittee on Commerce and Oversight, United States trade representative Carla A. Hills announced that these sanctions would not be lifted until PMA drug patents were honored. Mossinghoff lauded her action and called for further international measures to "achieve harmonization of patent systems at an appropriately higher level of protection drawing upon the most advanced patent systems of the world."[21]

Over the course of the 1990s, a set of regional initiatives like the 1994 North American Free Trade Agreement (NAFTA) and global initiatives by the World Intellectual Property Organization and the WTO were several means by which sovereign states directed their own intellectual property law regarding drugs. Before NAFTA, for example, Canada had reserved the right to issue compulsory licenses on patented drugs when issues of drug supply were crucial to the health of the public; these measures were weakened considerably by the trade agreement. Likewise in Mexico the implementation of NAFTA brought with it further external enforceability of American-style product patents.[22]

The same year that NAFTA went into effect witnessed the Uruguay Round of the GATT that led to the formation of the WTO and the ratification of TRIPS. With the implementation of TRIPS, Mossinghoff's dream of extending the pursuit of international patent pirates would grow from a regional to a global project. As country after country signed the treaty under bilateral pressure from the United States and multilateral pressure from the WTO, even former patent pirates like Brazil agreed to recognize product patents. TRIPS was alternately praised and criticized as a key milestone of global neoliberalism in the post–Cold War era. To its advocates, TRIPS eliminated antiquated forms of nationalistic variation that slowed the wheels of global commerce. To its critics, TRIPS imposed a new empire of intellectual property, dominated by nodes of commerce in the global North, which would have calamitous implications for public health in the global South.

A tangible demonstration of the public health impact of TRIPS arrived the same year that Brazil became a signatory to the treaty, at the eleventh annual International AIDS Conference in Vancouver, Canada. Delegates gathered under the conference slogan "One World, One Hope" indeed found hope in the announcement by virologist David Ho that combination antiretroviral therapies could render HIV infection undetectable and transform an invariably fatal disease into a manageable chronic condition. Yet to those delegates in the audience conscious of the impact of TRIPS, the one world of harmonized global intellectual property regimes threatened to snuff the one hope of a global response to the global epidemic. Under TRIPS, these new drugs would be patent protected *everywhere* and generically available nowhere. HIV/AIDS threatened instead to bifurcate into a manageable condition in those countries of the global North that could afford to provide antiretroviral cocktails to their citizens—and an ever-growing deadly scourge in those countries in the global South that could not. The events in Vancouver pushed the bilateral and multilateral contestation over pharmaceutical intellectual property to a grim paradox. "One World, One Hope" meant one thing for humanitarian groups interested in global health equity and quite a different thing for the harmonization of global intellectual property.

As the medical anthropologist João Biehl has discussed, Brazil found a third path that enabled it to continue its civil disobedience in the realm of pharmaceutical intellectual property and act as an "activist state." Even in a post-TRIPS era, Brazil continued to resist the imposition of global intellectual property regimes in the interest of providing generic versions of antiretrovirals for its citizens. As TRIPS went into effect, the Health Ministry of Brazil invoked HIV/AIDS as a public emergency and declared a policy of universal access to antiretroviral drugs produced generically in Brazil. A private Brazilian firm had already announced in 1993 its ability to manufacture a high-quality reproduction of the antiretroviral AZT. In 1997, the state-owned pharmaceutical laboratory in Farmanguinhos (part of the Oswaldo Cruz Foundation) was given a twentyfold infusion of funding to begin producing copies of other antiretroviral drugs, with a stated aim of making the AIDS treatment program of the Health Ministry feasible and cost effective. As the sociologists Marilena Correa and Maurice Cassier note, in the five years from 1997 to 2002, the output of the Farmanguinhos drug production facility increased sevenfold, and accounted for 40 percent of all Brazilian production of antiretrovirals.[23]

The result was the formation of an innovative sector of Brazilian public-private collaboration that privileges imitation as a form of innovation. Farmanguinhos has become exceptionally efficient at reverse engineering antiretroviral drugs through a "proven reverse engineering methodology" formalized on technical, methodological, and managerial levels. In some instances, the techniques reverse engineered by Brazilian chemists are superior to those developed by the originator firms. When attempting, for example, to reproduce Bristol-Myers Squibb's antiretroviral ddI (Videx), the formulation group produced a product that they declared was not just bioequivalent to Videx but "substantially better." Upon disclosing their formula at a conference, the Farmanguinhos team received a letter from Bristol-Myers Squibb congratulating them on this achievement. By 2007, the Farmanguinhos laboratory was working to expand its ability to transfer knowledge acquired in its own research to Brazilian companies and universities.[24]

The activities of the activist state in a post-TRIPS era were nonetheless limited. TRIPS still bound Brazil in unanticipated ways. While Farmanguinhos could still flaunt pre-1996 product patents on antiretrovirals, the production of on-patent drugs in other sectors shifted toward a tidier, more American model of generic industry. In October of 1999, the Brazilian government approved Law 9787, the Generic Medicines Law, which recognized a new kind of legitimate pharmaceutical copy in a regime that newly respected pharmaceutical product patents—and followed American standards of bioequivalence as well. By 2000, the first of these globalized generic drugs had been approved by the Agência Nacional de Vigilância Sanitária.

As Brazil expanded this new, legitimate generics market, it participated in the harmonization of not only on-patent drugs but off-patent drugs as well. Many of these globalized generic drugs could be produced more cheaply in other parts of the world, especially India. As the markets for generic drugs effectively globalized, Brazilian domestic pharmaceutical manufacturers soon faced new competition from overseas manufacturers themselves. Paradoxically, the globalization of the generic drug market weakened local production—exactly the opposite of the WHO's initial interest in expanding generic production of essential medicines some twenty-five years earlier.[25]

GENERICS AS AN EXPORT MARKET: PHARMACEUTICAL EXPANSION ON THE INDIAN SUBCONTINENT

Like Brazil, India was a frequent malfeasant in the eyes of the US trade representative and jockeyed for position on the watch list for pharmaceutical intellectual property violations over the course of the 1980s and early 1990s. Yet the development of the Indian generics sector traces a very different history than the Brazilian case.[26] Looking back from the vantage point of 2005, a US International Trade Commission report marveled that the Indian pharmaceutical industry was now the fourth-largest in the world, chiefly on the strength of its export market for generic drugs. "Over the last 30 years," the working paper noted, "India's pharmaceutical industry has evolved from almost nonexistent to a world leader in the production of high quality generic drugs."[27] Much of this growth had come not in spite of TRIPS but explicitly because of India's early engagement with the new regime of harmonized intellectual property. When India joined the WTO in 1995, its exports of pharmaceuticals were valued at less than $600 million; by 2005 they had exploded to $3 billion.[28]

Several Indian pharmaceutical firms like Cipla—now prominent exporters of generic antiretroviral drugs for global treatment of HIV/AIDS across the global South—got their start in the 1930s as private ventures in a colonial context. A public Indian pharmaceutical sector also expanded after independence, after the government of Jawaharlal Nehru recognized that the young postcolonial nation was economically dependent on imported brand-name drugs from Euro-American multinational corporations. The Indian state founded five public pharmaceutical companies and instituted a series of policies to favor local production. Although many multinational pharmaceutical firms fled the Indian market, the state-owned companies were soon joined by thousands of medium and small private Indian pharmaceutical firms. This arena was further shaped by the government of Indira Gandhi, whose Patent Act of 1970 formally ended product patents and granted seven years of process patent protection to each innovator able to successfully reverse engineer the production of a drug. The Hathi Committee Report of 1975, which built a strategy for domestic pharmaceutical self-sufficiency around a series of 117 essential drugs, should be viewed as the culmination of a long series of parliamentary interventions to align local pharmaceutical production with therapeutic access.[29]

India became pharmaceutically self-sufficient far earlier than many other former European colonies, but its expanding industry competed for a domestic pharmaceutical market whose rates of consumption would still rank among the lowest in the world by 2005 ($4.50 per person, as compared with $820 per person in the United States). As the private Indian pharmaceutical sector grew, it grew as an export industry, often at the expense of the public sector industry. Already by 1979 India—along with Argentina, Brazil, Mexico, South Korea, and Taiwan—was recognized as an exporter of formerly "transferred technologies" to other Southern locales. A consultant to the UN Centre on Transnational Corporations, the Oxford economist Sanjaya Lall held up the Indian pharmaceutical industry as a challenge to the received notion that technology transfer was a unidirectional vector from innovative and productive Northern donors to a passive and consuming geography of Southern recipients. In the three years between 1975 and 1978, Indian exports to countries in the Middle East, sub-Saharan Africa, and the Soviet Union doubled and were poised to double again in the next few years, with plans to set up a "turn-key" plant for Indian pharmaceutical production in Cuba. Lall pointed to the Indian pharmaceutical nexus as an example of an imitative industry that provided crucial innovations for further technological development in the developing world.[30]

With low barriers to entry, limited costs of formulating drug products, and a minimal regulatory environment, a field of two thousand Indian pharmaceutical companies in 1970 expanded to more than twenty thousand by 2005. Internally, the industry segmented into a three-tiered pyramid: a large base of small firms with almost no investment in quality control, a smaller pool of midsized firms with good reputations for domestic pharmaceutical use but whose products did not meet requirements for export markets, and a concentrated top tier of large firms with international brand recognition and export capacity. Export firms met the FDA's Good Manufacturing Practices and were regularly inspected for quality by FDA agents; it was the interests of these firms that largely drove India's engagement with the WTO-TRIPS process as a means of accessing broader and still broader export markets. India was at the table already in 1986 for the Uruguay Round, and while the nation initially approached the WTO with some ambivalence, key elements within the export sector stood to gain access to broader markets through TRIPS. By the time that the final TRIPS provisions took effect in 2005; India had a $5.2 billion export market (which accounted for nearly 40 percent of all drugs produced by Indian firms and pushed India's overall trade surplus to $3.8 billion). Having fully shed its public origins, the Indian pharmaceutical company under TRIPS had become a private entity poised to become the world's key geographic site for production and export of global generics.[31]

Indian generics became a cause célèbre for global public health activists when large export firms like Cipla announced the reduction of the annual cost of antiretrovirals from $15,000 in 1995 to $200 by 2005, which helped to finally make HIV control economically feasible in the least-developed countries. Shortly afterward, the WHO announced a global scheme to provide technical expertise in certifying the quality and bioequivalence of generic antiretrovirals for use in global HIV/AIDS rollout campaigns funded by the UN Global Fund and other humanitarian bodies. These joint actions of Cipla and the WHO helped establish the reputation of Indian generic firms as key actors in an emerging landscape of global health that revolved around forms of South-South pharmaceutical provision.[32]

Cipla continues to innovate in global deliverables for HIV/AIDS care—an August 2012 press release proudly introduced Qvir, a new branded-generic entry of a combination kit packaged at roughly $3 per day as a second-line treatment.[33] Yet Indian export firms must still be understood to be creatures of the private sector. Even as early as 1984, a report lauding the Indian pharmaceutical industry as "the most vertically-integrated of all the developing countries" nonetheless cautioned that "there is one area in which both local and foreign enterprises behave similarly. The product mix of both kinds of firms tends not to reflect social priorities. Both focus on remedies for diseases that afflict the middle and upper income strata at the expense of diseases affecting the vast majority of the population."[34] By the early twenty-first century, the three largest Indian firms—Ranbaxy Laboratories, Dr. Reddy's Laboratories, and Cipla, Ltd.—dominated the Indian export industry and much of the global generic drug market. These firms had taken on the form of complex multinational corporations in their own right. Ranbaxy, for example, was a fully vertically integrated firm selling bulk active pharmaceutical ingredients, commodity generics, and branded generics in more than a hundred countries, with manufacturing installations in eight. By 2005, exports accounted for nearly 80 percent of Ranbaxy's sales, with the US market taking up the lion's share.[35]

Yet in spite of a series of moves calculated to legitimate membership of Indian generics in the global pharmaceutical economy—including early participation in TRIPS, submission for WHO certification, and regular FDA inspections—a residue of piracy and counterfeit status still adheres to these pharmaceuticals of Southern origin. Consider, for example, these two reports from 2006:

> On 4 December 2006, the Dutch authorities seized a cargo of generic medicines en route from India to Brazil. The cargo consisted of 570 kilograms of losartan potassium, an active pharmaceutical ingredient used in the production of medicines for arterial hypertension. It was sent by Indian company, Dr. Reddy's Labs to the Brazilian importer EMS. The cargo was held back by Dutch authorities for 36 days after which, it was released and directed back to India. Losartan potassium does not enjoy IP rights in India, the country of origin, or in Brazil, the country of destination . . .

The latest confiscation has been a UNITAID funded shipment consisting of 49 kilograms of abacavir sulfate tablets at Schiphol Airport
by Dutch customs authorities under the misleading claim that it contained counterfeit goods. UNITAID issued a statement on its website
that the drugs were not counterfeit nor did the shipment infringe
any intellectual property rights. UNITAID clarified that these were
medicines manufactured by Indian company Aurobindo and were
approved by WHO as well as temporarily qualified by USFDA."[36]

Who was the pirate here? The kinds of arguments Mossinghoff and the
PMA had used in labeling Indian firms patent pirates no longer applied:
Indian firms were conforming to US-style pharmaceutical intellectual
property agreements and had received the quality approval of both the
FDA and the WHO. It was a group of Northern pirates from the Netherlands, Indian authors argued, who had effectively boarded the ships of
reputable Southern merchants and held their cargo ransom. Ranbaxy,
Cipla, Dr. Reddy's, Aurobindo—these Indian firms constituted a new
breed of multinational corporation and represented the new face of pharmaceutical globalization: the generic giants.

GENERIC GIANTS

Sketching out the trajectory of the generic drug from the United States
to the WHO, from Brazil to India, from the activist state to the Southern
multinational only begins to illustrate the series of vital paradoxes inherent in the globalization of generics. Once the darling local industries,
favored in radical critiques of exploitative multinational corporations,
generic exporters like Ranbaxy, Cipla, and Dr. Reddy's now increasingly
resemble the multinational pharmaceutical companies they replaced.
WHO programs meant to encourage local production of generic pharmaceuticals have collided with WTO policies that now place local generic manufacturers in competition with powerful generic giants from
overseas.

Nor was the geography of the generic giant limited to South Asia. By
the mid-1980s, as many Northern multinational firms began to expand
their own generic-production wings through acquisition of smaller generic manufactures around the world, one potential target for consumption, the Israeli-based Teva Pharmaceuticals, opted (in the words of its

CEO) to eat as well, rather than be eaten. Teva had already survived the first contractures of the Israeli pharmaceutical market by merging with two smaller firms, Assia and Zoria, and acquired the second-largest Israeli drug firm in the late 1970s. As American firms were acquiring local generic companies around the world, Teva made a bid in the opposite direction and acquired in 1985 the unfortunately named American generics manufacturer Lemmon Pharmaceuticals. By 1991, Teva controlled more than one-third of Israel's pharmaceutical market and continued to acquire generic manufacturing overseas over the course of the 1990s and early 2000s, including Biogal, Biocraft, Copley, and Novopharm, among others. In 2001, Teva was able to gain full ownership of the prominent American generics firm Marion Pharmaceuticals. By 2004, Teva announced its plans to enter the Indian subcontinent as well, opening a research and development facility and an active pharmaceutical ingredients and processing facility in the state of Uttar Pradesh.[37]

By the turn of the twenty-first century, Teva was the largest supplier of generic drugs to the United States. Teva's acquisition of Sicor for $3.4 billion in 2004 expanded its range to the injectable market; the purchase of the Ivax Corporation for $7.4 billion in 2005 consolidated Teva's position as the world's largest maker of lower-cost drugs.[38] Before the Teva-Ivax merger, the largest generic manufacturer in the world had been Sandoz, the repurposed Swiss pharmaceutical firm that—following the 1996 merger of Sandoz and Ciba-Geigy—had become rebranded as the branded generic division of the multinational megafirm Novartis. Like Teva, Sandoz grew by acquisition in the 1990s and 2000s. But the purchase of Ivax allowed Teva to surpass even Sandoz in scope and scale. By 2006, Teva could boast of operations in more than fifty markets worldwide, with production capacity for thirty-six billion tablets and capsules, forty-four manufacturing sites, and a roster of seven hundred compounds in more than 2,800 doses and formulation. In the ten years between 1999 and 2009, Teva's profits increased by more than an order of magnitude from $135 million to $2 billion; its global revenue increased from $1.3 billion to $14 billion—with another doubling projected for a revenue of $31 billion by 2016. As this book goes to press, Teva is the largest producer and distributor of generic drugs for both the US market and the world. As the largest company in Israel, Teva has become so crucial to the Israeli economy that when a wave of layoffs were announced for the firm in October 2013, the story made front-page news for weeks and fanned widespread fears that the now-multinational firm might shift its headquarters out of the country.[39]

Teva was not merely aggressive in its consumption of other companies; it was also aggressive in its legal challenges to products still under patent protection. Teva's CEO frequently boasted that the agility of its legal department was what set Teva apart from other pharmaceutical firms. "Teva's legal department is big," he laughed in a 2006 interview. "We have many, many, top notch lawyers that work for us around the country. What do you think? That Pfizer puts against us the low-key lawyers?" At the time of the interview, Teva was simultaneously engaged in fifteen patent challenges, with 160 Abbreviated New Drug Applications filed at the FDA, 88 with potential paragraph IV patent challenges. Teva learned to promote generic competition earlier and earlier in the patent life of a brand-name drug—launching generic versions of the popular Zithromax (azithromycin), Neurontin (gabapentin) and Allegra (fexofenadine) while each was still under patent. "The balance of power," *Pharmaceutical Executive* noted, "has changed, and punishments resulting from at-risk launches no longer frighten today's giant generic companies."[40]

David has become Goliath. Generic manufacturers can no longer meaningfully be characterized as a local "Little Pharma" in contrast to a multinational "Big Pharma." Along with Indian generic firms like Cipla, Ranbaxy, Dr. Reddy's, and European multinational generic firms like Sandoz, generic firms now dominate the global pharmaceutical market by volume—and are rapidly swallowing up the local pharmaceutical industries that WHO and UN Conference on Trade and Development policies had tried so hard to build up in the 1970s and 1980s. By 2010, one out of every seven prescriptions filled in the US was filled with Teva products.[41] With the company's growth as a multinational firm, it had acquired and expanded innovative facilities that went beyond reverse engineering patented drugs to designing its own drug-delivery systems and licensing and marketing new compounds in its own right. In July of 2006, Teva launched Azilect (rasagiline), an innovative anti-Parkinson drug, on the US market, having already experienced substantial success with its first innovative, patent-protected drug, Copaxone (glatiramer), an immunomodulatory drug used in the treatment of multiple sclerosis. And already by 2006, Teva had dipped a toe into the realm of "biogenerics"—marketing new versions of off-patent biotech blockbusters such as human growth hormone, interferon beta, and gram-colony stimulating factor.

These newer biological generics represent the latest challenge for the global industry of imitative pharmaceuticals. Unlike "small molecule" drugs (which characterized most brand-name and pharmaceutical products over the twentieth century), the "large molecule" drugs that have emerged from the biotech industry have posed new challenges to an imitative industry even as they go off patent in the early twenty-first century. Teva's early investigations into biogenerics have largely bypassed the American generics manufacturing, regulating, and consumption sphere: Teva manufactured them in places like Lithuania, Mexico, and China and sold them in other markets with looser regulatory regimens than the United States and Europe. "The concept of 'Let's first be careful' with biogenerics is reasonable," Amir Elstein, Teva's vice president for biogenerics told *Pharmaceutical Executive* in 2006. "But within two or three years, there will be good proof in the marketplace that the products delivered by the biogeneric community are not only essentially similar, but equivalent."[42]

The future growth of generic giants, as Elstein suggests, now depends on breaching the world of generic biotech. *Pharmaceutical Executive* recently predicted that without penetration into the biological sector, the "generics bubble will burst in 2015." What would be the future of a multinational generics firm like Sandoz, the editors mused "as the last small-molecule megablockbusters lose their patents"?[43]

THE CRISIS OF SIMILARITY

Concepts are not spontaneously created but are determined by their ancestors. That which occurred in the past is a greater cause of uncertainty—rather, it only becomes a cause of insecurity—when our ties with it remain unconscious and unknown.

—LUDWIK FLECK, *GENESIS AND DEVELOPMENT OF A SCIENTIFIC FACT*, 1935

In the fall of 2012, Teva Pharmaceuticals announced the launch of a new kind of pharmaceutical copy for the American market. Already the largest generic manufacturer in the world, Teva had already been marketing copies of off-patent drugs for decades. Yet what was different about this newest old drug, TevaGrastim, was that it was explicitly not chemically equivalent to the innovator product it copied. Indeed the European Medicines Agency (EMA) had approved it a few years earlier as the first "biosimilar" drug in its class that could be marketed as a legitimate copy of a large-molecule biotech drug. But TevaGrastim could not be approved as a biosimilar drug yet in the United States, because a formal pathway for approval of such drugs by the Food and Drug Administration did not yet exist.[1]

Unlike the small-molecule drug products of the mid-twentieth century, the wave of innovative pharmaceuticals that emerged from speculatively capitalized biotech firms in the 1980s and 1990s were molecularly larger and more complicated by orders of magnitude. The price tags on these large-molecule drugs were also bigger: twenty-two times larger, on average, than those of their small-molecule counterparts.[2] As the patents on these interferons, recombinant human insulins, human growth factors, epoetins, colony-specific bone marrow stimulators, TNF-alpha blockers, monoclonal antibodies, and other therapeutic products of recombinant genetic engineering began to expire in the first decades of the twenty-first century, enterprising manufacturers eagerly anticipated a new frontier in generic drug making. But copying large-molecule drugs raised a host of new challenges for the would-be marketers of generic biotech. Unlike small-molecule drugs, the substance of these large-molecule drugs cannot simply be mapped out, atom by atom, and replicated. Unlike small-molecule drugs, there is no way to prove that two versions of a protein are, from a molecular perspective, the same. The promise of generic biotech is opening up a new crisis of similarity.

ALL MOLECULES, LARGE AND SMALL

Yet these problems are not *entirely* new. Though it has become commonplace to speak of biotech as a recent hybrid of science and commerce that began in the late 1970s, a looser definition of the term would incorporate the history of standardized "biologics" and "biological" drug products from at least the late nineteenth century. After all, the first federal law to regulate the approval of therapeutic products in the United States was not the Pure Food and Drug Act of 1906 but the Biologicals Act of 1902. It is often forgotten that one of the first wonder drugs of the interwar era, insulin, was itself a biological technology—a biotech drug avant la lettre.[3]

The history of insulin illustrates many of the broader challenges of copying large-molecule therapeutics. Even though insulin was first patented in 1921, we still see almost no generic insulins on the market nearly a century later. This is partly due to the unique history of the insulin patent, one of the first therapeutic patents awarded to an academic institution. From the early 1920s until 1941, each batch of insulin marketed in North America was tested (and certified) by its patent holder, the University of Toronto. So extraordinary was the insulin patent that the US Congress, concerned about the public health consequences of the loss of such stewardship after patent expiry, passed the Insulin Amendments of 1941 commissioning the United States Pharmacopoeial Convention—in concert with the sole US licensee of the insulin patent, Eli Lilly—to develop precise assays to determine insulin quality control and mandated continued batch certification of all insulin products by the FDA to certify identity, strength, quality, and purity.[4] These additional barriers to market entry for would-be imitators of biological drugs would be extended with the establishment of the Division of Biological Standards in the 1950s.

Would-be imitators of insulin were also confounded by the shifting web of process patents and regulatory structures that surrounded insulin as a therapeutic object. These legal protections were focused as much on *how* insulin was made as on *what* it was from a molecular perspective. Eli Lilly began early on to innovate with modes of insulin delivery: slow insulins, fast insulins, neutral protamine hagedorn insulins, lente, ultralente forms, forms based on depot formulation. Insulin became a difficult drug to copy not just because it was a large molecule but because it was a moving target, an intercalated mesh of things developed at different moments in time.[5] Likewise many subsequent innovative biological and synthetic biological drugs—from Wyeth's conjugated estrogen preparation Premarin to Boots's levothyroxine preparation Synthroid—have remained "hard-to-copy drugs" long after the expiry of their original patents.

Most histories of the modern biotech industry nonetheless begin their narratives with the technological and legal developments of the late twentieth century that made large-molecule drugs into more easily patentable objects. Technological developments from recombinant DNA techniques to rapid protein sequencing enabled the manufacture of new large molecules using newly engineered biological systems; legal developments such as the court ruling in *Diamond v. Chakrabarty* and the 1980 Bayh-Dole Act recognized the patentability of these objects in the United States and made it easier for academia and industry to create new collaborative ventures to capitalize on them. These coproduced technological and legal developments are often invoked to narrate the new emergence of the biotech industry in the 1980s and 1990s.[6]

But the biotech boom of the last decades of the twentieth century recapitulated—often on a molecular basis—key aspects of the development in biological therapeutics seen in the first decades of the twentieth century. As new firms like Genentech took shape at the intersection of powerful new molecular biology techniques and new forms of speculative capital, their therapeutic targets often focused on older biological products that could now be reproduced de novo as patentable biotechnological artifacts: human growth hormone, interferon, and—most importantly—recombinant human insulin. Indeed Lilly's Humulin was one of the first commercially viable drugs from the biotech sector, and recombinant insulins have gradually replaced all older forms of insulin on the market.[7]

Even in their infancy, these new biotech firms paid close attention to the formation of regulatory pathways for generic competition laid out in the Hatch-Waxman Act of 1984. The FDA's initial interpretation of Hatch-Waxman declared that biotech drugs could not be copied like small-molecule drugs: anyone wishing to market a copy of a biotech drug would have to file a full New Drug Application. When, in 1990, several patients suffered therapeutic failure when switching from brand-name Pancrease to a supposedly generic preparation of enzymes for patients in pancreatic failure, an FDA spokesperson quickly announced that "different formulations of pancreatic enzyme products are not true generics in the sense of multisource products listed in the FDA's *Orange Book*" since their exact chemical or therapeutic equivalence could not be confirmed. "Physicians and patients must realize," he concluded, "that these products differ and are not generic equivalents."[8]

There was no clear path to producing generic biotech, but not for lack of interest on the part of generic manufacturers. By the turn of the twenty-first century, the generic pharmaceutical industry began to lobby lawmakers and pressure the FDA to establish a clear protocol of equivalence as the patents on the first blockbuster drugs of the biotech boom, such as Amgen's Epogen (epoetin alpha), began to expire. In turn members of the Biotechnology Industry Organization (BIO) lobbied Congress to block these efforts. A representative from Amgen denounced the entire concept of biogenerics as "a direct threat to patient health and safety," pointing to the fact that a competitor's brand of recombinant erythropoetin, Johnson & Johnson's Eprex, had been linked to the severe side effect of pure red cell aplasia in forty patients, compared to only one incident with Amgen's Epogen.[9] Though the problem seems to have been linked to Johnson & Johnson's European packaging (since both versions of epoetin alfa were manufactured in the same plant), advisors to the European Medicine Agency stressed that "the concept of generics cannot be extrapolated to biopharmaceutical products" and recommended that off-patent biopharmaceutical products continue to be evaluated on a case-by-case basis. The following year, the BIO lobby submitted a citizen's petition to the FDA to expressly protest the concept of a protocol for the approval of generic versions of biological drugs.[10]

Like their counterparts in the 1960s and 1970s, lawmakers and regulators in the early twenty-first century grapple once again with economically interested claims of therapeutic difference and forge new sciences of similarity to adjudicate these claims. The FDA began to research protocols to prove biocomparability or biosimilarity when bioequivalence could not be tested. On one level, if a protein-based drug could be considered as a sequence of amino acids, one could develop a protocol of similarity based on protein sequencing, as early FDA comparability protocols suggested for would-be generic versions of the small-peptide hormone glucagon. For larger, more complicated proteins, however, even identical amino-acid sequences do not ensure identical three-dimensional protein folding or glycosylation (a finishing stage in the production of many proteins in which the protein is covered with different carbohydrate moieties). Yet each of these new means of measuring differences between biosimilar drugs simultaneously offers a means of standardizing them. New regulatory protocols of equivalence might, for example, use fluorescence tagging, chromatography, or electrophoresis to prove identical levels of glycosylation between two protein products. And yet, while these new tests promise increased standardization of similarity, none remove the possibility of further differences that have not yet been detected or are not yet detectable.

The prospects for generic biotech are further complicated by the changing structures of intellectual property. If the molecular structure of follow-on biotech drugs is not exactly the same as the innovator molecule, why should competitors even wait for the patent to expire on the original version in the first place? If a competitor product could be similar enough to be exchangeable but not similar enough to be a patent infringement, what would a biotech patent be worth? Partly as a result, biotech firms such as Amgen and Genentech have successfully argued for long-term market exclusivity independent of patents, creating an even more complex set of legal barriers to the copying of large-molecule drugs.[11]

An additional hurdle to the production of biogenerics has been the private nature of the molecular and cellular techniques employed to make the innovator drug equivalent even to itself. Quality assurance in biotech firms is linked, in part, to tight control over the means of pharmaceutical production through engineered biological systems. But, as with earlier considerations of pharmaceutical know-how in the production of bioequivalent drugs, only a limited part of these techniques of production are publicly disclosed in the drug patent or the NDA. The rest lie half hidden in the realm of trade secrets. On the rare occasions that FDA regulators have tried to ease the burden on manufacturers of would-be biosimilar drugs by providing proprietary data on the production of innovator drugs, they have received sharp rebukes from BIO's lawyers and pulled back.[12]

Much as bioequivalence was written into law with Ronald Reagan's signature on the Hatch-Waxman Act of 1984, biosimilarity was signed into law with Barack Obama's pen as the Biologics Price Competition and Innovation Act of 2009 (BPCIA), a component of the broader Affordable Care Act known colloquially as Obamacare. In the hearings leading up to the BPCIA, generic manufacturers repeatedly pointed out that biotech drugs were now dose for dose, pound for pound, the single most expensive part of the US health-care system, and they called for a stable pathway for manufacturing cheaper generic versions. Members of BIO and the Pharmaceutical Research and Manufacturers of America (PhRMA), in turn, argued before Congress that safe, exchangeable biogeneric drugs simply did not exist. Interest groups played a powerful role in this political process: statements read by at least twenty-two Republican and twenty Democratic lawmakers in hearings on the BPCIA included language that had been directly drafted by Genentech.[13]

Although the enactment of the BPCIA provided a pathway to market for copies of large-molecule drugs, it is not at all clear that biosimilars will have the same economic and public health benefits as small-molecule generics. The architects of BPCIA assumed that the same market structures that worked for generic drugs will work for inexpensive protein copies as well. Subsequent estimates of the cost reductions achievable through biosimilars have been disappointing, however, because the structures of risk and reward that were balanced so well for small-molecule generics work in opposite directions for follow-on biologic drugs.[14]

Like the Hatch-Waxman Act of 1984, the BPCIA ostensibly provided a compromise measure to both sides in the biosimilarity debates. For brand-name manufacturers, the bill offered a twelve-year marketing exclusivity pathway—independent of patents—to help assure the biotech industry that its products would be protected from generic competition long enough to satisfactorily recoup investment costs. For generics manufacturers, the bill authorized the FDA to oversee two abbreviated pathways for approving production of legitimate copies of biotech drugs, known as (1) biosimilars and (2) interchangeables. The biosimilar pathway requires less data for FDA approval, but products approved as biosimilar are not considered equivalent to or exchangeable with innovator drugs. Products approved through the interchangeable pathway, in turn, are considered to be substitutable in much the same fashion as bioequivalent generic drugs. Yet, for their drugs to be approved as interchangeable, would-be manufacturers need to submit more extensive (and expensive) clinical trials data to demonstrate that patient response does not vary when switched between the original and the new drug. Economists fear that since the incentive structures of both pathways cross, the field of biogenerics will do little to affect skyrocketing health-care costs.[15]

Ironically, the most likely beneficiaries of these new approval pathways appear to be larger, established firms that have the funding and technical ability to organize comparability and interchangeability trials. As the *Economist* noted in 2010, storied drug houses like Pfizer were poised to become "one-stop shops" for branded biosimilars. In the years since, Pfizer executives have elaborated that their own entries into the field of follow-on biologicals will be similar but *better* than the original compounds—a new take on the me-too drug optimistically called "biobetters."[16] There is no simple opposition between brand and generic here: rather, large manufacturers are finding new ways to *rebrand* pharmaceutical copies at the same time that they work to *unbrand* the patent monopoly of the innovator drug. As with the broader movement toward selling "authorized generics"—that is, generic drugs that are licensed under the original NDA and often authorized by the innovator company to imitate their trademark pill size, shape, and color for a fee—this move may reflect an early twenty-first-century return to the early twentieth-century structures of the ethical pharmaceutical industry mapped at the beginning of this book. Then, as now, the pharmaceutical industry was dominated by large firms, each offering roughly comparable lines of products differentiated primarily by claims of superior quality vouchsafed by trademarks, institutional brands, and service promises of the manufacturers.[17]

Curiously, as the FDA and the European Medicines Agency prepare to regulate a new industry of imitative products, they face challenges in deciding even what these compounds should be named. Contemporary debates over the existence of a single biogeneric name evoke earlier twentieth-century debates about the role of the generic name in designating therapeutic objects as similar or different. As analyst Jessica DeMartino notes, the palatability or unpalatability of a cheaper copy of a biotech drug among its potential consumers will partly be determined by what it is be called—interchangeable, biosimilar, biocomparable—because each name itself suggests a different degree of similarity.

This is already true of many consumer products. As we have seen with the formation of generic drug names in the mid-twentieth century, a name denotes the degree to which an object can be considered the same or different from other similar objects. Calling a French wine a *grand cru*

Margaux designates a high degree of similarity to other highly valued wines from a specific territory in the Bordeaux wine region; calling it a Margaux slightly less so, calling it a Bordeaux even less, although all bottles of Margaux are Bordeaux wines. These expectations are valid, at least in theory, because the *appellation controlée* system of nomenclature for French wines is associated with exacting standards of product specificity at different gradations of quality control.

Could the FDA and the United States Adopted Names Council create a system that allowed physicians and patients to have the same trust in American biotech products that the oenophile has in French wines? De Martino is optimistic. The precise naming and labeling of biosimilars, she argues, would greatly affect the ability of practitioners to prescribe and dispense them because they need to be able to differentiate biosimilars from innovator products. In contrast to those who claim that all biosimilar products should have the same generic name as their originator products, De Martino argues for *similar* names that differentiate the biosimilar from the innovator drug while still suggesting *some* relationship between the two. A prefix of *sim-* or *neo-* added onto the original generic name could suggest the difference between original and copy while retaining the possibility of interchangeability. A biosimilar version of erythropoetin would not be called erythropoietin but *sim*erythropoetin; subsequent versions of interferon-alpha would be *neo*interferon-alpha. This proposition has also been advocated by BIO and PhRMA, who argue that drug nomenclature should properly reflect the differences between biogenerics and the innovator drugs they copy.[18]

Yet as Robert Ulin reminds us in his ethnohistory of French viniculture, wine appellations are as much about branding, histories of consumption, and national debates over what constitutes a quality wine product as they are about designating standards or similarity as such. Whatever nomenclature system emerges for biosimilars, in turn, will reflect the political economy of pharmaceutical regulation as much as it does any underlying similarity or difference between molecular forms.[19]

GENERIC HISTORIES, GENERIC FUTURES: THE SAME BUT NOT THE SAME

We have returned, full circle, to the question posed at the beginning of this book: What's in a name? Indeed every episode we have described in these debates over biosimilar drugs—which continue as this book goes to press—echoes with resonances of the history traced in these chapters. History cannot predict what will happen next or tell us unequivocally what to do today, but it can reorient our understanding of twenty-first-century problems, even in a realm as relentlessly future oriented as modern biomedicine.

Consider the story of biosimilarity from this perspective: recent developments in the innovative structure of the pharmaceutical industry, shifting concerns over intellectual property, and the production of new kinds of me-too drugs extend prior logics of therapeutic similarity in unorthodox ways (chapters 12, 13, and 14). These new forms of similarity produce new market opportunities—and new crises of confidence for physicians, pharmacists, and consumers (chapters 10 and 11). New forms of therapeutic similarity require new regulatory practices of comparability, interchangeability, and exchangeability that are geographically variable and expose discontinuities in the different possible ways of making things the same (chapters 8 and 9). These regulatory deliberations expose the insufficiency of our current sensors—and structures—for detecting significant differences between objects that we would like to call the same, and they demand a novel suite of sciences of similarity (chapters 6 and 7). These sciences of similarity and difference, in turn, alternately destabilize and stabilize markets for these newly similar things (chapters 3, 4, and 5). New debates erupt over what these things are, and our uncertainty over what, exactly, they are is reflected in our continued inability to know even how to name them (chapters 1 and 2).

The past, then, helps to orient us to the present. The stakes and stakeholders are not identical but they are similar. They are the same but not the same. Indeed, "the same but not the same" is a useful concept for understanding not just the problem of generic drugs in history but the relevance of historical perspective in social, medical, public health, and policy spheres more generally. It is not exactly true, as poor George Santayana is endlessly trotted out to state, that those who cannot remember the past are condemned to repeat it. We *are* constantly repeating our pasts, but what returns is never exactly the same as what came before, only similar. The historian's task is to articulate *these* protocols of similarity: to define from our prior experience what is relevant to understanding the returning forces of the present.

In so doing, we must deal with the challenges posed by a biomedical enterprise that moves very rapidly and operates almost exclusively in the future tense. The pill that you take today, you take for the promise of transformations in your body in days, weeks, or years to come. In board-rooms and legislative hearings, the profits that this pill might generate for its manufacturer are measured against the costs of some future un-named innovation yet to come. Patent-protected brand-name drugs of today, expensive as they seem now, are already calculated as the generic bargains they will become at some future date of patent expiry. The many forms of quantification, documentation, and analysis that constitute our understanding of the pharmaceutical market—forecasts for investors, policy briefs, competitive analyses for industry executives—likewise seem to traffic exclusively in future tenses.

In recent years, many of these future tenses have converged into a catastrophic imperative regarding the inevitably generic future of the in-dustry. Take, for example, the expanding genre of "patent cliff" reporting, explored in one of its earliest manifestations in chapter 4 with the antic-ipation of the expiry of the blockbuster Chloromycetin patent in 1966. Patent cliff journalists had a field day in 2011 predicting the impact of the expiry of Pfizer's patent on its blockbuster cholesterol drug Lipitor, the single most lucrative pharmaceutical in the history of the industry—and Lipitor was just the tip of the iceberg. "This year," the *New York Times* re-ported, "the industry will lose control over more than 10 megamedicines whose combined annual sales have neared $50 billion." Kenneth I. Kaitlin, the director for the Center for the Study of Drug Development, agreed that it was "panic time for the industry" and lamented, "I don't think there's a company out there that doesn't realize they don't have enough products in the pipeline or the portfolio, don't have enough revenue to sustain their research and development."[20]

As Kaitlin's comment suggests, the patent cliff was linked to another feature in the perilous landscape of twenty-first-century pharmaceuti-cals: the "dry pipeline." At least a decade's worth of debate and doomsday augury has now revolved around charts and graphs that illustrate the drying of the pharmaceutical pipeline or the sludging of pharmaceutical innovation. The graph (figure 16) accompanying a 2012 study published in *Nature Reviews Drug Discovery* entitled "Diagnosing the Decline in Phar-maceutical R&D Efficiency" is only a snapshot of one of the more recent entries into this new field of industry jeremiads.[21]

The downward slope of the graph is quite visible in the executive boardroom, as are its tidings of a bearish future. The ratio of marketable new drugs per dollar (or more accurately, per billion dollars spent on

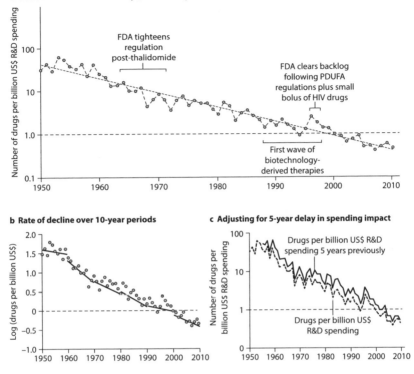

a Overall trend in R&D efficiency (inflation-adjusted)

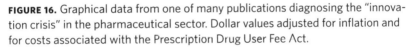

FIGURE 16. Graphical data from one of many publications diagnosing the "innovation crisis" in the pharmaceutical sector. Dollar values adjusted for inflation and for costs associated with the Prescription Drug User Fee Act.

Jack Scannell, Alex Blanckley, Helen Boldon, and Brian Warrington, "Diagnosing the Decline in Pharmaceutical R&D Efficiency," Nature Reviews Drug Discovery 11, no 3 (2012): 191–200, p. 193.

R&D) has been declining at a relatively steady rate for the past sixty years. With just a bit of cheek, the study authors named this constant rate of decline Eroom's Law, since it inverted the well-known model of innovative capacity in technology-based economic sectors known as Moore's Law. Moore's Law predicts that the increase in innovation is a constant: in the tech sector, the number of transistors that can be placed into an integrated circuit for roughly the same price has doubled every two years in the forty years between 1970 and 2010. Eroom's Law, in turn, observes that in the pharmaceutical sector the *decline* in innovation is itself constant: the yield of FDA-approved drugs per billion dollars spent has halved every nine years between 1950 and 2010.[22]

This crisis of innovation has been a bitter pill for PhRMA to swallow and an enduring riddle for those attempting to predict technological and market futures. Eroom's Law is all the more puzzling given the remarkable gains in efficiency that have characterized the pharmaceutical research process. Combinatorial chemistry has increased the number of drug-like compounds to be tested eight-hundred-fold in the two decades of the 1980s and 1990s alone. DNA sequencing, largely impossible in the 1960s, proceeds a billion times faster in 2010 than in 1970. It takes three orders of magnitude fewer hours to calculate a three-dimensional protein structure now than it did using the X-ray crystallographic techniques of the 1960s. Even since the 1990s, high throughput screening has reduced by ten thousand times the cost of testing libraries of compounds for potential therapeutic activity. Entirely new research objects—like transgenic mice—now make possible R&D programs undreamed of in the mid-twentieth century. In spite of these innovations we are developing drugs less efficiently now and will likely develop new drugs even less efficiently after another decade.[23]

Why? According to the framers of Eroom's Law, generic drugs are one of the primary causes for the decreased efficiency in pharmaceutical innovation:

> Imagine how hard it would be to achieve commercial success with new pop songs if any new song had to be better than the Beatles, if the entire Beatles catalogue was available for free, and if people did not get bored with old Beatles records. We suggest something similar applies to the discovery and development of new drugs. Yesterday's blockbuster is today's generic. An ever-improving back catalogue of approved medicines increases the complexity of the development process for new drugs and raises the evidentiary hurdles for approval, adoption, and reimbursement. It deters R&D in some areas, crowds R&D activity into hard-to-treat diseases and reduces the economic value of as-yet undiscovered drugs. The problem is progressive and intractable.[24]

The authors continue to argue that this "better than the Beatles" problem differentiates the pharmaceutical industry from other innovative industries that can rely more readily on planned obsolescence. As older, formerly innovative drugs live out their retirement on permanent display as generic entities, they crowd out the possibilities for the new. The generic is at once a past and future entity, but in both roles it serves as antagonist of innovation.

Not all generic futures are incompatible with innovation. IMS Health, a leading vendor of pharmaceutical market intelligence, has recently published a string of reports with more optimistic titles like *Generic Medicines: Essential Contributors to the Long-Term Health of Society* and *Brand Renewal: Maximizing Lifecycle Value in an Ever More Generic World*. These reports credit the generic sector as a driver of innovations in pharmaceutical delivery: "The development expertise within the generic medicines industry is recognized as being innovative. Indeed, it is the level of skill in chemistry and process development that has driven the successful introduction of so many generic medicines. Furthermore, there is an understanding within the industry of patient and pharmacist needs in relation to patient-oriented packaging, arising from the production of a wide range of products across many countries."[25] There are, then, at least two generic futures, one cautionary, one promissory. Generic medicines are either the death of biomedical innovation or its fullest translation into a tool for global public health.[26]

Like the crisis of similarity, the crisis of the generic future is not new. Alternating promissory and cautionary visions of a generic future have recurred throughout this book and have been eliciting hope and fear among policymakers, practitioners, and consumers for quite some time. In the expansive rhetoric of Senator Estes Kefauver and his allied therapeutic reformers in 1960, the generic future was promissory: a pledge to liberate the "captive consumer" of pharmaceuticals and produce a more egalitarian free market of therapeutics. In the more paranoid rhetoric of the Pharmaceutical Manufacturers Association, the generic future was a terminal condition for the innovative sector of the American free market economy. One speaker at the PMA's 1960 meetings asked the audience of executives gathered at the Boca Raton Hotel to travel in time twenty years forward into an imagined generic dystopia:

> Today—April 5, 1980—The citizen desiring medical attention merely dials the nearest government health center which arranges for his transportation to the nearest government diagnostic center . . . The electronic brains, which have been government programmed, will prescribe the "correct" drug. There is no problem of competing brands

to be concerned with because everything today is the "United States" brand, which, in the field of pharmaceuticals, is readily identified by the letter "K."[27]

K, one can assume, stands for Kefauver (or perhaps for Kafka?). This generic future features the loss of individual choice, market incentives, innovation, and creativity—all replaced by the grey tentacular embrace of the State.

Neither Kefauver's nor the PMA's visions of the generic future have fully come to pass. As early as 1960, however, their hopes and fears of a generic future animated key debates on contemporary decisions in American health policy and practice. We have, in other words, clearly been living in the generic future for at least half a century. Regardless of whether our visions of that future at this point feature blue skies or dark clouds, they continue to reflect and repeat prior futures imagined in the past.

These similarities between past and present visions of the generic future reveal some of the hidden political, economic, and moral stakes that are folded into seemingly technical debates about the commensurability of generic drugs. It is equally important, however, to attend to the differences between past and present anxieties about generic drugs. The significant differences between bioequivalence and biosimilarity, between the small Zeniths and Bolars of the early generic industry and the generic giants of the early twenty-first century and between the local market of early generics and global generic markets of today, should remind us not to assume that policy solutions that worked in the past (such as the framing of the Hatch-Waxman Act of 1984) will produce similar results in the present (such as the Biologics Price Competition and Innovation Act of 2009).

Markets for generic drugs are now global in reach, and the manufacturers of generic drugs are now structurally and functionally difficult to distinguish from the multinational giants they were once so dramatically compared against. And yet, as this book has documented, generic drugs have local histories. They have been and continue to be a more important solution to problems of access and cost in America than in many other parts of the world. It is precisely because the public sector has so little leverage in negotiating the prices of brand-name drugs in the United States that generic drugs—a private sector solution—became an operative mechanism for health-care cost containment. As the recent history of generic pay-for-delay settlements, limited liability for manufacturers of generic drugs, and increasing generic drug shortages has illustrated, that private sector solution has come with costs as well as benefits.[28]

The history of the generic drug challenges us to take a closer look at everyday categories we hold to be self-evident. Dichotomies that we treat as commonsensical—innovation versus imitation, small business versus multinational corporation, public health versus private markets—begin to untangle and fray. By historicizing how the generic drug became a commonplace object, we have discovered a site of intense personal, professional, economic, and political interest. The changing fortunes of generic forms in modern medicine reveal just how much is at stake in making, exchanging, and consuming these vital objects that are the same but not the same.

ABBREVIATIONS

MANUSCRIPT COLLECTIONS

AEP   Alfred Engelberg Papers (privately held), New York, NY

AJK   Alan J. Klawans Collection, 2005.144, Chemical Heritage Foundation, Philadelphia, PA

AMAA   Archives of the American Medical Association, Chicago, IL

APhAA   Archives of the American Pharmaceutical Association, Washington, DC

CEP   Charles Edwards Papers, MSS 447, University of California at San Diego, La Jolla, CA

DES   Drug Efficacy Study of the National Research Council's Division of Medical Sciences, 1966-1969, NAS 234 B-3-1 & 2, National Academy of Sciences, Washington, DC

EPP   Esther Peterson Papers, MC 450, Schlesinger Library, Radcliffe Institute for Advanced Study, Cambridge, MA

FDAAF   AF Correspondence Files of the Food and Drug Administration, White Oak, MD, accessed by Freedom of Information Act

FDAR   Records of the Food and Drug Administration, RG 88, National Archives and Records Administration, College Park, MD

FMIP   Felix Marti-Ibanez Papers, MS 1225, Yale University, New Haven, CT

GUC   Glenn E. Ullyot Collection, 2006.502.001, Chemical Heritage Foundation, Philadelphia, PA

HFDP   Harry Filmore Dowling Papers, MS C 372, National Library of Medicine, Bethesda, MD

JAP   John Adriani Papers, MS C 453, National Library of Medicine, Bethesda, MD

KRF   Kremers Reference Files, American Institute for the History of Pharmacy, Madison, WI

LGP   Louis Goodman Papers, ACCN 937, University of Utah, Salt Lake City, UT

LLP   Louis C. Lasagna Papers, D 302, University of Rochester, Rochester, NY

NYSGDIF New York State Generic Drug Investigations Files, L0136-80, New York State Archives, Albany, NY

PDC   Parke, Davis Collections, AC001, National Museum of American History, Washington, DC

PLP   Philip Lee Papers, MSS 91-1, University of California at San Francisco, San Francisco, CA

RBP   Richard Burack Papers (privately held), Jackson, NH

RFP   Robert Fischelis Papers, MSS 619, Wisconsin Historical Society, Madison, WI

SDCA   Sterling Drug Company Archives, AC 772, National Museum of American History, Washington, DC

SOAM   Records of the Subcommittee on Antitrust and Monopoly, Committee on
        the Judiciary, United States Senate, RG 46, National Archives, Washington,
        DC
SOM   Records of the Subcommittee on Monopoly, Select Committee on Small
        Business, RG 46, United States Senate, National Archives, Washington, DC
USPC   United States Pharmacopoeia Convention Archives, MSS 149, 2007 Addi-
        tions: 1888–2000, Wisconsin Historical Society, Madison, WI
WHOA   World Health Organization Archives, Geneva, CH
WPIB   Wisconsin Pharmacy Investigation Board, MSS 2638, Wisconsin Historical
        Society, Madison, WI
WSMP   William S. Middleton Papers, MS C 206, National Library of Medicine,
        Bethesda, MD

CONGRESSIONAL HEARINGS

Administered Prices   *Administered Prices in the Drug Industry, Hearings Before the
                          Subcommittee on Antitrust and Monopoly of the Select Com-
                          mittee on the Judiciary*, United States Senate
Competitive Problems   *Competitive Problems in the Drug Industry, Hearings Before
                          the Subcommittee on Monopoly of the Select Committee on
                          Small Business*, United States Senate
DIAA   *Drug Industry Antitrust Act, Hearings Before the Subcommittee
                          on Antitrust and Monopoly of the Select Committee on the
                          Judiciary*, United States Senate
FGDAP   *FDA's Generic Drug Approval Process, Hearings Before the
                          Subcommittee on Oversight and Investigation of the Com-
                          mittee on Energy and Commerce*, United States House of
                          Representatives
PDLPA   *Prescription Drug Labeling and Price Advertising, Hearings
                          Before the Subcommittee on Consumer Protection and Finance
                          of the Committee on Interstate and Foreign Commerce*, United
                          States House of Representatives
SPDA   *Substitute Prescription Drug Act, Hearings Before the Subcom-
                          mittee on Consumer Protection and Finance of the Committee
                          on Interstate and Foreign Commerce*, United States House of
                          Representatives

COMMONLY CITED PERIODICALS

FDCR   F-D-C Reports *"Pink Sheet"*
JAMA   *Journal of the American Medical Association*
JAPhA   *Journal of the American Pharmaceutical Association*
NEJM   *New England Journal of Medicine*
NYT   *New York Times*
WSJ   *Wall Street Journal*

NOTES

INTRODUCTION. THE SAME BUT NOT THE SAME

Epigraph. Nelson M. Gampfer, "The case for brand name specification," C 46 (p) I f 3, KRF, p. 1.

1. IMS Health, "IMS Health Reports U.S. Prescription Sales Grew 5.1 Percent in 2009, to $300.3 Billion," 1 April 2010 (press release), www.imshealth.com/portal/site /imshealth/menuitem.a46c6d4df3db4b3d88f611019418c22a/?vgnextoid=d690a27e9 d5b7210VgnVCM100000ed152ca2RCRD&cpsextcurrchannel=1.

2. IMS Health, *Exploiting Protection Expiry: Optimizing Off-Patent Opportunities in an Ever More Generic World*, accessed 24 June 2010, at http://amcp.org/WorkArea /DownloadAsset.aspx?id=11647.

3. Natasha Singer, "That Pill You Took? It May Well Be Teva's," *NYT*, 9 May 2010.

4. William F. Haddad to Don Hewitt, 18 April 1977, box 7, f 24 NYSGDIF, p. 3.

5. "New Drug Law: 'Safe and Effective Drugs at the Lowest Possible Cost,'" *Generics Magazine*, January 1985, 57–59.

6. The most detailed account of the generic drugs scandal of the late 1980s–early 1990s remains William C. Cray and C. Joseph Stetler's *Patients in Peril? The Stunning Generic Drug Scandal* (Washington, DC: PMA, 1991).

7. *FDCR*, 13 February 1989, p. T&G2; *FDCR*, 1 May 1989, T&G1; *FDCR*, 15 May 1989, pp. 10–12.

8. Dingell, as quoted in Cray and Stetler, *Patients in Peril?*, p. 41.

9. Vitarine admitted that it had "placed the active ingredients from the SK&F product into Vitarine gelatin capsules for bioequivalence studies in support of an ANDA." *FDCR*, 3 July 1989, pp. 13–15.

10. Testimony of Marvin Seife, *FGDAP*, vol. 2, pp. 2–43; Cray and Stetler, *Patients in Peril?*, p. 59.

11. In a "smoking gun" letter to Bolar, PharmaKinetics wrote that their "technical staff here has concluded that the tablets you sent for the bioequivalence study are not of your production," adding that "PharmaKinetics is not the keeper of morality in the pharmaceutical industry, but rather a small company which really needs your business." *FDCR*, 18 September 1989, p. 11; *FDCR*, 4 September 1989, p. 7.

12. Although an exhaustive review by the FDA found that most generic drugs on the market had been approved based on valid data, for Dingell the FDA's study showed too little, too late. The representative spent the next few years working to pass the Generic Drug Enforcement Act to allow more speedy prosecution of fraudulent generic manufacturers. *FDCR*, 8 January 1990, pp. T&G6-7; *FDCR*, 11 June 1990, 30 July 1990; Cray and Stetler, *Patients in Peril?*, p. 104; *FDCR*, 9 October 1989 [1990], p. 8; Tamar Nordenberg, "Inside FDA: Barring People from the Drug Industry," *FDA Consumer*, March 1997, n.p.

13. *FDCR*, 29 October 1990, p. 22.

14. 15 November 1990, pp. 3–6; *FDCR*, 28 December 1991.

15. Unger, "A Lunch That Cost Him Dearly," undated news clipping in KRF C 46(p) I f 6 batch 3; *FDCR*, 23 March 1991, p. TG1; Phil McCombs, "The Bungled Punishment of Prisoner Seife," *Washington Post*, 3 April 1992, A1.

16. McCombs, "Bungled Punishment of Prisoner Seife."

17. Betty J. Dong, "The Nonequivalence of Thyroid Products," *Drug Intelligence and Clinical Pharmacy* 20, no. 1 (1986): 77; W. A. Kehoe, B. J. Dong, F. S. Greenspan, "Maintenance Requirements of L-thyroxine in the Treatment of Hypothyroidism," *Western Journal of Medicine* 140, no. 6 (1984): 907–9; B. J. Dong, V. R. Young, B. Rapoport, "The Nonequivalence of Thyroid Products," *Drug Intelligence & Clinical Pharmacy* 20, no. 1 (January 1986): 77–78.

18. R. T. King Jr., "Bitter Pill: How a Drug Firm Paid for University Study, Then Undermined It," *WSJ*, 12 April 1996.

19. Ibid. p. 4.

20. Gilbert Mayor, T. Orlando, and N. M. Kurtz, "Limitations of Levothyroxine Bioequivalence Evaluation: Analysis of an Attempted Study," *American Journal of Therapeutics* 2 (1995): 417–32.

21. Drummond Rennie, "Thyroid Storm," *JAMA* 277, no. 15 (1997): 1238. Lawrence K. Altman, "Drug Firm, Relenting, Allows Unflattering Study to Appear," *NYT*, 16 April 1997, A1; B. J. Dong, W. W. Hauck, J. G. Gambertoglio, L. Gee, J. R. White, J. L. Bubp, F. S. Greenspan, "Bioequivalence of Generic and Brand-Name Levothyroxine Products in the Treatment of Hypothyroidism," *JAMA*, 277, no. 15 (1997): 1205–13; Sheryl Gay Stolberg, "Gifts to Science Researchers Have Strings, Study Finds," *NYT*, 1 April 1997, p. A17.

22. Dorothy S. Zinberg, "A Cautionary Tale," *Science* 273 (July 26, 1996): 411; Lawrence K. Altman, "Experts See Bias in Drug Data," *NYT*, 29 April 1997, p. C1; Carey Goldberg, "Urging a Freer Flow of Scientific Ideas," *NYT*, 6 April 1999, p. F3; Sheldon Rampton and John Stauber, "Research Funding, Conflicts of Interest, and the 'Meta-Methodology' of Public Relations," *Association of Schools of Public Health* 117, no. 4 (2002): 331–39.

23. $800 million figure estimated from Public Citizen, as quoted in Altman, "Drug Firm, Relenting, Allows Unflattering Study to Appear." See also "Lawsuit Says Companies Suppressed Drug Study," *NYT*, 19 May 1997, p. B7.

24. On hero/villain narratives in pharmaceutical studies, see Anne Pollock, "Transforming the Critique of Big Pharma," *BioSocieties* 6 (2011): 106–18.

25. A number of scholars have begun to attend to the problem of similarity and exchangeability in science, clinical medicine, and health policy. Anthropologists Cori Hayden, Stefan Ecks, and Kaushik Sunder Rajan have developed ethnographic projects examining generic pharmaceuticals in Mexico, Argentina, Brazil, and India, respectively; see Hayden, "A Generic Solution? Pharmaceuticals and the Politics of the Similar in Mexico," *Current Anthropology* 48, no. 4 (2007): 475–95; Hayden, "No Patent, No Generic: Pharmaceutical Access and the Politics of the Copy," in *Making and Unmaking Intellectual Property: Creative Production in Legal and Cultural Perspective*, ed. Mario Biagioli, Peter Jaszi, and Martha Woodmansee (Chicago: University of Chicago Press, 2011), pp. 285–304; Stefan Ecks and Soumita Basu, "The Unlicensed Lives of Antidepressants in India: Generic Drugs, Unqualified Practitioners, and

Floating Prescriptions," *Transcultural Psychiatry* 46, no. 1, 86–106; Kaushik Sunder Rajan, "Pharmaceutical Crises and Questions of Value: Terrains and Logics of Global Therapeutic Politics," *South Atlantic Quarterly* 111, no. 2 (2012): 321–46. Historian Dominique Tobbell (both alone and in collaboration with political scientist Daniel Carpenter) has explored the political, interprofessional, and regulatory skirmishes over generic drugs in the United States; see Tobbell, " 'Eroding the Physician's Control Over Therapy': The Post-War Politics of the Prescription," in *Prescribed: Writing, Filling, Using, and Abusing the Prescription in Modern America*, ed. Elizabeth Watkins and Jeremy Greene (Baltimore: Johns Hopkins University Press, 2012), pp. 66–90; Tobbell, *Pills, Power, and Policy: The Struggle for Drug Reform in Cold War America and its Consequences*, Milbank Series on Health and the Public (Berkeley: University of California Press, 2012); Dominique Tobbell with Daniel P. Carpenter, "Bioequivalence: The Regulatory Career of a Pharmaceutical Concept," *Bulletin of the History of Medicine* 8, no. 1 (2011): 93–131; Anne Pollock likewise devotes a chapter to the complex semiotics of generic thiazide use in her recent book, *Medicating Race: Heart Disease and Durable Preoccupations with Difference* (Durham, NC: Duke University Press, 2012).

26. Referring to the chronology of patent life as part of the life cycle of a prescription drug was already an actor's term in the 1960s; see William E. Cox Jr., "Product Life Cycles as Marketing Models," *Journal of Business* 40 (1967): 376–81. Jonathan Liebenau, *Medical Science and Medical Industry: The Formation of the American Pharmaceutical Industry.* (Baltimore: Johns Hopkins University Press, 1987); Jean-Paul Gaudilliere, "How Pharmaceuticals Became Patentable: The Production and Appropriation of Drugs in the Twentieth Century," *History and Technology* 24, no. 2 (2008): 99–106; Joseph Gabriel, *Medical Monopoly: Intellectual Property Rights and the Origins of the Modern Pharmaceutical Industry.* (Chicago: University of Chicago Press, 2014).

27. On the opposition of ontological and physiological notions of disease, see Owsei Temkin, "The Scientific Approach to Disease: Specific Entity and Individual Illness," in *Scientific Change: Historical Studies in the Intellectual, Social and Technical Conditions for Scientific Discovery and Technical Invention from Antiquity to the Present*, ed. A. C. Crombie, 629–47 (New York: Basic Books, 1963); for a later adaptation, see Robert Aronowitz, *Making Sense of Illness: Science, Society, and Disease* (Cambridge: Cambridge University Press, 1997); for more on the technological mobilization of biomedical objects, see Hannah Lanndecker, *Culturing Life: How Cells Became Technologies* (Cambridge: Harvard University Press, 2007).

28. Nicholas Jewson, "The Disappearance of the Sick Man from Medical Cosmology," *Sociology* 10, no. 2 (1976): 225–44; Arthur Kleinman, "What Is Specific to Biomedicine?," in *Writing at the Margin: Discourse between Anthropology and Medicine* (Berkeley: University of California Press, 1997), pp. 21–40; Byron Good, "How Medicine Constructs Its Objects," in *Medicine, Rationality, and Experience* (Cambridge: Cambridge University Press, 1994), pp. 65–87; Margaret Lock, *Encounters with Aging: Mythologies of Menopause in Japan and North America* (Berkeley: University of California Press, 1997).

29. Jeff Baker, *The Machine in the Nursery: Incubator Technology and the Origins of Newborn Intensive Care* (Baltimore: Johns Hopkins University Press, 1996); George Weisz, *Divide and Conquer: A Comparative History of Medical Specialization* (Oxford: Oxford University Press, 2005).

30. Julie Livingston, *Improvising Medicine: An African Oncology Ward in an Emerging Cancer Epidemic* (Durham, NC: Duke University Press, 2012); Annemarie Mol, *The Body Multiple: Ontology in Medical Practice* (Durham, NC: Duke University Press, 2003); Margaret Lock and Vinh-Kim Nguyen, *An Anthropology of Biomedicine* (New York: Wiley, 2010).

31. The subject of similarity and difference of alchemical gold is treated in William Newman, "Technology and Alchemical Debate in the Late Middle Ages," *Isis* 80, no. 3 (1989): 423–45; Newman, *Promethean Ambitions: Alchemy and the Quest to Perfect Nature* (Chicago: University of Chicago Press, 2004); Tara Nummedal, "On the Utility of Alchemical Fraud," in *New Perspectives on Alchemy*, ed. Lawrence Principe (Canton, MA: Science History, 2007); Lawrence Principe, *The Secrets of Alchemy* (Chicago: University of Chicago Press, 2012). On tables of similarity and difference in Paracelsian and early modern chemistry before Lavoisier, see Owen Hannaway, *The Chemists and the Word: The Didactic Origins of Chemistry* (Baltimore: Johns Hopkins University Press, 1975); Mi Gyung Kim, *Affinity, That Elusive Dream: A Genealogy of the Chemical Revolution* (Cambridge, MA: MIT Press, 2010).

32. Maurice P. Crosland, *Historical Studies in the Language of Chemistry* (Cambridge, MA: Harvard University Press, 1962); Ursula Klein, *Experiments, Models, Paper Tools: Cultures of Organic Chemistry in the Nineteenth Century* (Palo Alto, CA: Stanford University Press, 2002); Allan J. Rocke, *Image and Reality: Representation, Science, and the Scientific Imagination*, Synthesis (Chicago: University of Chicago Press, 2010).

33. Roald Hoffman develops this argument in *The Same and Not the Same* (New York: Columbia University Press, 1997); see also John Parascandola, "The Evolution of Stereochemical Concepts in Pharmacology," in *Van't Hoff-LeBel Centennial*, ed. O. Bertrand Ramsay (Washington, DC: American Chemical Society, 1975), 143–58; Arthur Cushny, Optical Isomerism and the Mechanism of Drug Action, *Journal of the History of Biology* 8, no. 2 (1975): 145–65.

34. W. F. Kean and C. J. L. Lock, "Chirality in Antirheumatic Drugs," *Lancet* 338, nos. 8782–83 (1991):1565–68; F. M. Khorfan, B. T. Ameredes, and W. J. Calhoun, "Levalbuterol versus Albuterol," *Current Allergy and Asthma Reports* 9, no. 5 (2009): 401–9; "Effects of Nebulized Bronchodilator Therapy on Heart Rate and Arrhythmias in Critically Ill Adult Patients," *Chest* 140 (December 2011): 1466.

35. Ken Alder, "Making Things the Same: Representation, Tolerance and the End of the Ancien Régime in France," *Social Studies of Science* 28, no. 4 (1998) 499–545; Merritt Roe Smith, *Harpers Ferry Armory and the New Technology: The Challenge of Change* (Ithaca, NY: Cornell University Press, 1977); David Hounshell, *From the American System to Mass Production, 1800–1932* (Baltimore: Johns Hopkins University Press, 1984).

36. Martin Zeiger, "The Generic Drug Industry: An Overview," *Vital Speeches of the Day* 54, no. 5 (15 December 1987): 142–47.

37. Examples of different works in recent history of medicine that situate disease as a zone of increasing commodification include Keith Wailoo, *Drawing Blood: Technology and Disease Identity in Twentieth Century America* (Baltimore: Johns Hopkins University Press, 1999); Charles Rosenberg, "What Is Disease? In Memory of Owsei Temkin," *Bulletin of the History of Medicine* 77 (2003): 491–505; Nicholas King, "Infectious Disease in a World of Goods" (PhD diss., Harvard University, 2001); Gregg

Mitman, *Breathing Space: How Allergies Shape Our Lives and Landscapes* (New Haven, CT: Yale University Press, 2008). In recent years, more popularized accounts have extended the analysis, e.g., Ray Moynihan, *Selling Sickness: How the World's Biggest Pharmaceutical Companies are Turning Us All into Patients* (New York: Nation Books, 2006). On the more general commodification of health, see Nancy Scheper-Hughes and Loic Vacquant, eds., *Commodifying Bodies* (London: Sage, 2002); for recent reconceptions of commodification see Martha M. Erdman and Joan C. Williams, *Rethinking Commodification: Cases and Readings in Law and Culture* (New York: New York University Press, 2005). On the critique of commodification in tandem with neo-liberalism, see Melinda Cooper, *Life as Surplus: Biotechnology and Capitalism in the Neoliberal Era* (Seattle: University of Washington Press, 2008), especially chapter 2.

38. After Arjun Appadurai, *The Social Life of Things: Commodities in Cultural Perspective* (Cambridge: Cambridge University Press, 1988).

CHAPTER 1. ORDERING THE WORLD OF CURES

Epigraph. T. S. Eliot, "The Naming of Cats," in *Old Possum's Book of Practical Cats* (London: Faber & Faber, 1939), pp. 11–12.

1. As Joseph Gabriel describes in *Medical Monopoly: Intellectual Property Rights and the Origins of the Modern Pharmaceutical Industry* (Chicago: University of Chicago Press, 2014) the apparent simplicity of these earlier names belies a taxonomical debate that took place during early nineteenth century over the multiple substances derived from botanical sources such as cinchona bark and the opium poppy.

2. This compound was also referred to by chemists as diacetylmorphine. The circulation of multiple names for this product suggests the extent to which even chemists themselves resisted systemic chemical names for many complex substances until after the Liege conference of 1930. See Evan Hepler-Smith, "Standards Bound to Disappoint: A Rational Chemical Nomenclature Defeated" (paper presented at the History of Science Society, Cleveland, OH, November 2011).

3. Both Adrenalin and Aspirin became the subject of complex interplay of common and trade names. Parke-Davis's Joichi Takamine proposed the name Adrenalin in part because he was unsure whether the substance he had isolated was the same as the epinephrine described by Johns Hopkins pharmacologist J. J. Abel, while Bayer's trademarking of the name Aspirin was not proposed until the company realized the prospect of competition after the patent expired on the drug (after the company lost control of the Phenacetin brand name in 1906). See Gabriel, *Medical Monopoly*; Jan R. McTavish, *Aspirin Wars: Money, Medicine, and 100 Years of Rampant Competition* (New York: Alfred A. Knopf, 1991).

4. For more on fears of trademark dilution or genericide, see Susan Strasser, *Satisfaction Guaranteed: The Making of the American Mass Market* (Washington, DC: Smithsonian, 1996); Jacqueline Stern, "Genericide: Cancellation of a Registered Trademark," *Fordham Law Review* 51 (1982): 666.

5. On the generic name in the 1920s, see Joseph Gabriel, "Trademarks and Generic Names in the 1920s," in *The Corporate Logic: Intellectual Property Rights and the Twentieth-Century Pharmaceutical Industry* (forthcoming); on sulfamethazine,

see R. Hazard to Expert Committee on the Unification of Pharmacopoeias, "Note concerning the Common Designation of Medicaments," 4 October 1948, WHO.IC/Pharm/35 WHOA, pp. 2–3.

6. The anthropological and sociological literature on the social forces inherent in all taxonomic structures is extensive. See, for example, Emile Durkheim and Marcel Mauss, *Primitive Classification*, trans. Rodney Needham (Chicago: University of Chicago Press, 1963 [1903]); Mary Douglas, *Purity and Danger* (New York: Routledge & Keagan Paul, 1966); Claude Lévi-Strauss, *The Raw and the Cooked* (New York: Harper & Row, 1969); Geoffrey Bowker and Susan Leigh Star, *Sorting Things Out: Classification and Its Consequences* (Cambridge, MA: MIT Press, 1999); Gilles Deleuze, *Repetition and Difference* (New York, Columbia University Press, 1999). On the relation of generics and taxonomic systems in the philosophy of language, see Francis Jeffry Pelletier, ed., *Kings, Things, and Stuff: Mass Terms and Generics* (New York: Oxford University Press, 2009).

7. On the dynamic role of the naming and shaping of objects of biomedical inquiry, see Ian Hacking, *Historical Ontology* (Cambridge, MA: Harvard University Press, 2004). The history of pharmaceutical nomenclature reform in the late twentieth century in some ways recapitulates the nomenclature reform of inorganic chemistry in the eighteenth century or of organic chemistry in the nineteenth and early twentieth centuries; see Maurice P. Crosland, *Historical Studies in the Language of Chemistry* (Cambridge, MA: Harvard University Press, 1962); Ursula Klein, *Experiments, Models, Paper Tools: Cultures of Organic Chemistry in the Nineteenth Century* (Palo Alto, CA: Stanford University Press, 2002); Allan J. Rocke, *Image and Reality: Representation, Science, and the Scientific Imagination*, Synthesis (Chicago: University of Chicago Press, 2010); Hepler Smith (forthcoming).

8. R. Hazard to Expert Committee on the Unification of Pharmacopoeias, pp. 2–3, 4.

9. Amy Sayward, *The Birth of Development: How the World Bank, Food and Agriculture Organization, and World Health Organization Changed the World* (Kent, OH: Kent State University Press, 2006). Javed Siddiqi, *World Health and World Politics: The World Health Organization and the UN System* (London: Hurst, 1995); John Farley, *Brock Chisholm, the World Health Organization, and the Cold War* (Vancouver: University of British Columbia Press, 2008).

10. Robert P. Fischelis to Leroy E. Burney, 1 October 1957, box 63, f 6, "World Health Organization 1945–1959," RFP. On the more general bureaucratization of diagnosis, see Charles E. Rosenberg, "The Tyranny of Diagnosis: Specific Entities and Individual Experience," *Milbank Quarterly* 80, no. 2 (2002): 237–61.

11. Jorge Luis Borges, "The Analytical Language of John Wilkins," in *Other Inquisitions*, trans. Ruth L. C. Simms, 1966 (Austin: University of Texas Press, 1964); John Wilkins, *An Essay towards the Real Character and the Philosophical Language* (London: John Martin, 1668).

12. Expert Committee on the Unification of Pharmacopoeias, *Report on the First Session Held at the Palais des Nations*, Geneva, 13–17 October 1947, WHO.IC/Pharm./2, WHOA; R. Hazard and L. Volckringer to the Expert Committee on the Unification of Pharmacopoeias, "Note Concerning the Latin Designations of Medicaments," 28 September 1948, WHO.IC/Pharm/33, quotation, p. 1, WHOA. See also Lembit Rago

and Sabine Kopp, "The International Pharmacopoeia in the Changing Environment," *Pharmaceuticals Policy and Law* 9 (2007): 357–68; E. Fullerton Cook, "Discussion of Synonyms in the International Pharmacopoeia," WHO/Pharm./39, WHOA, pp. 1, 4.

13. P. Hamill, "Names of New Drugs," *Practitioner* 156 (1946): 61–64; List of Synonyms of Drugs Included in Volume I of the First Edition of the International Pharmacopoeia, 10 April 1951, WHO/Pharm/113 Rev.1 WHOA. For more on the contested history of the Aspirin trademark, see Jan R. McTavish, *Aspirin Wars: Money, Medicine, and 100 Years of Rampant Competition* (New York: Alfred A. Knopf, 1991).

14. D. M. Dunlop and T. C. Denston, "The History and Development of the British Pharmacopoeia," *British Medical Journal* 2 (22 November 1958): 1250–62, quotation p. 1255. On the shifting geography of the languages of science, see Michael Gordin, *Scientific Babel* (Chicago: Chicago University Press, forthcoming 2015).

15. At the first meeting of the expert committee, Professor E. Fullerton Cook, of the USP, sent copies of the USP stylebook to all committee members. With the exception of the inclusion of maximal doses—a nonnegotiable point Hazard insisted must be derived from the French *Codex*—the initial framing of the *Pharmacopoeia Internationalis* was largely modeled on the *United States Pharmacopoeia*. Expert Committee on the Unification of Pharmacopoeias, *Report on the First Session Held at the Palais des Nations*, p. 4.

16. United States Pharmacopoeial Convention, *The Pharmacopoeia of the United States of America*, 1st ed. (facsimile) (Madison, WI: American Institute of the History of Pharmacy, 2005), p. iii. As Michael Flannery reminds us, in the Civil War there was one set of compounds that formed the pharmacopoeia for the Union army and another, entirely different, materia medica list created for the Confederate army. Michael Flannery, *Civil War Pharmacy: A History of Drugs, Drug Supply and Provision, and Therapeutics for the Union and Confederacy* (Cleveland, OH: CRC Press, 2004). And when the *British Pharmacopoeia* was first published in 1864, it already represented an attempt to harmonize the much older conflicting *Pharmacopoeia Londoniensis*, the *Edinburgh Pharmacopoeia*, and the *Dublin Pharmacopoeia*; D. M. Dunlop and T. C. Denston, "The History and Development of the British Pharmacopoeia," *British Medical Journal* 2 (22 November 1958): 1250–52. See also Lembit Rägo and Budiano Santoso, "Drug Regulation: History, Present, and Future," in *Drug Benefits and Risks: International Textbook of Clinical Pharmacology*, ed. C. J. van Boxtel, rev. 2nd ed. (Uppsala: IOS Press, 2008). For the nineteenth-century origins of the US generic drug name, see Gabriel, *Medical Monopoly*.

17. This myth of ethical marketing concealed, however, the fact that many ethical pharmaceutical firms also sold proprietary products and increasingly began to sell ethical products under proprietary names. Tom Mahoney, *The Merchants of Life: An Account of the American Pharmaceutical Industry* (New York: Harper & Brothers, 1959). Harry M. Marks, *Progress of Experiment: Science and Therapeutic Reform in the United States, 1900–1990* (Cambridge: Cambridge University Press, 1997).

18. Council on Pharmacy and Chemistry of the AMA, *New and Non-official Remedies* (Chicago: American Medical Association, 1953), p. xviii. The council officially began acting as a sounding board for generic names in 1951, vetting questions by drug manufacturers about the suitability of potential common terms used to describe new substances. Within a few years, all drugs in *New and Non-official Remedies*

were listed by nonproprietary names in the title and text of the monograph. Memorandum from H. D. Kautz to W. Wolman, "Comparative Display of Proprietary with Non-proprietary (Generic) Names," 16 December 1959, COD 3324, box 15-13, 1959, vol. 6, AMAA, p. 1.

19. Though the USP remained a private body, the quasi-official status it assumed after recognition by the 1906 Pure Food and Drug Act did nonetheless render it subject to public debate. Lee Anderson and Greg Higby, *The Spirit of Voluntarism: A Legacy of Commitment and Contribution* (Washington, DC: United States Pharmacopoeial Convention, 1995).

20. *USP Board Committee on General Principles of International Cooperation*, February 13, 1953, box 209, f 1, USPC, p. 1; Glenn Sonedecker, "Contributions of the Pharmaceutical Profession toward Controlling the Quality of Drugs in the Nineteenth Century," *Safeguarding the Public: Historical Aspects of Medicinal Drug Control*, ed. John B. Blake (Baltimore: Johns Hopkins University Press, 1970), 97–111. From the first meeting in 1947, the WHO expert committee had been concerned that any project of naming drugs would quickly get bogged down in the patchwork of international intellectual property law. The WHO counsel had hoped that by ruling only on *nonproprietary* drug names—and avoiding the third rail of international regulation of brand names—the WHO would both avoid problems of intellectual property and critiques of overstepping its role as a scientifically based international advisory body. Expert Committee on the Unification of Pharmacopoeias, *Report on the First Session Held at the Palais des Nations*, p. 5.

21. *USP Board Committee on General Principles of International Cooperation*, p. 4.

22. Miller was dismissive of all Latin American pharmacopoeial attempts, national or transnational: "this movement, however ill or well conceived, was practically doomed to endless delay if for no other reason than the fact that there is so little talent for drafting a really workable pharmacopoeia among the pharmacists of our neighbors to the south." *USP Board Committee on General Principles of International Cooperation*, pp. 4–6, 9.

23. Ibid., p. 6. Though French had long been the language of key medical textbooks throughout much of Latin America and Asia, American texts were steadily replacing French texts by the early 1950s. See also Lloyd Miller, "U.S.P. Policy on Non-Proprietary Name," 20 February 1953, box 209, f 1, USPC, pp. 15–16.

24. Miller's cooperation with industrial interests should not surprise us, as the major pharmaceutical houses represented a significant part of the United States Pharmacopoeial Convention. The USP's Committee on General Principles of International Cooperation included Carson P. Frailey (executive VP of the American Drug Manufacturers Association), Chester S. Keefer (who served as FDR's "penicillin czar" during the Second World War and the dean of the Boston University School of Medicine thereafter), and Theodore G. Klumpp (president of Winthrop/Sterling Pharmaceutical Company and the president of the proindustry National Pharmaceutical Council), and was chaired by Austin Smith, at the time the editor of *JAMA* and a member of the AMA's Council of Pharmacy and Chemistry (later to become the first full-time director of the new pharmaceutical lobbying group, Pharmaceutical Manufacturers Association (PMA), and then chair of the board at Parke, Davis Pharmaceutical Company). "Draft Report to the Board of Trustees of the U.S.P., U.S.P.

Board Committee on General Principles of International Cooperation," 2 March 1953, box 209, f 1, USPC, p. 18.

25. APMA, press release, 17 June 1953, C 34 (d), KRF; John Horan, "Report of Action Taken by Sixth World Health Assembly in Regard to Non-proprietary Name Program," C 34(d), KRF, p. 4; see also "Draft Minutes of Meeting, June 30, 1953, Non-proprietary Names," box 6, f 6, "World Health Association: 1945–1959," RPF.

26. Memorandum from H. D. Kautz to W. Wolman, "Comparative Display of Proprietary with Non-proprietary (Generic) Names," pp. 1–2. For more detail on the role of the Business Division of the AMA in the dissolution of the Seal of Acceptance program, see Jeremy A. Greene and Scott H. Podolsky, "Keeping Modern in Medicine: Pharmaceutical Promotion and Physician Education in Postwar America," *Bulletin of the History of Medicine* 83, no. 2 (2009): 331–77.

27. John McDonnell to H. D. Kautz, 22 March 1956, COD 3324, box 15-13, 1956 vol. 5, AMAA. Kautz, in his reply, gently reproaches Schering for the error: "The Council no longer gives formal consideration to trade names, although we appreciate the spirit of cooperation which motivates your request for consideration of your name, Miradon. Without exact knowledge of the chemical terminology and structural formula of your new compound, we cannot undertake formal consideration of your proposal, methophinindione. At this stage, it is our offhand opinion that your proposal might well be shortened to something, like, mephinadione." H. D. Kautz to John McDonnell, 28 March 1956, COD 3324, box 15-13, 1956, vol. 5, AMAA.

28. H. D. Kautz to Torald Sollmann, 14 December 1955, COD 3324, box 15-13, 1955, vol. 9, AMAA.

29. Memorandum from H. D. Kautz to W. Wolman, 16 December 1959, COD 3324, box 15-13, 1959, vol. 6, AMAA, p. 2.

30. Robert B. Clark to World Health Organization, 31 July 1959, COD 3324, box 15-13, 1959, vol. 6, AMAA. Clark noted that mepazine (and the variant vernacular, mepazin) had been published already in journals in the United States, Australia, Canada, Germany, United Kingdom, and Soviet Union and marketed in addition in Brazil, Colombia, Costa Rica, Cuba, the Dominican Republic, Ecuador, Guatemala, Honduras, Netherlands, West Indies, New Zealand, Panama, Peru, the Philippines, Portugal, Puerto Rico, and Venezuela. Chemical Abstracting Services to this day retains both mepazine and pecazine as common names for 10–1(1-methyl-piperid-3-ylmethyl) phenothiazine.

31. World Health Organization, *Chronicle of the World Health Organization* (WHO: Geneva, 1956), 10:28. Other rules in the emerging protocols of generic naming were intended to differentiate the nonproprietary name from the chemical name, avoiding pesky variables as numbers, capitalized letters, and hyphens wherever possible. Names were to be distinctive in sound and spelling; they "should not be inconveniently long and should not be liable to confusion."

32. Peter G. Forster, *The Esperanto Movement* (The Hague: Mouton, 1982); Arika Okrent, *In the Land of Invented Languages: Esperanto Rock Stars, Klingon Poets, Loglan Lovers, and the Mad Dreamers Who Tried to Build a Perfect Language* (Philadelphia: Spiegel & Grau, 2009).

33. Closer evaluation of the table draws out the arbitrariness or "heterotopic organization" that scholars from Jorge Louis Borges to Michel Foucault to Susan

Leigh Starr have noted can creep into any taxonomic system. Borges recounts the example of an apocryphal Chinese text, *Celestial Emporium of Benevolent Knowledge*: "On those remote pages it is written that animals are divided into (a) those that belong to the Emperor, (b) embalmed ones, (c) those that are trained, (d) suckling pigs, (e) mermaids, (f) fabulous ones, (g) stray dogs, (h) those that are included in this classification, (i) those that tremble as if they were mad, (j) innumerable ones, (k) those drawn with a very fine camel's hair brush, (l) others, (m) those that have just broken a flower vase, (n) those that resemble flies from a distance." Foucault celebrates this example as an illustration of the heterotopic character of all taxonomic projects. Michel Foucault, *The Order of Things: An Archaeology of the Human Sciences*, trans. by the publisher (New York: Pantheon Books, 1971 [1966]), preface; Jorge Luis Borges, "The Analytical Language of John Wilkins," in *Other Inquisitions*, trans. Ruth L. C. Simms (Austin: University of Texas Press, 1966 [1964]). For a closer analysis of the taxonomies of the Chinese encyclopedias referred to by Foucault via Borges, see Carla Nappi, *The Monkey and the Inkpot: Natural History and Its Transformations* (Cambridge: Harvard University Press, 2009). See also Bowker and Starr, *Sorting Things Out*.

34. E. C. S. Little, "'Donomen—A Nomenclature System for Pesticides and Pharmaceuticals," *Pharmaceutical Journal* 183, no. 5003 (1959): 131–32, p. 132. Little's argument here echoes the midcentury suggestion by Leopold Gmelin, professor of chemistry at the University of Heidelberg, to rationalize the nomenclature of organic chemistry with a system that broke completely from established usage and incorporated terms instead such as vinak, vinek, vinik, and vinok. Maurice Crosland, in his account of Gmelin, is more dismissive, concluding that "he could hardly have expected such an arbitrary synthetic language which made no concessions to euphony, etymology or current usage to win general acceptance." See Crosland, *Historical Studies in the Language of Chemistry*, pp. 310–11.

35. Little, "'Donomen,'" p. 132. Little submitted his system to an experimental analysis of five hundred Donomen words coined at random, studied by a team of linguists in seven European languages. Only eight had strong similarity to any existing words, with thirty having slight similarity—yielding less than 10 percent similarity or more than 90 percent conflict-free words. "From this test," Little concluded, "it can thus be estimated that about 10 per cent of Donomen words might have to be rejected on grounds of similarity with other words leaving the other 90 per cent available for use because, at present, they are meaningless" (p. 132). Waxing grandly, Little suggested his Donomen typology might eventually supplant (or at least enhance) Linnaeus's own nomenclature: "Perhaps it is asking too much, but if a system such as Donomen were internationally accepted, would there be a chance of biologists being persuaded gradually to revert to simplified nomenclature for organisms? In a busy biological world conversion of, say, *Lepinotarsa decemlineata* to *Lepotar dekemat* would, over the years, save a massive amount of time, paper, and energy. A change from (say), Rhododendron to Rodenon would upset conservative opinion but future generations would benefit and be thankful for the change."

36. "Kemikal Names Computed," *New Scientist* 6, no. 153 (22 October 1959): 738.

37. Ibid.

38. Statement of Charles O. Wilson, *Administered Prices*, vol. 21, p. 11500.

39. Paul G. Stecher, "Generic Names of Drugs," *Journal of Chemical Education* 34 (September 1959): 454–56, quotation 456.

40. Ibid., p. 455.

41. Lloyd Miller to T. C. Denston, 6 June 1961, box 249, f 40, USPC.

CHAPTER 2. THE GENERIC AS CRITIQUE OF THE BRAND

Epigraph. George Clifford, "Brand and Generic Names" n.d., group 7, box 52, f "Industries 5-2: Generics," SOM.

1. Opening statement of Sen. Estes Kefauver, 10 May 1960, *Administered Prices*, vol. 21, p. 11494.

2. "Resolution of the American Pharmaceutical Association," Miami Beach, FL, 6 May 1955, as cited in Ashok K. Gumbhir and Christopher A. Rodowskas Jr., "The Generic–Brand Name Drug Controversy: A History" *Medical Marketing and Media*, November 1971, 27–33. "The Name of a Drug," *Di Cyan & Brown Monthly Bulletin*, New York, September 1959, found in box 251, f 6, USPC. For more on the anticommercial aspirations for the twentieth-century uses of the generic drug name, see Harry M. Marks, *Progress of Experiment: Science and Therapeutic Reform in the United States, 1900–1990* (Cambridge: Cambridge University Press, 1997); Joseph Gabriel, *The Corporate Logic: Intellectual Property Rights and the 20th Century Pharmaceutical Industry* (forthcoming).

3. *Administered Prices*, vol. 21. The political history of this dispute is most thoroughly laid out in Richard Edward McFadyen, "Estes Kefauver and the Drug Industry" (PhD diss., Emory, 1973); also McFadyen, "Estes Kefauver and the Tradition of Southern Progressivism," *Tennessee Historical Quarterly* (Winter 1978): 430–43. For earlier accounts, see Estes Kefauver with Irene Till, *In a Few Hands: Monopoly Power in America* (New York: Pantheon Books, 1965); Richard Harris, *The Real Voice: The First Fully Documented Account of Congress at Work* (New York: Macmillan, 1964). For more recent treatments, see Daniel Carpenter, *Reputation and Power: Organizational Image and Drug Regulation at the FDA* (Princeton, NJ: Princeton University Press, 2010); and Dominique Tobbell, *Pills, Power, and Politics: Drug Reform in Cold War America and Its Consequences* (Berkeley: University of California Press, 2012).

4. Different advertisements for Robins & Co.'s popular antihistamine Dimetane in 1960 alternately referred to the drug generically as brompheniramine (in *JAMA*) and as parabromdylamine (in *Modern Medicine*). Nor was the chemical name any more stable: Robins & Co. named it 1-(p-bromophenyl-1-(2-pyridyl))-3-dimethylamino-propane maleate, the AMA's *New and Non-official Drugs* named it 2(p-bromo-alpha-(2-diethylaminoethyl)benzyl)pyridine maleate, and the WHO referred to it as (3-p-bromophenyl-3-pyrid-2'-ylpropyl)dimethylamine maleate. *Administered Prices*, vol. 14, p. 11526.

5. Statement of Charles O. Wilson, 10 May 1960, *Administered Prices*, vol. 21, pp. 11494–96; see also Wilson, "Inconsistency in Pharmaceutical Names," *American Journal of Hospital Pharmacy*, 16 (1959): 433. Jack Drury, the executive secretary of the Nashville Academy of Medicine, had complained to the Council on Drugs four years earlier about limitations of the AMA-WHO system: while the partnership had been

influential in shaping the naming practices of several major drug firms, especially those interested in international markets, the AMA no longer had any real means of enforcing their recommendations, and manufacturers remained free to make up their own generic names as they saw fit. Jack Drury to H. D. Kautz, 7 July 1956, COD 3324, box 15-13, 1956, vol. 4, AMAA.

6. Statement of Charles O. Wilson, 10 May 1960, *Administered Prices*, vol. 21, p. 11499.

7. E. Dunham, "Generic Jawbreakers" *NEJM* 280, no. 15 (1969 April 10): 841.

8. Chauncy Leake, "Names of New Drugs," *JAMA* 171, no. 14 (5 December 1959): 182; Statement of Dr. Russell Cecil, *Administered Prices*, vol. 14, p. 7986. Quotation from Frank C. Ferguson Jr., MD, chair of the Department of Pharmacology of Albany Medical College, found in Edgar F. Mauer, *Report of Sub-committee on Generic Terms*, 6 June 1960, box 9, f 8, HFDP, p. 3.

9. Testimony of Charles O. Wilson, *Administered Prices*, vol. 21, pp. 11507, 11513–15. Advertisements referred to *JAMA* 172, no. 9 (27 February 1960): p. 37; *Modern Medicine*, March 15, 1960. On earlier problems of nominal plurality, see Maurice P. Crosland, *Historical Studies in the Language of Chemistry* (Cambridge: Harvard University Press, 1962), pp. 294–95.

10. John Blair, cross-examination of Walter Modell, *Administered Prices*, vol. 21, p. 11628.

11. Cross-examination of Kurt Weilburg, *DIAA*, vol. 6, p. 3230, emphasis in original.

12. Testimony of Charles O. Wilson, *Administered Prices*, vol. 21, p. 11503.

13. "In regard to brand names," he continued, "there is no such thing as a generic equivalent." Testimony of Charles O. Wilson, *Administered Prices*, vol. 21, pp. 11521–22.

14. Testimony of Walter Modell, *Administered Prices*, vol. 21, p. 16067.

15. Statement of Hugh Hussey, *DIAA*, vol. 1, pp. 42–43.

16. Walter Hartung to Lloyd C. Miller, 9 September 1961; Lloyd C. Miller to Walter H. Hartung, 6 September 1961, box 248, f 1, USPC.

17. "Current Procedures of the AMA-USP Cooperative Program in the Consideration of Proposed Nonproprietary Names for Drugs," box 248, f 1, USPC, p. 5.

18. Unlike the language of chemistry, in which structure and function were considered to be immutable, the functional world of pharmacology was far more plastic, evolving alongside changing conceptions of disease, therapeutics, and physician practice patterns. Joseph B. Jerome to AMA-USP Committee on Nomenclature, 29 September 1961, box 248, f 1, USPC; Windsor Cutting to Hans Møller, 17 September 1961, box 248, f 1 USPC, pp. 1–2.

19. D. R. Zimmerman to Joseph B. Jerome, 19 September 1961, box 248, f 1, USPC.

20. Nelson M. Gampfer, "The Case for Brand Name Specification," C 46 (p) I f 3, KRF, p. 5.

21. National Pharmaceutical Council, *24 Reasons Why Rx Brand Names Are Important to You* (New York: National Pharmaceutical Council, 1957), p. 1.

22. Linwood Tice, editorial, *American Journal of Pharmacy*, January 1960, as cited in *Administered Prices*, vol. 21, pp. 11648.

23. Testimony of Charles O. Wilson, *Administered Prices*, vol. 21, pp. 11511–12.

24. Ibid., p. 11519.

25. "V-53" (memorandum), n.d., box 1, f "Drugs: Generic vs. Trade Name: Recommendations, Sept.–Dec. 1960," SOAM.

26. Harold O'Keefe, "The FDA and Nonproprietary Names for Dugs," C 46 (p) I f 3, KRF, quotation p. 1. In 1963, thirty-seven major manufacturers and the PMA had filed suit in district federal court in Delaware that new regulation requiring the generic name of a drug to be listed every time a brand name appears in labeling and advertising was "unauthorized by and contrary to law." According to the *AMA News*, this action was the first time that the prescription drug industry was unified in a lawsuit against the FDA. It would not be the last. "Generic Labeling Rule Contested," *AMA News*, 16 September 1963, p. 6; "Court Upholds PMA's Generic Labeling Plea," *Drug Trade News*, 11 May 1964.

27. *United States Adopted Names: A Compilation of the United States Adopted Names Selected and Released from June 15, 1961, through December 31, 1964* (New York: USPC, 1965).

28. "Guiding Principles for Coining United States Adopted Names for Drugs. 1315-8," in *2009 USP Dictionary of USAN and International Drug Names* (Rockville, MD: USPC, 2009), quotation p. 1315. Another early function of the unified USAN was to reduce the risk of "lookalike names" as a source of medication error. In 1962 *Drug Topics* prepared a list of thirteen "dangerously similar" generic names (e.g., digoxin/digitoxin; quinine/quinidine) that could lead to inadvertent substitution of the wrong compound. USAN did not succeed in reducing this problem: by 1971, *Pharmacy Times* published a list of seven hundred names now considered dangerously similar. Frederick D. Lascoff, "A List for Double-Checking Similar Drug Names," *Drug Topics*, 12 February 1962, p. 40; Benjamin Teplitsky, "Caution! 700 Drugs Whose Names Look-Alike or Sound-Alike," *Pharmacy Times*, 27–29 March 1971, n.p.

29. *FDCR*, 12 December 1966, pp. 13–14.

30. Ibid.

31. Lloyd C. Miller to Harvey Richards, 31 July 1961, box 251, f 6, USPC.

CHAPTER 3. DRUGS ANONYMOUS

Epigraph. Gabriel Tarde, *Psychologie economique* [1902], as translated and cited in Bruno Latour and Vinent Antonin Lepinay, *The Science of Passionate Interests: An Introduction to Gabriel Tarde's Economic Anthropology* (Chicago: Prickly Paradigm Press, 2009).

1. Quotation in text is from the opening statement of Sen. Gaylord Nelson, *Competitive Problems*, 1967, vol. 1, p. 3, emphasis mine. In a later hearing, Nelson equated the differential pricing of equivalent products with consumer fraud. "I can only say about that that the ordinary, hard-working little consumer having trouble meeting the grocery bill and paying the taxes ought to be shocked out of his shoes, and would be, if he knew that he is paying $17.90 plus a markup for a drug when right there in the marketplace there is an equivalent item, with scientific evidence to support its therapeutic value, at one eighth the price, and one-tenth, and one-twentieth the price, as a matter of fact at $0.59 . . . This is a shocking business." *Competitive Problems*, 1967, vol. 4, p. 1287. For the use of *generic drug* as a negative term in the Kefauver

hearings, see E. Gifford Upjohn, *Administered Prices*, vol. 14, p. 8299. The case in point regarded the differences between Upjohn's brand of hydrocortisone, Cortef, and a drum of the unfinished chemical hydrocorstisone acetate. All parties agreed that although the generic name was known in both instances, no clear concept of a generic drug existed in this case. Joseph Gabriel narrates a fascinating if brief-lived effort by the FTC in the 1920s to create markets for generic equivalents of German pharmaceuticals after their patents had been seized by the US government in World War I, in *The Corporate Logic: Intellectual Property Rights and the 20th Century Pharmaceutical Industry* (forthcoming).

2. Durward G. Hall, "Anyone Suggesting That One Drug Firm Is as Good as Another Is a Fool or Is Naïve, or Both," *Rhode Island Medical Journal* 10 (1967): 691–95, quotation p. 695.

3. Drew Pearson, "LBJ Says No," *New York Post*, 25 May 1967; Joseph Stetler, *The Nelson Hearings on Drugs: A Study in Distortion and Omission* (Washington, DC: PMA, 1968); pamphlet, C 36 (e) I, KRF.

4. Harry Wiener, *Generic Drugs: Safety and Effectiveness* (New York: Pfizer, 1973), p. 1. Benjamin Wells of the NPC would likewise claim in 1974 that "every drug product, regardless of the name we give it, has been made by someone." Benjamin B. Wells, "Generic Nomenclature," *JAMA* 229, no. 5 (1974): 527; see also William C. Cray, "If Your Files on the Prescription Drug Industry Don't Include the Information in This Folder, Your Files May Be Incomplete," C 36 (e) I PMA-1976, KRF.

5. Recent scholarship on the history of industrial standards has opened up a lively interface between business history and the history of science and technology; e.g., Amy Slaton and Janet Abbate, "The Hidden Lives of Standards: Technical Prescriptions and the Transformation of Work in America," *Technologies of Power: Essays in Honor of Thomas Parke Hughes and Agatha Chipley Hughes*, ed. Michael Tadd Allen and Gabrielle Hecht (Cambridge, MA: MIT Press, 2001), 95–144; Amy Slaton, "As Near as Practicable: Precision, Ambiguity, and the Social Features of Industrial Quality Control," *Technology and Culture* 42 (2001): 51–80; Andrew Russell, "Industrial Legislatures: The American System of Standardization," *International Standardization as a Strategic Tool* (Geneva: International Electrotechnical Commission, 2006). Historians of the pharmaceutical industry have recently become interested anew in this intersection. See for example Christian Bonah, Christophe Masutti, Anne Rasmussen, and Jonathan Simon, eds., *Harmonizing Drugs: Standards in 20th Century Pharmaceutical History* (Paris: Éditions Glyphe, 2010).

6. Richard M. Burack, *The Handbook of Prescription Drugs*, 1st ed. (New York: Pantheon Books, 1967), p. 72. Cambridge City's use of generically named drugs echoed earlier explorations by academic physicians working with municipal health assistance programs in Baltimore, New York, and other cities. See, for example, "Cost of Drugs: Baltimore Formulary 1956–1958," box 11, f 15, SOAM.

7. Morton Mintz, "Drug Store Holdup," *Washington Post Book Week*, 7 May 1967. Gaylord Nelson, "The Need for a Great Awakening on Prescription Drugs," Cong. Rec., 26 April 1967, pp. 10894–96. Edward Kennedy, "New Book on Drugs," Cong. Rec., 10 May 1967, p. 12263. Durward Hall, "A Bad Prescription for Drugs," Cong. Rec., 9 August 1967, pp. 12980–81; C. Joseph Stetler to William S. Middleton, 3 May 1967, box 6, f "DRB Correspondence and Miscellaneous Data—May 1967," WSMP.

8. Ed Wallace, "Rackets Feed the Gusher: Savage, Deadly U.S. Pill Orgy," *New York Sunday News*, 17 December 1967, p. 24.

9. Margaret Kreig, *Black Market Medicine* (New York: Prentice-Hall), p. 127; Morton Mintz, "Drug Store Holdup," *Washington Post Book Week*, 7 May 1967.

10. The *Post* reviewer's praise for Burack's book was matched with his disdain for Kreig's "attempt to equate counterfeit drugs made by hoodlums with generic drugs made by small businessmen." The reviewer in question, Morton Mintz, had authored a popular account of the Kefauver hearings, *The Therapeutic Nightmare*, and portrayed Kreig in turn as a public relations stooge for the prescription drug industry. Morton Mintz, "Drug Store Holdup," *Washington Post Book Week*, 7 May 1967. Kreig denied these allegations in a statement before the Legal and Monetary Affairs Subcommittee, House Committee on Government Operations, June 13, 1967, RG 46, Nelson, box 208, f 2, SOAM.

11. "It cannot be denied," the pamphlet read, "that substandard drug products do find their way to market, and these are often sold under generic names." Pharmaceutical Manufacturers of America, *Drugs Anonymous?* (New York: NPC, May, 1967), quotation p. 8. Other PMA pamphlets bore titles like *Compulsory Generic Prescribing—a Peril to Our Health Care System*. "PMA Issues Pamphlet Restating Stand on Generic Prescriptions," *Drug Trade News*, 19 June 1967, p. 6. Dominique Tobbell also writes of *Drugs Anonymous?* in *Pills, Power, and Policy: The Struggle for Drug Reform in Cold War America* (Berkeley: University of California Press, 2012).

12. *A Historical Retrospective of NPC* (Reston, VA: NPC, 2003). The NPC's twelve founding firms included Abbott Laboratories; Ciba Pharmaceutical Products, Inc.; Hoffmann–La Roche, Inc.; Lederle Laboratories (a division of American Cyanamid Company); McNeil Laboratories; the William S. Merrell Company; Charles Pfizer & Co., Inc.; G. D. Searle & Co.; Smith, Kline & French; and E. R. Squibb. Two of three former and present heads of the NPC present at the Kefauver hearings in 1960 were pharmacists, one of whom was the outgoing president of the APhA.

13. Frederick Accum, *A Treatise on Adulterations of Food, and Culinary Poisons* (Philadelphia, 1820); Mitchell Okun, *Fair Play in the Marketplace: The First Battle for Pure Food and Drugs* (DeKalb: Northern Illinois University Press, 1986); James Harvey Young, *Pure Food: Securing the Federal Food and Drugs Act of 1906* (Princeton, NJ: Princeton University Press, 1989); Harry M. Marks, "What Does Evidence Do? Histories of Therapeutic Research," in *Harmonizing Drugs: Standards in 20th-Century Pharmaceutical History*, ed. Christian Bonah, et al. (Paris: Éditions Glyphe, 2010), 81–100.

14. Robert A. Hardt, "Prescription Brands and Substitution," *Journal of American Pharmaceutical Association* 18, no. 2 (1957), as included as exhibit 385, *Administered Prices*, vol. 21, p. 11759. On the vertical integration of the pharmaceutical industry, see Alfred Chandler, *Shaping the Industrial Century: The Remarkable Story of the Evolution of the Modern Chemical and Pharmaceutical Industries* (Cambridge, MA: Harvard University Press, 2009).

15. Jack Anderson, "Counterfeit Drugs," *Parade*, 26 October 1960, p. 2. On an earlier celebration of mass manufacture as a remedy to local adulterative practices in pharmaceuticals, see Okun, *Fair Play in the Marketplace*, pp. 169–84. On drugs and moral panics, see Nicholas Rasmussen, "Goofball Panic: Barbiturates, "Dangerous," and "Addictive Drugs, and the Regulation of Medicine in Postwar America," in *Pre-*

scribed: Writing, Filling, Using, and Abusing the Prescription in Modern America, ed. Jeremy Greene and Elizabeth Watkins (Baltimore: Johns Hopkins University Press, 2012).

16. John T. Connor, "Responsibilities of the Pharmaceutical Industry," *Northwest Medicine*, July 1961, reprinted in *DIAA*, vol. 3, p. 1895.

17. *Administered Prices*, vol. 21, p. 11624; Robert A. Hardt, "Prescription Brands and Substitution," *Journal of American Pharmacists Association* 18, no. 2 (1957), as included as exhibit 385, *Administered Prices*, quotation vol. 21, p. 11759.

18. *Parade*, 23 October 1960, pp. 1–3.

19. "A New Program to Protect You against . . . Counterfeit Drugs," *Parade*, 15 January 1961; Paul Rand Dizon to T. C. Williams, "Counterfeit Drugs," 20 February 1961, box 1, f 1, SOAM.

20. Lucile B. Wendt to M. R. Fensterald, "Counterfeit Drugs," 3 April 1961, box 1, f "Drugs: "Counterfeit," Counterpart 1960-1, SOAM.

21. *Hearings on H.R. 6245*, 17–24 May 1962, *DIAA*, vol. 1, p. 47. See also statement of Newell Stewart, *Administered Prices*, vol. 21, p. 11696.

22. Wallace, "Rackets Feed the Gusher," p. 24.

23. Irwin di Cyan, quoted by Margaret Kreig in *Black Market Medicine*, p. 127.

24. K.E. to FDA, 2 September 1966, AF30-841, vol. 6, FDAAF.

25. *FDCR*, 3 March 1968, p. 10.

26. Leonard W. Cannon to John Adriani, 6 March 1969, box 34, f 5, JAP.

CHAPTER 4. ORIGINS OF A SELF-EFFACING INDUSTRY

Epigraph. "Book Club Selection Boosts Generic Drugs," *Drug News Weekly*, 13 March 1967, clipping found in f 5, News Clippings & Letters about the "Handbook," RBP.

1. Statement of William Apple, *Competitive Problems*, 1967, vol. 5, p. 1290. On Apple's early identification of generic substitution as a defining issue for his presidency of APhA, see "Substitution Committee," 21 September 1966, f 33.46.1.1, "Drug Products Selection," APhAA.

2. *Competitive Problems*, vol. 21, p. 1292.

3. Coleman & Company, *An Appraisal of Parke-Davis & Company—May 1965*, box 65, f 1, SOM; Frost & Sullivan, *The US Generic Drug Market* (New York: Frost & Sullivan, 1976).

4. Estes Kefauver, introduction of Herman E. Nolen, *DIAA* v. 5, p. 2648; statement of Herman E. Nolen, *DIAA*, quotation v. 5, p. 2652. Barbara Yuncker, "Major Firm to Cut Price of Rx Drugs," *New York Post*, 5 Oct 1961. Market researchers hired by the PMA contested the imminence of the generics market. A survey of 190 wholesale companies presented at the PMA Annual Meeting in December 1961—just days before Nolen's Senate testimony—suggested that 88 percent of forty-nine thousand pharmacists serviced had "practically no interest in generic drugs." Wayne Luther (Druggists' Service Company), "Generics—A D.S.C. Survey" (remarks given at the PMA Annual Meeting, 11 December 1961), box 8, f "Drugs—Companies—McKesson-Robbins, 1963–64," SOM.

5. Marvin Seife, the first head of the FDA's Generic Drug Division, linked the birth

of the Abbreviated New Drug Application (ANDA) and the Generic Drug Division to the implementation of DESI in 1970. *FGDAP*, part 2, p. 4. See also Daniel Carpenter, Jeremy Greene, Susan Moffitt, and Jonathan Warsh, "Therapeutic and Economic Effects of Efficacy-Based Drug Withdrawals: The Drug Efficacy Study Initiative and Its Manifold Legacies" (paper presented at the Standard Exchanges Programme: Workshop international sur la standardization en histoire et de la medicine, Strasbourg, France, 7 December 2012).

6. It was unethical, they argued, to expose research subjects to placebo-controlled trials for a drug already known to work; it was economically wasteful to require a full NDA—which could reach thirty or forty volumes of paper for a single submission. For more on the role of the FDA in market making and unmaking, see Daniel Carpenter, *Reputation and Power: Organizational Image and Pharmaceutical Regulation at the FDA* (Princeton, NJ: Princeton University Press, 2010).

7. George C. Harlan, "Drugs' Branded Battle," *Sunday Herald-Tribune*, 30 January 1966, 2E.

8. Shortly after the passage of Medicare and Medicaid in 1965, HEW issued circulars favoring the use of generic names for drug purchasing "whenever it is practicable and economical." "HEW Advises Agencies: Use Generic Drugs," *Drug Trade News*, 17 January 1966, p. 1. For figures of estimated cost savings of generic substitution, see United States Task Force on Prescription Drugs, *Final Report* (Washington, DC: US Government Printing Office, 1969), 25; also discussed in Milton Silverman and Philip Randolph Lee, *Pills Profits and Politics* (Berkeley: University of California Press, 1974), p.145.

9. "Basically," Seymour Blackman told Kefauver's subcommittee, "the firm was founded with the premise of selling pharmaceuticals under generic names." Seymour Blackman, *Administered Prices*, vol. 14, p. 8212.

10. By September 1959, Premo could boast that more than a hundred of its products had been approved by the Council on Pharmacy and Chemistry (now Council on Drugs) of the AMA. "Premo Hospital Net Price List," exhibit 58, *Administered Prices*, vol. 14 (appendix), p. 8691.

11. "Looking Backward: Highlights in the History of an American Institution of Service," exhibit 59, *Administered Prices*, vol. 14 (appendix) pp. 8721–22.

12. Thomas Mahoney, *The Merchants of Life: An Account of the American Pharmaceutical Industry* (New York: Harper, 1959).

13. "Looking Backward," p. 8721.

14. William J. Barbour, "Inspection Report: Premo Pharmaceutical Laboratories, Inc.," 10 June 1958, AF13-610, vol. 7, FDAAF, p. 2; Ralph C. Smith to Premo Pharmaceutical Laboratories, Inc., 9 November 1960; "Memorandum of Telephone Conversation: Leslie Harrup, J. K. Kirk," 29 June 1961; F. L. Lofsvold to Los Angeles District, "Investigation of Certain Drug Firms," 9 February 1960, AF 13-610, v. 8, FDAAF.

15. "In order to convince a physician that he is getting an equivalent product, merely writing the generic name is not sufficient. In the hospitals, when they get a formulary together, for example, at Lennox Hill Hospital, we had our plant and facilities examined by R. Bogash, their hospital pharmacist, and our facilities had to first be approved before our generic products could be accepted as equal." Cross-examination of Seymour Blackman, *Administered Prices*, vol. 14, pp. 8213, 8223.

16. Statement of Seymour Blackman, *Administered Prices*, vol. 14, p. 8212.

17. "We would like to know," the director of a community hospital in San Antonio, Texas, asked in a letter to the FDA, "if you consider these two firms ethical and reliable houses, and whether or not you have any information that would cause us to be cautious in dealing with them. We are a tax supported institution on a limited budget, and savings such as their prices offer will be most welcome. However, we do not desire to deal with houses that are not considered reputable." John W. Simpson to FDA, 25 January 1960, AF13-610, vol. 8, FDAAF. Many similar letters from other pharmacies and hospitals found their way to the FDA in the early 1960s, which generally voiced its support of small firms with reputations for high quality. The FDA's standard reply to these queries would repeat a catechism: "a review of our files over the past six years indicates no record of our having proceeded in the Federal courts against Premo Pharmaceutical Laboratories, Inc. or its products, for violation of the Act." H. W. Chaddack to H. R. Williers, 30 September 1964. All letters found in AF 13-610, vol. 9, FDAAF.

18. In November of that year, Premo's bid to supply tolbutamide for military and Veterans Administration hospitals was rejected. After the military reiterated that it would only purchase tolbutamide produced and sold by Upjohn as Orinase, Premo filed a complaint to the Department of Defense and to Senator Estes Kefauver. Though this complaint did not help Premo's tolbutamide find a military or VA market—and though a patent infringement suit by Upjohn helped shut down the possible private sector market for Premo's copy—it indicates a company aware of new market possibilities on the horizon. Premo's NDA for tolbutamide was approved 17 January 1963; the drug, however, was never marketed, and NDA approval was withdrawn in 1971. See Ralph Smith to W. B. Rankin, "Tolbutamide, NDA 12-678," 8 January 1963, AF13-610, vol. 9, FDAAF; George P. Larrick to Frank B. Berry, 29 January 1963, AF 13-610, vol. 9, FDAAF, p. 1; Donald R. Martin to Richard McDermaid, 2 July 1973, AF 13-610, vol. 13, FDAAF; Frank B. Berry to George P. Larrick, 10 January 1963, AF 13-610, vol. 9, FDAAF; "Memorandum of Meeting: Jerry Thomas, Attorney, the Upjohn Company, Kalamazoo, Michigan; Franklin D. Clark, Deputy Director, Bureau of Regulatory Compliance," 24 August 1964, AF 13-610, vol. 9, FDAAF.

19. A 1968 inspection of expanded facilities showed "deplorable conditions were discovered in every area of plant operations from the first day of the inspection"; many of these issues were still not resolved a year later. Frances J. Flaherty, "IDI Inspection: Premo Pharmaceutical Labs," 5 December 1969; Bruce Byer to Dick Anderson, "IDI Review: Premo Pharmaceutical Labs," 10 December 1969; Bruce Byer to Dick Anderson, "IDI Review: Premo Pharmaceutical Labs," 29 December 1969. All found in AF13-610, vol. 11, FDAAF.

20. For Premo's illegal distribution of chloriazepoxide/clinidium, see "Seizure" (memorandum), 1 August 1977, AF13-610, vol. 15, FDAAF; for tolbutamide, see Gene Knapp to Carol Byone, 3 March 1978; for chlorpropamide, see Prescription Drug Compliance Branch (HFD-31), "Request for Inspection Regarding the Shipment of Chlorpropamide without an Approved NDA" (memorandum), 19 May 1978; both found in AF13-610, vol. 16, FDAAF. Carlos Dixie to Carleton Sharp, "Drug Substitution w/ No Approved ANDA," 3 January 1979; affidavit, 28 December 1978, AF13-610, vol. 17, FDAAF.

21. Morton M. Schneider, *Inspection Report*, 23 June 1959; M. Boyle and Frank Bruno, memo to Accompany EIR of Bolar Pharmaceutical Co., 10 December 1962, vol. 1; EIR of Bolar Pharmaceutical Co., 10–11 April 1967, vol. 2; AF10-156, FDAAF.

22. Lawrence Raisfeld to FDA, 11 October 1966, vol. 2, AF10-156, FDAAF.

23. "Report of Intensified Inspection, 4 November 1968–20 March 1969," vol. 2, AF10-156, FDAAF.

24. Edward Warner, "Violative Inspection—for Information Purposes," 14–16 May 1968, vol. 2, AF10-156, FDAAF.

25. Robert Shulman to Henry Simmons (FDA), 30 October 1970, vol. 3, AF10-156, FDAAF. The FDA was initially unwilling to play such a role, as noted by Albert Lavender to Robert Shulman, 23 November 1970, vol. 3, AF10-156, FDAAF.

26. Ibid.

27. "ANDA Inspection," 17, 19, 22, 23, and 26 January 1973, vol. 4 AF10-156, FDAAF.

28. Coleman & Company, *Appraisal of Parke-Davis & Company—May 1965*; Donaldson, Lufkin, & Jenrette, Inc., *The Ethical Drug Industry: Stock Prices Ignore Long Term Outlook for Slower Growth*, November 1965, group 7, box 65, f 1, SOM.

29. Quotation *FDCR*, 5 Oct 1965, p. 7; see also AF30-841, vol. 6-20, FDAAF.

30. Benjamin Wiener to Oregon State Pharmaceutical Association, August 1965, reprinted in *FDCR*, 6 September 1965, p. 16.

31. Benjamin Wiener to Wallace Werble, 1 September 1965, reprinted in *FDCR*, 6 September 1965, p. 17. Oregon's proposed Seal of Approval for generic drug products was immediately denounced by APhA—at the time still very much aligned with the NPC—who issued a sharp reprimand that a professional society "should not compromise its reputation and integrity by 'selling' its endorsement of specific firms or their products." William Apple to Speckman, reprinted in *FDCR*, 23 August 1965, p. 35. Two years later, however, Apple would testify in front of Nelson's subcommittee that generic drugs should be considered to be equivalent with their brand-name counterparts.

32. "Catalog & Price List: Zenith Laboratories, 1967," AF30-841, vol. 6, FDAAF. Quotation from *FDCR*, 29 May 1967, pp. 16–17.

33. *FDCR*, 29 May 1967, pp. 16–17.

34. The company hoped publicly that the possibility of an antitrust suit would likewise draw support of Senator Gaylord Nelson and the Subcommittee on Monopoly and Antitrust. The Department of Justice ultimately filed a civil antitrust complaint against Beecham and Bristol-Myers in 1970. Frost & Sullivan, *U.S. Generic Drug Market*, p. 47. "Zenith's Generic Ampicillin Price," *FDCR*, 14 April 1969, p. TG-1. On the analogies to tetracycline, "NJ Federal Court & 3rd Circuit Refuse to Enjoin Trial in Fla. Court," *FDCR*, 25 January 1965, p. 13.

35. Frost & Sullivan, *U.S. Generic Drug Market*, pp. 42–43.

36. Quotation from International Resource Development, Inc., *Generic Drugs in the 1980s* (Norwalk, CT: IRD, 1979), p. 58. The top fifteen generic drugs by prescription in 1976 were (in alphabetical order) ampicillin, tetracycline, phenobarbital, prednisone, thyroid, erythromycin, digoxin, meprobamate, penicillin VK, penicillin G depot nitroglycerin, quinidine, nicotinic acid, paregoric, and reserpine. Frost & Sullivan, *U.S. Generic Drug Market*. John P. Curran, "Major Thrust Expected in the Promotion of Generic Products," 7 April 1977, *Wood Gundy Progress Report*, box 7 f 24, NYSGDIF. On the rising specter of generic drug shortages in the early twenty-first century, see

Margaret Clapp, Michael A. Rie, and Phillip L. Zweig, "How a Cabal Keeps Generic Scarce," *NYT*, 2 September 2013.

37. International Resource Development, *Generic Drugs in the 1980s*, p. 1.

CHAPTER 5. GENERIC SPECIFICITY

1. International Resource Development, Inc., *Generic Drugs in the 1980s* (Norwalk, CT: IRD, 1979), p. 87.

2. Ibid., pp. 89–90.

3. Cori Hayden has also developed the concept of generic specificity in her comparative analysis of contemporary Argentinian and Brazilian generics markets; see Hayden, "No Patent, No Generic: Pharmaceutical Access and the Politics of the Copy," in *Making and Unmaking Intellectual Property: Creative Production in Legal and Cultural Perspective*, ed. Mario Biagioli, Peter Jaszi, and Martha Woodmansee (Chicago: University of Chicago Press, 2011), pp. 285–304.

4. If Purepac had distinguished itself from other firms in the eyes of FDA officials, it was mostly as a firm that resisted regulation. By 1941, a senior regulator, Charles N. Lewis, authored a thirty-three-page-long memorandum entitled "Purepac: Chronological Listing of Firm's Contact with the Food and Drug Administration: Evidence of Bad Faith," AF17-705, FDAAF, vol. 1. In spite of the company's early motto ("Fine Remedies for Many Years") consumer letters documented frequent lapses in quality control of early Purepac products, as did continued investigations through the 1950s and 1960s. See, for example, Wayne B. Adams to Robert S. Roe, 28 May 1952; AF17-705, vol. 4, FDAAF.

5. Purepac was acquired by a pharmaceutical house in 1980 and eventually folded into the multinational branded generics firm Actavis. Ian R. Ferrier to Marion Finkel, 25 February 1980. AF17-705, vol. 11, FDAAF.

6. Frost & Sullivan, *The US Generic Drug Market* (New York: Frost & Sullivan, 1976), pp. 4, 16.

7. By December of 1977, SKF offered physicians and pharmacists a "Uni-Price" program that guaranteed a stable price to all clinicians, pharmacists, and patients. International Resource Development, Inc., *Generic Drugs in the 1980s*, p. 41.

8. Jeff Feldman, "No-Name Drugs," *Orange Coast Magazine*, February 1982, pp. 126–32, p. 132. See also Barrie G. James, *The Marketing of Generic Drugs: A Guide to Counterstrategies for the Technology Intensive Pharmaceutical Companies* (London: Associated Business Press, 1984).

9. Pfizer followed suit with a series of medicolegal symposia on the issue of pharmacist liability, the implication being that branded generics offered pharmacists, as well as patients and physicians, an ease of mind not possible with lesser-known generics; "Pharmacy and the Law—Final Script." 11 August 1977, box 4 f 2, Generic Drug Investigation Files, NYGDIF; for more on Pfipharmecs, see International Resource Development, Inc., *Generic Drugs in the 1980s*, p. 40.

10. *FDCR*, 11 September 1978, p. 13.

11. Kefauver confronted the head of Schering with this issue in 1959, noting that Schering did not even produce its own branded prednisone. "Let's get it very clear.

You buy this material from Upjohn. You bought 440 pounds of prednisone in 1958. All you do is put in in a capsule, add your brand name to it, and sell it"; John Blair exhibit, *Administered Prices*, vol. 21; Kefauver, cross-examination of Frank Brown, *Administered Prices*, vol. 14, p. 7859.

12. *Patent Term Extension and Pharmaceutical Innovation: Hearing before the Subcommittee on Investigations and Oversight of the Committee on Science and Technology, U.S. House of Representatives*, Ninety-Seventh Congress (statement of William Haddad), p. 206; *FDCR*, 11 September 1978, p. 13.

13. "Memorandum: Lederle Laboratories Material," David Langdon to Susan Guthrie, 21 March 1978, box 6, f 25, NYGDIF. As Lederle noted in its own internal documents, "Mylan Pharmaceutical is a supplier of the V.A., the United Nations, the World Health Organization, and major pharmaceutical companies such as Abbott, Bristol, Mallinckrodt, Parke-Davis, SKF, Squibb, Wyeth, and A.H. Robbins." "Contract Manufacturer's Profiles Update," 18 May 1976, box 6, f 25, NYGDIF, p. 3. Lederle claimed it imposed an extra quality assurance step on Mylan products that made them independently eligible for the Lederle label. See also "Quality Control of Products Which Are Not Manufactured by Lederle," 23 November 1976, box 6, f 25, NYGDIF, p. 1.

14. April 1961 catalog of Milan Pharmaceuticals, 13 November 1961. By 1963, Milan had set up its own raw manufacturing plant in Pennsauken, NJ, that manufactured drugs, with what FDA inspectors noted to be subpar quality control procedures. FDA inspection report of Milan Pharmaceuticals, 4 February 1963, inspection report of Milan's Pennsauken plant, from f 1, AF17-986, FDAAF. By 1965, however, twenty-seven of the fifty products in the Milan catalog were produced in house, and they had begun to venture into the riskier but potentially more profitable realm of injectable drugs as well. Catalog of Milan Pharmaceutical Co., 11 September 1965, f 1, AF17-986, FDAAF.

15. Marvin Seife to Allen Dines, 5 December 1973, folder 4, vol. AF17-986, FDAAF; Paul Bryan to Smith, Kline, & French Laboratories, 24 May 1973, folder 4, AF17-986, FDAAF; Mary McEniry to Marvin Seife, December 15, 1976, folder 7, vol. AF17-986, FDAAF.

16. "As in the past," he continued, "we will select only those products which offer the greatest potential for return on investment. In most cases, competition among these drugs is less intense because they demand the expertise and quality production on which Mylan has established its reputation." *FDCR*, 8 October 1979, p. 9; *FDCR*, May 31 1982, p. 6; *FDCR*, May 31 1982, p. 6.; James, *Marketing of Generic Drugs*, pp. 60–63.

17. Seymour Blackman to Ted Byers, 27 March 1975, AF13-610, vol.15, FDAAF.

18. Paul Hyman and Herman Rosenstein, "Memo of Telecon," 4 August 1975, f 14, AF 30-841, FDAAF. HEW counsel William Vodra, however, was uncomfortable with the proposition and wrote to the FDA's Marion Finkel that, until the parameters of a key legal case filed by the Swiss manufacturer Roche was fully resolved, a drug marketed for the first time after 1962 under an approved New Drug Application could be marketed by a second firm only after the approval of a full new drug application for that product or by submission of a "paper NDA." William Vodra to Marion Finkel, 11 August 1975, f 14, AF 30-841. Zenith promptly submitted a literature review plus bioequivalence studies and proof of Good Manufacturing Practices as part of its "paper

NDA" for generic diazepam. Marion Finkel to [dir., Bureau of Drugs], 12 December 1975, f 14, AF 30-841, FDAAF.

19. These requests to petition for old drug status made little headway with FDA lawyers, who feared (with good reason) that once a "new drug" was no longer new the FDA might lose authority over its regulation altogether. William Vodra to Marion Finkel, 11 August 1975, f 14, AF 30-841, FDAAF; Marion Finkel to [dir., Bureau of Drugs], 12 December 1975, f 14, AF 30-841, FDAAF. The FDA declared its intent to extend ANDA protocols to post-1962 drugs in 1978 but did not propose any specific regulations to make this happen. 43 Fed. Reg. 39128 (1 September 1978).

20. J. Kevin Rooney to Theodore E. Byers, 17 March 1978, f 16, AF 30-841, FDAAF.

21. Richard Chastonay to J. Kevin Rooney, 15 September 1978, f 16, AF30-841, FDAAF.

22. Larsen continued, "Either the agency decided that once a pioneer drug was demonstrated to be safe, there was no real need to have 'me too' versions duplicate that task, or the agency decided that once safety was demonstrated for a particular drug, subsequent versions of it were no longer new drugs. (We do know that literally thousands of 'not new drug' letters were issued by the FDA) . . . This of course, was the encouragement for the beginning of the generic drug industry." See also Seymour Blackman to Richard Crout, 10 January 1980, vol. 18, FDAAF.

23. If Zenith was one of the few "responsible" generic firms, Crout viewed other firms less favorably—a view supported by correspondence files filled with evidence of evasions, deceptions, and shoddy manufacturing and quality control. News leaked somehow to generic competitors who were not invited, such as Bolar, who complained bitterly of their exclusion from this meeting. Richard Crout to Kenneth Larsen, 13 March 1980, f 20, AF30-841, FDAAF.

24. Kenneth Larsen to Henry Waxman, 30 June 1980, f 21, AF30-841, FDAAF. Larsen also wrote to Republicans, including Richard Schweiker, Ronald Reagan's newly designated appointee as head of HEW. Kenneth Larsen to David Winston, 9 January 1981, f 22, AF30-841, FDAAF; quotation from statement of Kenneth Larsen, chair of the GPIA and president of Zenith Laboratories, Inc., "Post-1962 Drug Approval," 10 March 1981, f 22, AF30-841, FDAAF.

25. William Haddad to GPIA members, 24 January 1984, f 4, AEP; memo to PMA board of directors, 15 July 1984, f 5, AEP.

26. Statement by William F. Haddad, president and CEO of the Generic Pharmaceutical Industry Association, on the Drug Competition Act of 1984, 12 June 1984, f 4, AEP.

27. "Bolar under Spotlight as Generics Become New Glamour Stocks on Wall Street," *FDCR*, 8 April 1985, p. 10.

28. Richard G. Frank, "The Ongoing Regulation of Generic Drugs," *NEJM* 357, no. 20 (2007): 1993–96. Greg Critser has described some of the deal making required in the production of the Hatch-Waxman Act in *Generation Rx* (New York: Houghton-Mifflin, 2005). The legislative history of the Hatch-Waxman Act is told elsewhere in greater detail; e.g., Allan M. Fox and Alan R. Bennett, *The Legislative History of the Drug Price Competition and Patent Term Restoration Act of 1984* (Washington, DC: Food and Drug Law Institute, 1987).

29. The industry journal *FDC Reports* effectively got it right when it reported on

the bill's passage that "for the generic industry the ANDA part of the bill may mark a coming of age." "ANDA/Patent Restoration Bill Culminates 10-Year Cycle of Process & Profits Hearings on Hill," *FDCR*, 10 September 1984, p. s2.

30. International Resource Development, Inc., *Generic Drugs in the 1980s*, p.10.

31. Claude Lévi-Strauss, *The Raw and the Cooked* (Chicago: University of Chicago Press, 1969).

CHAPTER 6. CONTESTS OF EQUIVALENCE

Epigraph. Max Sadove, "What Is a Generic Equivalent?," *American Professional Pharmacist*, February 1965, p. 6.

1. For an engaging narrative of the role of the state in producing bioequivalence as a new form of regulatory science in the years leading up to Hatch-Waxman, see Daniel Carpenter and Dominique Tobbell, "Bioequivalence: The Regulatory Career of a Pharmaceutical Concept," *Bulletin of the History of Medicine* 85, no. 1 (2011): 93–131.

2. *FDCR*, 24 June 1985, p. TG3-4.

3. *FDCR*, 23 September 1985, p. TG2.

4. *FDCR*, 23 September 1985, p. TG2, emphasis mine.

5. The relevance of "paper technologies" in ordering the world of medical practice has been developed most thoroughly by Volker Hess and J. Andrew Mendelsohn, "Case and Series: Medical Knowledge and Paper Technology, 1600–1900," *History of Science* 47 (2010): 287–314; it elaborates the concept of paper tools developed by historians of science such as Ursula Klein, *Experiments, Models, Paper Tools: Cultures of Organic Chemistry in the Nineteenth Century* (Palo Alto, CA: Stanford University Press, 2002). The broader concept—that innovations in forms, lists, and ways of representing affect medical theory and practice as much as X-ray machines do—can also be found in earlier works such as Joel Howells, *Technology in the Hospital: Transforming Patient Care in the Early Twentieth Century* (Baltimore: Johns Hopkins University Press, 1996). As Adrian Johns has noted, early English pharmacopoeiae were technologies that powerfully linked words and things, connecting texts of medicine to material medicines. Adrian Johns, *Piracy: Intellectual Property Wars from Gutenberg to Gates* (Chicago: University of Chicago Press, 2010).

6. United States Pharmacopoeial Convention, *The Pharmacopoeia of the United States*, 8th ed. (Philadelphia: P. Blakiston, 1907), p. 294.

7. United States Prescription Drug Task Force, *The Prescribers: Background Papers* (Washington, DC: Government Printing Office, 1968), p. 26.

8. Tom Mahoney, *The Merchants of Life: An Account of the American Pharmaceutical Industry* (New York: Harper & Brothers, 1959). Lee Anderson and Greg Higby, *The Spirit of Voluntarism: A Legacy of Commitment and Contribution* (Washington, DC: USPC, 1995).

9. Amy Slaton, "As Near as Practicable: Precision, Ambiguity, and the Social Features of Industrial Quality Control," *Technology and Culture* 42 (2001): 51–80; Andrew Russell, "Industrial Legislatures: The American System of Standardization," *International Standardization as a Strategic Tool*, 71–79 (Geneva: International Electrotechnical Commission, 2006).

10. Nichols recalls the concoction of the *United States Pharmacopoeia Digitalis Reference Standard* as a landmark moment in the ability of the USP to issue therapeutic standards for pharmacological equivalence. To conform to the recently developed international digitalis unit, the *United States Pharmacopoeia* standard lot was concocted from "a blend of three lots of high quality digitalis, two thirds from an American source, one-sixth from England, and one-sixth from Germany." A. B. Nichols, "U.S.P. Reference Standards for Biologic Assay," *Journal of American Pharmacists Association* 42 (1953): 215–25, quotation p. 218.

11. A. B. Morrison, D. G. Chapman, and J. A. Campbell, "Further Studies on the Relation between in Vitro Disintegration Time of Tablets and the Urinary Excretion Rates of Riboflavin," *Journal of American Pharmacists Association* 49 (1959): 634. As cited in Gerhard Levy, "Therapeutic Implications of Brand Interchange," *American Journal of Hospital Pharmacy* 17 (1960):756–59, quotation p. 756. Levy's speech provoked a spirited discussion. Donald Francke, an early pioneer of the hospital formulary movement, interpreted the Canadian evidence as a justification for a new role of the pharmacist in the late twentieth century: "if, as Dr. Levy points out, there are therapeutically ineffective brands on the market, it is the responsibility of the hospital pharmacist and the medical staff to sift out these products and eliminate them."

12. D. G. Chapman, R. Crisafir, and J. A. Ampbell, "The Relation between *in Vitro* Disintegration Time of Sugar-Coated Tablets and Physiological Availability of Sodium P-Aminosalicylate," *Journal of American Pharmacists Association* 45 (1956): 374.

13. D. G. Chapman, L. G. Chatten, and J. A. Campbell, "Physiological Availability of Drugs in Tablets," *Canadian Medical Association Journal* 76 (1957): 102–7.

14. K. G. Shenoy, D. G. Chapman, and J. A. Campbell, "Sustained Release in Pelleted Preparations as Judged by Urinary Excretion and in Vitro Methods," *Drug Standards* 27 (1959): 77; A. B. Morrison, C. B. Perusse, and J. A. Campbell, "Physiologic Availability and in Vitro Release of Riboflavin in Sustained-Release Vitamin Preparations," *NEJM* 253, no. 3 (1960): 115–19, quotation p. 118.

15. E. Lozinski, "Physiological Availability of Dicumerol," *Canadian Medical Association Journal* 83, no. 4 (1960): 177–78, quotation p. 178.

16. Enteric coating refers to an extra layer applied to the outside of an oral medication to delay absorption. Gerhard Levy, "Therapeutic Implications of Brand Interchange," *American Journal of Hospital Pharmacy* 17 (1960): 756–59, quotation p. 759; John G. Wagner, William Veldkamp, and Stuart Long, "Correlation of in Vivo with in Vitro Disintegration Times of Enteric Coated Tablets," *Journal of the American Pharmaceutical Association* 47, no. 9 (1958): 681–85, quotation p. 681.

17. John G. Wagner, "Biopharmaceutics: Absorption Aspects," *Journal of Pharmaceutical Sciences* 50, no. 5 (1961): 359–87.

18. Gerhard Levy and Eino Nelson, "Pharmaceutical Formulation and Therapeutic Efficacy," *JAMA* 177, no. 10 (1961): 689–91.

19. Eino Nelson, "Kinetics of Drug Absorption, Distribution, Metabolism and Excretion," *Journal of Pharmaceutical Sciences* 50 (1961):181–92; John G. Wagner, "Pharmacokinetics," *Annual Review of Pharmacology* 8 (1968): 67–94. John G. Wagner, "History of Pharmacokinetics," *Pharmaceutical Therapeutics* 12 (1981): 537–62, quotation p. 537. The term *pharmacokinetics* had earlier been proposed by the German pharmacologist F. H. Dost in 1953, and some of its driving questions can be found in the writings of laboratory-based pharmacologists since the mid-nineteenth century.

20. Wagner, "History of Pharmacokinetics," p. 537.

21. Wagner, "Biopharmaceutics," p. 376.

22. Charles M. Mitchell to Edward G. Feldmann, 23 March 1962, box 251, f 25, USPC. "Inter-Tablet Dosage Variation Committee" (memorandum), 22 March 1962, box 251, f 25, USPC, p. 1. For more on the origins of tolerance in industrial quality control, see Walter A. Shewhart, *Economic Control of Quality of Manufactured Product* (New York: D. Van Nostrand, 1931).

23. "Inter-Tablet Dosage Variation Committee," p. 2

24. C. M. Mitchell, "Memo, PMA Contact Section Tablet Committee," 22 March 1962, box 43251, f 25, USPC.

25. "Procedure for Dissolution Rates for Tablets," March 1962, 251, f 25, USPC.

26. Jerome Bodin to Edward Feldmann, "Memo: PMA Contact Section, Tablet Committee Report," 24 April 1962, b 251, f 25, USPC, p. 2.

27. C. M. Mitchell to Edward Feldmann, 7 February 1963, box 43, f 25, USPC; "Erweka Tester Type AT-3," 11 September 1963, box 251, f 25, USPC, pp. 1–2.

28. Ibid., pp. 4–5.

29. Jerome Bodin to Edward Feldmann, "Memo: PMA Contact Section, Tablet Committee Report"; "Memo, PMA Contact Section Tablet Committee," 18 October 1962, 2 box 251, f 25, USPC, p. 1.

30. Edward G. Feldmann to C. Leroy Graham, 27 June 1968, box 251, f 25, USPC; C. Leroy Graham, "Pharmaceutical Manufacturers Association Memo to Representatives on the Quality Control Section," 19 June 1968, box 251, f 25, USPC.

31. Notably, acetohexamide, hydrochlorothiazide, meprobamate, methandrostenolone, indomethacin, methylprednisolone, nitrofurantoin, prednisone, sulfamethoxazole, sulfisoxazole, theophylline, ephedrine, phenobarbital, and tolbutamide. John Colaizzi, "Pharmacy's Responsibility and the Antisubstitution Law Controversy" (memorandum), 27 August 1970; "Correspondence in Favor of Substitution," box 18.11, f 33.46.1.1, APhAA.

32. Jaime N. Delgado and Frank P. Cosgrove, "Fallacies of Generic Equivalence Thesis," *Texas State Journal of Medicine* 59 (1963):1008–12.

33. Dale G. Friend, "Pharmaceutic Preparation and Clinical Efficacy of Drugs," *Clinical Pharmacology and Therapeutics* 3, no. 3 (1962): 417–20, quotations pp. 418–20.

34. Ibid.

35. Nelson's hearings on the prescription drug industry would fill more than thirty-four volumes and form the basis for his most successful re-election campaign ever in 1968, a year in which president Lyndon Johnson was likewise induced to state—in his health message—that "the taxpayer should not be forced to pay $11 if a $1.35 drug is equally effective. To do this would permit robbery of private citizens with public approval." Milton Silverman and Philip Lee, *Pills, Profits, & Politics* (Berkeley: University of California Press, 1974), p. 147

36. Testimony of Edward Feldmann, *Competitive Problems*, vol. 1, p. 410; Lloyd C. Miller, *Competitive Problems*, vol. 2, p. 508.

37. Testimony of Edward Feldmann, *Competitive Problems*, vol. 1, p. 410.

38. Ibid., pp. 413–14. Full quote: "From a technical standpoint there really is no such thing as complete 'drug equivalence'—whether we compare two drug products sold under their nonproprietary, or generic name; or whether we compare one generic name drug product to a brand name drug product; or whether we compare to

brand name drug products, or even if we compare two batches of the very same drug product of a single firm."

39. Nelson sent the document to Feldmann, who replied that this evidence still contained fewer than five true cases of clinically important difference. United States Prescription Drug Task Force, *Prescribers*, p. 27.

40. Harry Wiener, *Generic Drugs: Safety and Effectiveness* (New York: Pfizer, 1973), p. 42. These arguments were echoed by academic pharmacologists such as Gerhard Levy, who noted that even if only five cases of *proven* difference had challenged the *United States Pharmacopoeia* standards, "there has not been a single study which shows that these standards *are* suitable for assuring bioavailability or therapeutic equivalence." Gerhard Levy, *Drug Intelligence and Clinical Pharmacy* 6 (1972): 18.

41. Testimony of Henry F. DeBoest, *Competitive Problems*, vol. 3, p. 971.

42. Thomas Maeder, interview with Benjamin Gordon, 6 May 1991, cited in Maeder, *Adverse Reactions* (New York: William Morrow, 1994), p. 333.

43. On the role of industry-funded science in the production and maintenance of useful controversies, see Allan Brandt, *The Cigarette Century: The Rise, Fall, and Deadly Persistence of the Product That Defined America* (New York: Basic Books, 2007); Naomi Orsekes and Erik M. Conway, *Merchants of Doubt: How a Handful of Scientists Obscured the Truth on Issues from Tobacco Smoke to Global Warming* (New York: Bloomsbury Press, 2010).

CHAPTER 7. THE SIGNIFICANCE OF DIFFERENCES

Epigraph. Smith, Kline, & French Laboratories, "An Indispensable Book in Perspective . . . The United States Pharmacopoeia" (pamphlet), 1961, group 7, box 52, folder 1, "Industries 5–2: Generics," SOM.

1. Smith was also the former editor of *JAMA* and the former Secretary of the AMA Council on Drugs. Quotation from Austin Smith to James L Goddard, 23 September 1966, AF12-757, f 28, FDAAF. Smith continued, "If you feel it would be helpful we would be glad to have our scientists meet with you or your staff for personal discussion of any questions that may arise." See also "Certification of Chloramphenicol" (position paper), 7 October 1966, f 28; memorandum of meeting, 31 October 1966, f 28; H. Summerson to James L. Goddard, 17 November 1066; Goddard to Smith, 25 November, 1966, f 28; Smith to Goddard, 15 December 1966, f 29; Goddard to Smith, 3 January 1967, f 29, all from AF12-757, FDAAF, all cited in Maeder, *Adverse Reactions*, pp. 320–22. All manufacturers were subject to the regular antibiotic certification process, which required batch-to-batch proofs of identity, strength, quality, and purity of the chemical product by the foreign manufacturing plant—four of which, Goddard pointed out, were owned and operated by Parke-Davis.

2. Sales volume discussed in *FDCR*, 24 October 1966, p. 6. The problem of a tablet failing to dissolve in water suggests that the McKesson product did not even conform to existing compendial standards—the kind of problem the USP disintegration test would have easily pointed out.

3. This initial trial was followed by a larger study of 500 mg oral doses distributed among "volunteer" inmates of the Lorton Reformatory. "Brand, Generic Drugs

Differ in Man," *JAMA* 205, no. 9 (1968): 23–36; Ronald T. Ottes and Robert A. Tucker, "History of the U.S. Food and Drug Administration: Herbert Ley," Oral History, 15 December 1999, Food and Drug Administration, White Oak, MD, p. 8; see also Maeder, *Adverse Reactions*, pp. 326–28.

4. *FDCR*, 11 Nov. 1967, p. TG-2.

5. Maeder, *Adverse Reactions*, p. 328.

6. Herbert L. Ley to James L. Goddard, 16 November 1967, as cited in ibid., pp. 329–30.

7. As Herbert Ley, then director of the FDA Bureau of Medicine, wrote to Zenith in 1968,

> Recent studies have shown that capsules of chloramphenicol with a dissolution rate in 30 minutes of at least 98 percent give adequate blood levels in humans, whereas capsules with slower dissolution rates may or may not correlate well with acceptable blood levels. We are very much interested in continuing to explore the relationship between dissolution rates or other in vitro test systems and blood level data, but in the meantime, for any new formulation, certification must be contingent on the presentation of data showing the material will produce adequate blood levels in humans. Subsequent batches may then be certified on the basis of similar dissolution rate the batch previously certified provided all other requirements are fulfilled.

Herbert L. Ley to Zenith Laboratories, 11 June 1966, AF30-841, vol. 7, FDAAF.

8. Comment of Senator Gaylord Nelson, *Competitive Problems*, vol. 11, p. 4527.

9. As the head of the PMA, C. Joseph Stetler, complained to Nelson, "The burden has always been shifted to the proof of lack of equivalence. Nobody has come up with proof of equivalency . . . It is a question of where the burden of proof lies in this matter." Statement of C. Joseph Stetler, *Competitive Problems*, vol. 4, p. 1368.

10. Alan B. Varley, "The Generic Inequivalence of Drugs," *JAMA* 206, no. 8 (1968): 1745–46.

11.Bean especially scoffed at Varley's claims of having objectively demonstrated that "drug availability," not *United States Pharmacopoeia* specifications, was "the present most sensible and feasible way of establishing generic equivalence of drugs." *FDCR*, 16 December 1968, p. 8; Comment of Ben Nelson, *Competitive Problems*, vol. 10, p. 3925.

12. *FDCR*, 17 March 1969, p. TG-10; *FDCR*, 12 January 1970, p. 18.

13. William M. Heller to Thomas M. Durant, 7 April 1969, box 250, f 32, USPC; Thomas M. Durant to William M. Heller, April 16 1969, box 250, f 32, USPC.

14. Quotation from John Adriani to Alan B. Varley, 18 June 1969, box 34, f 5, JAP. Although Upjohn and others referred to the proprietary "dispersant" VeeGum as an "inert binding gum," this seemed a paradox if the problem at hand was absorption— presumably, a dispersant is not inert in its effects on absorption. Dale Friend, "Generic Drugs and Therapeutic Equivalence," *JAMA* 206, no. 8 (1968): 1785. See also Thomas M. Durant to William M. Heller, 16 April 1969, box 250, f 32, USPC.

15. Lloyd C. Miller to Benjamin Gordon, 10 December 1968, reprinted in *Competitive Problems*, vol. 10, p. 3966. Lloyd C. Miller to Alan Varley, 4 March 1969, *Competitive Problems*, vol. 10, p. 3970.

16. Statement of John Adriani, *Competitive Problems*, vol. 12, p. 5128.

17. United States Task Force on Prescription Drugs, *The Drug Prescribers: Background Papers* (Washington, DC: Government Printing Office, 1968).

18. Milton Silverman and Philip Lee, *Pills, Profits, & Politics* (Berkeley: University of California Press, 1974) p. 152.

19. United States Task Force on Prescription Drugs, *Drug Prescribers*, p. 24.

20. Emphasis mine. The report concluded that only generic versions of critical or life or death drugs with known solubility/absorption problems should be made to provide evidence of "essentially equivalent" biological availability prior to FDA approval; all other drugs needed only to provide proof of chemical equivalence. United States Task Force on Prescription Drugs, *Drug Prescribers*, 21, 31. The original twenty-four compounds tested were aminophylline (adrenergic agent), bishydroxycoumarin (an anticoagulant), chloramphenicol (broad-spectrum antibiotic), chlortetracycline (broad-spectrum antibiotic), diethylstilbestrol (estrogen replacement), diphenydramine (antihistamine), diphenylhydantoin (anticonvulsant), erythromycin (antibiotic), ferrous sulfate (iron), griseofulvin (antifungal), hydrocortisone (steroid), isoniazid (anti-tuberculosis), meperidine (minor tranquilizer), meprobamate (minor tranquilizer), oxytetracycline (antibiotic), para-amino-salicylate (antituberculosis), penicillin G (antibiotic), penicillin V (antibiotic), prednisone (steroid), quinidine (antiarrhythmic), reserpine (antidepressant/antihypertensive), secobarbital (major tranquilizer), sulfisoxazole (antibiotic), tetracycline (antibiotic), thyroid (hormone), tripelennamine (antihistamine), warfarin (anticoagulant).

21. When the DES final report was issued in 1969, it was accompanied by a set of comments solicited from the nearly two hundred panel members regarding their reflections on the problem of therapeutic equivalence of generic and brand-name drugs. Generic equivalence was a question that "continually intruded into the work of the panels." Duke Trexler to DES panelists, 24 January 1969, in *Drug Efficacy Study: A Report to the Commissioner of Food and Drugs* (Washington, DC: National Academy of Sciences, 1969). As the DES reviews of effective drugs were completed and phased into regulatory implementation, the members of the review panels voiced discomfort that DES standards of efficacy would be attached to generically named drug products. Paul A. Bryan and Lawrence H. Stern, "The Drug Efficacy Study: 1962–1970," *FDA Reports*, September 1970, 14–16, p. 15. For more on the divisions between these advisory panels on generic equivalence, see Daniel Carpenter and Dominique Tobbell, "Bioequivalence: The Regulatory Career of a Pharmaceutical Concept," *Bulletin of the History of Medicine* 85, no. 1 (2011): 93–131.

22. United States Task Force on Prescription Drugs, *Final Report* (Washington, DC: Government Printing Office, 1969); "Review Committee of the Task Force on Prescription Drugs, 1967–1969," box 36, f 6–8, JAP; Drug Bioequivalence Study Panel, US Office of Technology Assessment, *Drug Bioequivalence* (Washington, DC: Office of Technology Assessment, 1974).

23. Statement of Robert Berliner before Senate Select Committee on Monopoly, 19 March 1975, *Competitive Problems*, vol. 26, p. 11656.

24. Remarks by Alexander M. Schmidt, commissioner of Food and Drugs, at the Quinquennial Meeting of the United States Pharmacopoeial Convention, Inc., Washington, DC, 22 March 1975, as published in *PDLPA*, 383.

25. *FDCR*, 5 July 1975, p. T&G5.

26. Jerome Philip Skelly, "Biopharmaceutics in the Food and Drug Administration 1968–1993," FDA Generics File, FDAAF, p. 47.

27. Ibid., p. 37. By January of 1978, Cabana claimed to have found the in vitro test that would solve the problem of bioavailability without need for human subjects— his lab had found proof, Cabana claimed to the NAPM, of an "important equation: dissolution can be equated with bioequivalence." Yet even those in the generic industry were not confident they could take Cabana's proclamation at its word. One industry consultant quoted in the pink sheet reported that Cabana's remarks "scared the hell out of me: I just don't think that the dissolution test itself has reached that high a point where you can replicate it." *FDCR*, 30 January 1978, p. 32.

28. Marvin Seife to Jean Callahan and David Langdon, 28 November 1977, box 5, f 56, NYSGDIF, p. 1.

29. "Bioavailability Game and Bernard Cabana, PhD," Marvin Seife to David Langdon, 19 December 1977, box 5, f 56, NYSGDIF, pp. 3–4.

30. Skelly, "Biopharmaceutics in the Food and Drug Administration," pp. 47–48. FDA Oral History, John E. Simmons, conducted by John P. Swann and Robert A. Tucker, 6 June 2006, Rockville, MD, p. 17.

31. The PMA criticized the FDA's approach as exceeding its proper role vis-à-vis the private sector: "this would seem to be an extraordinary method of operation by FDA, to take unto itself a very important activity of this sort and freeze out the industry involved and academic medicine." *FDCR*, 19 August 1984, p. 1, 5. See also Dan Ermann and Mike Millman, "The Role of the Federal Government in Generic Drug Substitution," in *Generic Drug Laws: A Decade of Trial, A Prescription for Progress*, ed. Theodoer Goldberg, Carolee A. DeVitto, Ira E. Raskin (Bethesda: National Center for Health Services Research and Health Care Technology Assessment, 1986), pp. 99–115, esp. p. 103; Carpenter and Tobbell, "Bioequivalence."

32. Abbott comments on Upjohn Studies, 31 October 1974, *Competitive Problems*, vol. 26, pp. 11815–19, quotation p. 11816.

33. Perhaps, Barrows continued, animal models might even produce superior results to in vivo human subjects. *FDCR*, 19 August 1984, p. 9.

34. Manufacturers of several specific drugs continued to fight the equation of bioequivalence with therapeutic equivalence. See, for example, M. B. Ross, "Status of Generic Substitution: Problematic Drug Classes Reviewed," *Hospital Formulary* 24, no. 4414 (1989): 447–49; P. H. Rheinstein, "Therapeutic Inequivalence," *Drug Safety* 5 (1990): 114–19; D. H. Rosenbaum, A. J. Rowan, L. Tuchman, and J. A. French, "Comparative Bioavailability of a Generic Phenytoin and Dilantin," *Epilepsia* 35 (1994): 656–80; R. T. Burkhardt, I. E. Leppik, K. Blesi, S. Scott, S. R. Gapany, and J. C. Cloyd, "Lower Phenytoin Serum Levels in Persons Switched from Brand to Generic Phenytoin," *Neurology* 63, no. 8 (2004): 1494–96; Yu-tze Ng, "Value of Specifying Brand Name Antiepileptic Drugs," *Archives of Neurology* 66 no. 11 (2009): 1415–16.

35. *FDCR*, 4 July 1983, p. 5.

36. *FDCR*, January 13 1986, p. T&G-10.

37. Tamar Lewin, "Drug Makers Fighting Back against Advance of Generics," *NYT*, 28 July 1987, p. A1.

38. Sharon Anderson and Walter Hauck, "Consideration of Individual Bioequivalence," *Journal of Pharmacokinetics and Biopharmaceutics* 18, no. 3 (1990): 259–73. As

Don Schuirmann, of the FDA's Office of Epidemiology and Biostatistics, told the committee, "We should be interested in the population equivalence in so far as it implies the individual equivalence. If the population equivalence does not imply individual equivalence, then population equivalence would be no basis for approval of generic products." All quotes (including in-text quotation from Sharon Anderson) found in *FDCR*, 15 February 1993, pp. 13–14. See also Mei Ling Chen and Lawrence Lesko, "Individual Bioequivalence Revisited," *Clinical Pharmacokinetics* 40, no. 10 (2001): 701–6; "Individual Bioequivalence Stricter Standards Suggested by Committee," *FDCR*, 63, 17 December 2001, 39; On theoretical solutions to theoretical problems, see Iris Pigeot et al., "The Bootstrap in Bioequivalence Studies," *Journal of Biopharmaceutical Statistics* 21 (2011): 1129.

39. "The One the Patient Takes Is Never Tested," Lilly advertisement, *American Druggist*, September 1977. Earlier versions of these advertisements can be found in the *Carolina Journal of Pharmacy* 56 (June 1976).

40. Eli Lilly advertisement, *American Druggist*, September 1977.

41. Eli Lilly, *Implications of Drug Substitution Laws: Analysis and Assessment* (staff report to the Indiana Joint Legislative Committee), June 1978, NYSGDIF, box 4, f 8, p. ii.

42. John Swann, "The 1941 Sulfathiazole Disaster and the Birth of Good Manufacturing Practices," *Journal of Pharmaceutical Science and Technology* 53, no. 3 (1999): 148–53.

43. United States Task Force on Prescription Drugs, *Drug Prescribers*, pp. 32–33.

44. Ibid., pp. 32–34.

45. Henry E. Simmons, "Brand vs. Generic Drugs: It's Only a Matter of Name" (speech presented before the California Council of Hospital Pharmacists, San Diego, 30 Sept 1972), reprinted in *FDA Consumer*, March 1973. J. Richard Crout "The FDA's View of Generic Equivalence," draft of speech given at the 17th Annual Ohio Pharmaceutical Seminar, box 5, f 4, NYSGDIF, p. 6.

46. "The digoxin example clearly implicates smaller manufacturers rather than the leading firms," he continues, "but this is by no means a universal pattern for all drugs." Crout, "FDA's View of Generic Equivalence," p. 6.

47. Statement by Marvin Seife before the Drug Formulary Commission, Department of Public Health, Boston, MA, 8 September 1977, text of speech in box 5, f 4, NYSGDIF, p. 2.

48. See, for example, "The Generic Scam," *Private Practice*, October 1978, p. 30. Lilly's researchers also sampled lots of thirty-five companies' propoxyphene HCl products and found that though all samples of Lilly's Darvon met all parameters, more than 14 percent of competitors failed USP weight variation or content uniformity tests, 72 percent of competitors exceeded USP standards of common acetoxy impurities, 80 percent of the prophyphene/aspirin combinations exceeded USP standards for free salicylic acid. Eli Lilly, *Implications of Drug Substitution Laws*, p. 5. A more condensed PMA study on 1977 data alone showing 73 percent (214) of recalls in that year came from non-PMA firms as compared to 27 percent (79) of PMA firms; 89 percent (81) of all seizures were in non-PMA firms, as compared with 11 percent (10). Pharmaceutical Manufacturers Association, " 'Real World' Factors That Affect What the Consumer Actually Pays at Retail" (backgrounder), 21 July 1978, box 12, f 13, NYSGDIF.

49. Eli Lilly, *Implications of Drug Substitution Laws*, p. 5.

50. Gregory Bateson, "Form, Substance, and Difference," *Steps to an Ecology of Mind* (New York: Ballantine Books, 1972).

CHAPTER 8. SUBSTITUTION AS VICE AND VIRTUE

Epigraph. Bruce M. Chadwick, "Physician-Controlled Source Selection—a Suggested Approach to Substitution," *Journal of Legal Medicine* 4, no. 3 (1976): 27–32.

1. Michigan State Board of Pharmacy v. Casden (Wayne Cnty. Ct. 1949) (No. 301,799), as printed in *Administered Prices*, vol. 21, p. 11761.

2. Ibid., p. 11762.

3. Joseph H. Stamler, "Some Aspects of the Substitution Problem," *Food Drug and Cosmetics Law Journal* 8 (1953): 643–55, quotation p. 645.

4. "Michigan Bill Would Permit Using 1 *USP, NF* Item for Another," *American Druggist*, Feb 18, 1952; 125:16.

5. Mark Prendergast, *For God, Country, and Coca-Cola: The Definitive History of a Great American Soft Drink and the Company That Makes It* (New York: Basic Books, 2000), p. 182.

6. The US Constitution reserved the regulation and licensing of professionals activities largely for the states: this is why, for example, architecture licensing examinations in California can require earthquake certification even though those in New York do not.

7. Schering's effective loss in the *Casden* ruling was one of a series of judicial defeats of brand-name companies seeking to prohibit substitution of chemically equivalent products at the pharmacy. Stamler, "Some Aspects of the Substitution Problem," p. 645, also cited in N. J. Facchinetti, and W. M. Dickson, "Access to Generic Drugs in the 1950s: The Politics of a Social Problem," *American Journal of Public Health* 72 (1982): 468–75; Statement of Newell Stewart, *Administered Prices*, vol. 21, p. 11692. For an extended account of the NPC's origin and early actions, see Dominique Tobbell, *Pills, Power, and Policy: How Drug Companies and Physicians Resisted Federal Reform in Cold War America*, Milbank Series on Health and the Public (Berkeley: University of California Press, 2012).

8. Facchinetti and Dickson, "Access to Generic Drugs in the 1950s," p. 470–71; Tobbell, *Pills, Politics, and Power*, pp. 105–11.

9. Jeremy Greene and David Herzfeld, "Hidden in Plain Sight: Marketing Prescription Drugs to Consumers in the 20th Century," *American Journal of Public Health* 100, no. 5 (2010): 793–803.

10. Statement of Newell Stewart, *Administered Prices*, vol. 21, pp. 11692–95.

11. John L. Hammer Jr., "Substitution on Prescription," *Food Drug and Cosmetics Law Journal* 6 (1951): 775–79.

12. Wilbur E. Powers to Secretaries and Members of Boards of Pharmacy, box 7, f 36, WPIB; Facchinetti and Dickson, "Access to Generic Drugs in the 1950s," p. 471, note that the NPC had by the end of 1955 paid formal visits to every board of pharmacy in the United States.

13. Wilbur E. Powers to Secretaries and Members of Boards of Pharmacy. Like-

wise the NPC's work with state boards of pharmacy began to have an influence on the pharmacy profession on a national level—APhA conventions in 1953, 1955, and 1956 proudly "condemned as unethical the dispensing of a pharmaceutical preparation or brand other than that ordered and prescribed." Facchinetti and Dickson, "Access to Generic Drugs in the 1950s," p. 471.

14. Stewart frequently cited an article from the *American Journal of Pharmacy* in 1897 stating that the pharmacist "has no right . . . to substitute his own or anybody else's preparation for the one specified, even if he is sure the substitute is as good, or, as he may think, better." Statement of Newell Stewart, *Administered Prices*, vol. 21, p. 11696; see also J. J. Galbally, "Substitution as Gross Immorality," *Food Drug & Cosmetic Law Journal* 12 (1957): 759–64.

15. Commenting on the *Casden* case in 1960, Cornell's Walter Modell was careful to distinguish the older sin of "substitution" from the benign action of generic dispensing. "Morally, I have no feeling about this at all," he told Kefauver's committee. "I don't think we are substituting when we give the same thing under another name." Statement of Walter Modell, *Administered Prices*, vol. 21, 11627.

16. George F. Arcambault, "The Formulary System versus the New Concept of 'Substitution,'" *Journal of the American Hospital Association* (*Hospitals*) 34 (1960): 71–73.

17. Tobbell, in *Pills, Power, & Politics,* locates the dissolution of pharmacy's relationship with the pharmaceutical industry in the 1971 speech of APhA president William Apple entitled "Pharmacy's Lib" and the corresponding furor surrounding the APhA white paper. Earlier evidence of this break can be dated to the mid-1960s, however, based on internal correspondence of Apple and colleagues within the archives of the APhA.

18. Quotation from statement of William Apple, *Competitive Problems*, 1967, vol. 5, p. 1294.

19. Almost immediately following the enactment of its Medicaid law, for example, the state of New York tried to amend its antisubstitution law to allow generic substitution for Medicaid patients. Sidney H. Wilig, "Ethical and Legal Implications of Drug Substituting," *FDCLJ* 23, no. 6 (1968): 284–305.

20. Ibid.; see also Dean Linwood Tice to Joseph V. Swintosky, "Drug Products Selection," 12 July 1966, f 18.11, APhAA.

21. Joseph V. Swintosky to Linwood Tice, 3 July 1966, DPS box 3, 18.11, APhAA, p. 1.

22. William Apple, "Special Committee on Substitution," and "Survey of Anti-Substitution Laws" (memoranda), 21 September 1966, DPS box 3, 18.11, APhAA; Turner & Wolin, an advertising agency, offered to develop the resolution as a public-interest advertisement. Irwin Wolin to William S. Apple, "Correspondence in Favor of Substitution," 26 May 1970, box 18.11, f 33.46.1.2, APhAA.

23. Dean Linwood Tice, "Anti-Substitution Laws," *American Journal of Pharmacy* 142, no. 3 (1970): 107–8.

24. A. E. Rothenberger, "He Rocks Our Ivory Tower," *Drug Topics,* 8 June 1970, 38.

25. Quotation from Edward W. Brady to William S. Apple, 2 September 1970, "Correspondence against Substitution," box 18.11, f 33.46.1.2, APhAA. See also American Pharmaceutical Association, "A White Paper on the Pharmacist's Role in Product Selection," *JAPhA,* n.s., 11 (1971): 181–99. Within the APhA the topic of substitution soon

became, in the words of one member, a "schismatic subject." George M. Scattergood to George B. Griffenhagen, 20 May 1970; Mark E. Barmak to William S. Apple, 5 May 1970, both in "Correspondence against Substitution," box 18.11, f 33.46.1.2, APhAA.

26. Edward O. Leonard to APhA, 21 June 1971, "Correspondence against Substitution," box 18.11, f 33.46.1.2 , APhAA.

27. James D. Hawkins to Edward O. Leonard, 22 July 1971, "Correspondence against Substitution," box 18.11, f 33.46.1.2, APhAA.

28. Edward O. Leonard to APhA, 26 October 1971, "Correspondence against Substitution," box 18.11, f 33.46.1.1, APhAA.

29. Albert C. Veid to William S. Apple, 22 May 1970, "Correspondence in Favor of Substitution," box 18.11, f 33.46.1.1, APhAA.

30. Quotation from Donald E. Francke, "The Formulary System: Product of the Teaching Hospital," *Hospitals* 41, no. 22 (1967): 110–16, p. 122.

31. Quotations from statement of August Groeschel, *Administered Prices*, vol. 21, pp. 11566, 11571. Quotation attributed to Robert Hatcher (professor of pharmacy at Cornell and head of the formulary) by Groeschel. See also Francke, "Formulary System." The abundance of new drugs in the decades of the 1950s made Groeschel despair of the future of the formulary as a regularly printed book. "More than anything else the accelerated pace of the changes in pharmacotherapy in the last decade has made it virtually impossible to publish a useful formulary which is bound between hard covers" (11571).

32. Statement of August Groeschel, *Administered Prices*, vol. 21, p. 11568, quotation p. 11571.

33. The formulary concept spread quickly among inpatient facilities in the postwar years and with it the agency of the hospital pharmacist. By 1959, the American Society of Hospital Pharmacists had set up the *American Hospital Formulary Service* which provided drug monographs to participating hospitals to help them create and update their own formularies. After the passage of the Social Security Amendments of 1965 (which enacted Medicare), the hospital formulary was rendered more important by the stipulation that Medicare would only reimburse costs of drugs for hospitalized patients if they were in a formulary adopted in an accredited hospital by the P&T committee or other acceptable compendium. "Statement on the Pharmacy and Therapeutics Committee," *American Journal of Hospital Pharmacy* 16 (1959).

34. Edward W. Brady to William S. Apple, 2 September 1970, "Correspondence against Substitution," box 18.11, f 33.46.1.2, APhAA. See also statement of August Groeschel, *Administered Prices*, vol. 21, p. 11573.

35. T. Donald Rucker, "The Role of Drug Formularies and Their Relationship to Drug Product Selection," in *Generic Drug Laws: A Decade of Trial—A Prescription for Progress*, ed. Theodore Goldberg, Carolee A. DeVitro, and Ira E. Raskin (Washington, DC: United States Department of Health and Human Services, 1986), pp. 465–86.

36. United States Task Force on Prescription Drugs, *The Drug Prescribers: Background Papers* (Washington DC: Government Printing Office, 1968), p. 43.

37. Ibid.

38. By the early twentieth century, Scandinavian countries had set up national formularies to focus the provision of essential medical supplies to citizens, and in the wake of World War II many countries that had set up unified single-payer national

health insurance programs followed suit. United States Task Force on Prescription Drugs, *Current American and Foreign Programs: Background Papers* (Washington, DC: Government Printing Office, 1968).

39. Ibid.; United States Task Force on Prescription Drugs, *Drug Prescribers*, p. 48. See also Pierre S. Del Prato to William F. Haddad, 31 March 1977, box 7, f 24, NYSGDIF.

40. Some early state programs (such as those in Illinois and Georgia) had considered "open" formulary designs, in which drugs not listed on the formulary would be reimbursable as long as there was no equivalent in the formulary. Other programs (such as California and Tennessee) used "closed" formulary systems in which drugs not on the list would only be reimbursed if physicians provided special authorization. Neighborhood health centers often followed formularies made by the Public Health Service and the Veterans Administration. United States Task Force on Prescription Drugs, *Current and American Foreign Programs*, pp. 42, 46.

41. New Mexico used the *Physicians' Desk Reference* as its formulary in 1967. Quotations from United States Task Force on Prescription Drugs, *Drug Prescribers*, p. 46.

42. Quotations from United States Task Force on Prescription Drugs, *Drug Prescribers*, p. 46. See also Judith Nelson, "What Ever Happened to the Pharmacy Class of '47?," *Private Practice*, September 1982, 61–66.

43. "Report of Reference Committee B," *JAPhA* 10, no. 6 (1970): 338–39; "A White Paper on the Pharmacist's Role in Drug Product Selection," *JAPhA* 11, no. 4 (1971): 181–99; American Pharmaceutical Association Academy of Pharmaceutical Sciences and Board of Trustees, *Critique and Response to "A White Paper on the Pharmacist's Role in Drug Product Selection"* (Washington, DC: American Pharmaceutical Association, 1972).

44. Fred Wegner, "A Consumer's Perspective on the Use and Costs of Prescription Drugs," in Goldberg, DeVitro, and Raskin, *Generic Drug Laws*, pp. 117–24, quotation pp. 117–18.

45. An internal survey of accredited schools of pharmacy conducted in 1975 showed that 92 percent of the seventy-one responding schools actively included drug product selection—or brand substitution—in the formal education of pharmacists. Edward Feldmann, "Drug Product Selection" (memorandum), 20 June 1975; "Correspondence in Favor of Substitution," both in box 18.11, f 33.46.1.1, APhAA. See also Louis M. Sesti, "DPS assessment report," *Michigan Pharmacist* 1 (July 1975): 7–8; see also C. Joseph Stetler, "Purchasing Drugs," *Wilmington, Delaware, Evening Journal*, 7 March 1975.

46. Statement of Robert A. Lewinter, *PDLPA*, pp. 69, 72.

47. Maryland had passed a bill favoring generic prescribing, but the PMA considered it a "narrow escape," since it had no teeth to force physicians or pharmacists to comply. William C. Cray, *The Pharmaceutical Manufacturers Association: The First 30 Years* (Washington, DC: Pharmaceutical Manufacturers Association, 1989), pp. 219–20; Christopher S. Harrison, *The Politics of the International Pricing of Prescription Drugs* (Westport, CT: Praeger, 2004), p. 54.

48. Both quotations from Robert C. Johnson, "The Changing Role of the Pharmacist," in Goldberg, DeVitro, and Raskin, *Generic Drug Laws*, pp. 58, 473. See also "Positive Formulary," n.d., box 4, f 10, NYSGDIF.

49. Allen White to William Haddad, 10 June 1977, box 4, f 10, NYSGDIF, p.1; see also

Rucker, "Role of Drug Formularies and Their Relationship to Drug Product Selection," in Goldberg, DeVitro, and Raskin, *Generic Drug Laws*, p. 475.

50. Dominique Tobbell, "Eroding the Physician's Control over Therapy," in *Prescribed: Writing, Filling, Using, and Abusing the Prescription in Modern America*, ed. Jeremy Greene and Elizabeth Watkins (Baltimore: Johns Hopkins University Press, 2012).

51. Weinberger, as quoted in Dan Ermann and Mike Millman, "The Role of the Federal Government in Generic Drug Substitution," in Goldberg, DeVitro, and Raskin, *Generic Drug Laws*, p. 102.

52. "Report on FDA List of Drugs Presenting Actual or Potential Bioequivalence Problem ('Blue Book')" (undated memorandum), box 5, f 4, NYSGDIF.

53. Ibid., pp. 2–3.

54. "Report on FDA List of Drugs Presenting Actual or Potential Bioequivalence Problem ('Blue Book')," p. 5.

55. Statement by Marvin Seife before the Drug Formulary Commission, Department of Public Health, Boston, MA, 8 September 1977, text of speech in box 5, f 4, NYSGDIF, p. 2; Fred Wegner to Donald Kennedy, 3 April 1978, box 5, f 4, NYSGDIF, p.1; Marvin Seife to William Haddad (memorandum), "Therapeutic Equivalence Categories for Approved New Drug Products," 22 September 1978, box 5, f 4, NYSGDIF.

56. Fred Wegner to Donald Kennedy, 3 April 1978, box 5, f 4, NYSGDIF, pp. 2–3.

CHAPTER 9. UNIVERSAL EXCHANGE

Epigraph. Felton Davis Jr. to William F. Haddad, 16 June 1977, box 13, f 4, NYSGDIF.

1. T. Donald Rucker, "The Role of Drug Formularies and Their Relationship to Drug Product Selection," in *Generic Drug Laws: A Decade of Trial—A Prescription for Progress*, ed. Theodore Goldberg, Carolee A. DeVitro, and Ira E. Raskin, pp. 465–86 (Washington, DC: United States Department of Health and Human Services, 1986), p. 475.

2. William C. Cray, *The Pharmaceutical Manufacturers Association: The First 30 Years* (Washington, DC: Pharmaceutical Manufacturers Association, 1989), p. 224; Morton Kondracke, "Long Pursuit of the Drug Price-Fixing Rebates," *Chicago Sun-Times*, 9 February 1969, pp. 7–9.

3. Kondracke, "Long Pursuit of the Drug Price-Fixing Rebates."

4. For comparisons with Kefauver, see "Battle of the Bulge," *Newsweek*, 29 May 1967, 103; memo to Bill Haddad, 17 March 1967, box 7, f 21, NYSGDIF; George Glotzer, "How Generic Drug Dispensing Can Be Accomplished," 19 January 1968, box 7, f 21, NYSGDIF; William F. Haddad to Mayor John V. Lindsay, 14 June 1967, box 7, f 21, NYSGDIF; Governor Nelson A. Rockefeller to William F. Haddad, 18 April 1967, box 7, f 21, NYSGDIF.

5. In the 1969–1970 legislative season, bill S. 3954 was submitted "to establish and maintain in office, current, official list of drugs generically equivalent to trade name drugs" to encourage generic substitution. When this failed, Strelzin resubmitted a new bill on generic substitution in 1971–1972 (A. 11428 / S. 9605). Lloyd Stewart to William F. Haddad, 5 June 1975, "Memo: Analysis of Pending Drug/Pharmaceutical Legislation," box 12, f 8, NYSGDIF.

6. Quotations from "NY State Law: Green Book" (MS), n.d., box 13, f 4, NYSGDIF. Strelzin believed that all prior attempts to pass a substitution law in New York had foundered on questions of bioequivalence. Statement of Harvey L. Strelzin, *PDLPA*, 249.

7. Going into the 1977–1978 legislative session, five bills were already in motion, representing three very different approaches to the issue. Two bills in the assembly (A. 6489 / A. 6512, Grannis/Harenberg) recommended generic prescribing. A pair of bills in the assembly and senate (S. 1623 / A. 1502, McFarland/Fremming) proposed permissive substitution. Another pair (S. 2659 / A. 6130; Pisani/Passannante, and A.5884, Harenberg) mandated generic substitution at the pharmacy. "Generic Drug Substitution for Prescriptions," *Senate Research Service Issues in Focus* (Albany: n.p., 1977), vol. 35, p. 1.

8. William F. Haddad to Don Hewitt, 18 April 1977, box 7, f 24, NYSGDIF, p. 5.

9. Both quotations from William F. Haddad to Don Hewitt, 18 April 1977, box 7, f 24, NYSGDIF, p. 3.

10. For example, for 50 mg tablets of hydrochlorothiazide, the FDA had cleared Bolar Pharmaceutical Company, Zenith Laboratories, and Cord Laboratories as manufacturers of therapeutically exchangeable versions of Merck's brand name Diuril. In the NYGDIF files, the original five requested source files are preserved: (a) ANDAs 1970–1975 (1200), (b) approved ANDAs 1976–1977 (250), (c) DESI project report 1938–1962 (1550), (d) computer printout 1962–1969), and (e) pre-1938 drugs (100). "FDA Drug Formulary" (MS), n.d.; "Record of the New York State Formulary for Safe, Effective, and Interchangeable Drugs," n.d., all in box 12, f 8, NYSGDIF.

11. William F. Haddad to Harvey Strelzin, 21 April 1977, box 7, f 24, NYSGDIF.

12. William F. Haddad to Don Hewitt, 18 April 1977, box 7, f 24, NYSGDIF, p. 2.

13. William F. Haddad to Representative Lester L. Wolff, 29 April 1977, box 7, f 24, NYSGDIF.

14. Memorandum of a telephone conversation: Paul DeMarco and Gene Knapp; memorandum of a telephone conversation: David Langdon and Gene Knapp, both 17 May 1977, box 12, f 8, NYSGDIF.

15. Felton Davis Jr. to William F. Haddad, 16 June 1977, box 13, f 4, NYSGDIF.

16. The delay was connected in part to deliberations over drugs not yet found to be proven effective under DESI review but for which some compelling medical need existed and adequate methodology was not yet available to establish substantial proof of efficacy. William Haddad, David Langdon, Donald Kennedy, J. Paul Hile, Carl M. Leventhal, Gene Knapp, Marvin Seife, and Allan Shurr (memorandum of meeting), 5 May 1977, box 12, f 8, NYSGDIF. Quotation from Gene Knapp to William F. Haddad, 23 May 1977, box 12, f 8, NYSGDIF. Premo Laboratories, for example, wrote the FDA early in 1979 after learning that its version of tolbutamide was not allowed for substitution in the state of New York. "Dear Sir," Premo's letter began,

> It has come to our attention from one of our customers that Premo's Tolbutamide Tablets 0.5g cannot be substituted for the Upjohn product Orinase in the State of New York because the Premo product does not appear listed in a little green book called, "Safe, Effective and Therapeutically Equivalent Drugs" which is a list of drugs published by the State of New York Department of Health and certified by the Food and Drug Adminis-

tration. In addition, this list does not include several other important Premo products. We were wondering on what basis they have been excluded from this certified list.

The FDA's reply was curt. Since no data on the bioavailability for Premo products had been provided prior to the formation of the *Green Book*, "FDA is not in a position to verify the therapeutic equivalence of your product to the innovator's until your study has been reviewed and approved." Steven T. Blackman to Gene Knapp, 26 April 1979; Paul A. Bryan to Steven T. Blackman, 7 May 1979, both in AF13-610, vol. 18.

17. "Record of the New York State Formulary for Safe, Effective, and Interchangeable Drugs."

18. William E. Haddad memo to editors, "FDA Certifies List of Interchangeable Drug," 27 May 1977, box 12, f 8, NYSGIF.

19. William F. Haddad to Edward Kosner, 3 June 1977; William F. Haddad to Robert Laird, 3 June 1977, both in box 7, f 24, NYSGDIF.

20. "New York State Backs Generics Bill," *Chain Store Age*, September 1977, p. 111.

21. Quotation from Gene Knapp to Harriet Morse, 6 July 1977, box 12, f 8, NYGDIF; see also Mary Doug Tyson, "Drug Monographs—Bureau Director Staff Meeting of June 23, 1977" (memorandum), 28 June 1977, box 12, f 8, NYSGDIF.

22. *Safe, Effective and Therapeutically Equivalent* Prescription Drugs* (Albany: New York State Department of Health), 1 October 1977, box 5, f 3, NYSGDIF, quotation p. 2; see also "New York Drug List Expected to Have Major National Impact," *Hospital Formulary*, August 1977, p. 498.

23. Op-ed, *NYT*, 9, 15 June 1977.

24. Quotations from C. Joseph Stetler to Donald Kennedy, 5 July 1977, box 12, f 5, NYSGDIF, p. 2. The PMA would later dismiss the *Green Book* as "an administrative compilation of the firms that had been granted market approval." William F. Haddad, 10 August 1978, box 12, f 4, NYSGDIF. Stetler complained to the FDA that Marvin Seife's "coercing" of the New York State Assembly to pass a substitution law was a violation of federal law (5 U.S.C. 7322)—a move that Kennedy dismissed as ridiculous. C. Joseph Stetler to Donald Kennedy, 9 November 1977, box 12, f 4, NYSGDIF; Donald Kennedy to C. Joseph Stetler, 14 March 1978, box 12, f 4, NYSGDIF. As Stetler wrote to Haddad in August of 1977, "How FDA can say that non-existent bioequivalence standards can show clinical comparability is still a mystery. Clinical comparisons demonstrate clinical comparability." C. Joseph Stetler to William F. Haddad, 18 August, 1977, box 12, f 4, NYSGDIF, p. 4. In an earlier letter Stetler had asked Kennedy for (1) records of FDA inspections for all firms mentioned, (2) proof of comparable clinical effectiveness for each product, (3) records of all bioavailability data, and (4) proof of compliance with standards. Joseph Stetler to Donald Kennedy, 5 July 1977, box 12, f 5, NYSGDIF, p. 2. Haddad would later write to Stetler in December 1977 that though the PMA claimed to have found fifty-three errors in the New York list, they had never specified any of them. "The purpose of this mailgram," Haddad concluded, "is to document for the record your failure to supply this information and to respectfully implore you to quote put up or shut up unquote." William F. Haddad to C. Joseph Stetler, 2 December 1977, box 12, f 4, NYSGDIF.

25. Quotations from Marvin Seife to Jean Callahan and David Langdon, 16 December 1977, box 5, f 56, NYSGDIF, pp. 4–5.

26. Marvin Seife to David Langdon, "Bioavailability Game and Bernard Cabana, PhD," 19 December 1977, box 5, f 56, NYSGDIF, p. 2.

27. Quotations from Marvin Seife to Jean Callahan and David Langdon, 16 December 1977, box 5, f 56, NYSGDIF, pp. 5–6, emphasis mine.

28. Donald Kennedy to Robert P. Whelan, 23 January 1978, p. 1; Kennedy to State Health Officers, State Boards of Pharmacy, and State Drug Program Officials, 31 May 1978, p. 1, both in box 4, f 15, NYSGDIF.

29. *FDCR*, 30 January 1978, quotation pp. 33–36; "FDA to Become Distributor for New York Drug List," *PMA Newsletter* 2, no. 5 (1978), quotation p. 1.

30. Statement of Donald Kennedy, *SPDA*, 191; also cited in "Out of Its Bioequivalence Depth?," *Pharmaceutical Technology*, October 1978, p. 72.

31. *What Is a Generic Drug?* (Montpelier, VT: Vermont Public Interest Research Group, 1969, box 4, f 5, NYSGDIF.

32. Kennedy to State Health Officers, State Boards of Pharmacy, and State Drug Program Officials, p. 1.

33. "Policy Statement: National Consumer Alliance on Prescription Drugs," 8 June 1978, box 9, f 8, NYSGDIF, p. 2. In its early documents the alliance was particularly concerned that the Drug Reform Act proposed by Sen. Edward Kennedy (D-MA) "is silent on the generic-trade name controversy." "Second Draft—National Consumer Alliance on Prescription Drugs," 12 May 1978, box 9, f 8, NYSGDIF, pp. 8–9.

34. Patricia S. Coyle, "Fear and Loathing and Generic Drugs," *Private Practice*, September 1978, 18–45, quotation p. 18.

35. C. Joseph Stetler, "New York State and Drug Lists: A History of Confusion," *Medical Marketing and Media*, September 1978, 36–43.

36. Box 7, f "Drugs: Latin America," NYSGDIF.

37. Quotation from *FDCR*, 18 November 1978, p. 11; see also "Industry Sues State of New York," *PMA Bulletin*, September 1978, p. 1. The PMA was joined by a similar case launched by the New York State Pharmaceutical Society, which was heard by the Second Circuit Appeals Court in the same week.

38. "Out of Its Bioequivalence Depth?," p. 16.

39. Ibid.

40. Abbott Laboratories' erythromycin Filmtabs—covered in a thin layer of lacquer—were inappropriately listed to be therapeutically equivalent with eight generic uncoated versions. Filmtabs reportedly could be taken with meals, while uncoated erythromycin could not (because of absorption issues—it was widely known at the time that erythromycin lost 90 percent of its bioavailability if taken with a meal). FDA moved swiftly to advise a change in labeling for erythromycin, but the move hurt the credibility of the *Green Book*. "Out of Its Bioequivalence Depth?," p. 16; "New Substitution Risk: Rx Labeling Differences," *Drug Topics*, 1 August 1978, p. 17.

41. Richard M. Cooper to HEW commissioner, "Model Generic Drug Substitution Bill," 2 May 1978, box 4, f 15, NYSGDIF. The FTC had been investigating the issue of prescription drug price disclosures for several years in the late 1970s and prepared its own model generic drug law; see also Federal Trade Commission, *Drug Product Selection* (staff report) (Washington, DC: Government Printing Office), 1978.

42. Statement of John Murphy, *SPDA*, pp. 1, 3, 4, 13–14.

43. Statement of Fred Wegner, *SPDA*, p. 70.

44. Statement of John H. Budd, *PDLPA*, p. 288.

45. Statement of Nicholas Ruggieri, *SPDA*, p. 82.

46. Other progeneric advocates in the federal government saw the bill as a form of either over- or underreach. Nelson criticized the House bill as not going far enough to "mitigate the evils caused by the extensive use of drug trade names," which he believed needed to be abolished outright. HEW and the FDA both objected that the federal government had no business regulating professional practice. HEW's efforts at supporting substitution, then, lay clearly with the states. Although the FTC had proposed a model generic substitution bill for use at the state level, it did not see a federal law as a viable solution. Statement of Sen. Gaylord Nelson, *PDLPA*, p. 84; statement of Gene R. Haislip, *PDLPA*, p. 154; statement of Michael Perschuck, *SPDA*, pp. 167–68.

47. Statement of Theodore Goldberg, *SPDA*, p. 14.

48. Of particular interest to Goldberg and his colleagues in this nascent field was the underwhelming uptake of generic substitution among pharmacists in most states studied—less than 3 percent of all prescriptions in available samples by 1978. Statement of Theodore Goldberg, *SPDA*, p. 17.

49. Carolee A. Devito and Theodore Goldberg, "Methods of Drug Product Substitution Analysis"; and "Prospects for a Second Generation of State Generic Drugs Substitution Laws," in *Generic Drug Laws: A Decade of Trial—A Prescription for Progress*, ed. Theodore Goldberg, Carolee A. DeVitro, and Ira E. Raskin (Washington, DC: United States Department of Health and Human Services), 1986, quotations pp. 199, 527.

50. W. H. Shrank, T. Hoang, S. L. Ettner, P. A. Glassman, K. Nair, D. DeLapp, J. Dirstine, J. Avorn, and S. M. Asch, "The Implications of Choice: Prescribing Generic or Preferred Pharmaceuticals Improves Medication Adherence for Chronic Conditions," *Archives of Internal Medicine* 166, no. 3 (13 February 2006): 332–37; W. H. Shrank, M. Stedman, S. L. Ettner, D. DeLapp, J. Dirstine, M. A. Brookhart, M. A. Fischer, J. Avorn, and S. M. Asch, "Patient, Physician, Pharmacy, and Pharmacy Benefit Design Factors Related to Generic Medication Use," *Journal of General Internal Medicine* 22, no. 9 (September 2007): 1298–304, e-pub 24 July 2007; W. H. Shrank, S. L. Ettner, P. Glassman, and S. M. Asch, "A Bitter Pill: Formulary Variability and the Challenge to Prescribing Physicians," *Journal of the American Board of Family Medicine* 17, no. 6 (November–December 2004): 401–7. W. H. Shrank, "Effect of Incentive-Based Formularies on Drug Utilization and Spending, *NEJM* 350, no. 10 (4 March 2004): 1057; W. H. Shrank, A. K. Choudhry, J. Agnew-Blais, A. D. Federman, J. N. Liberman, J. Liu, A. S. Kesselheim, M. A. Brookhart, and M. A. Fischer, "State Generic Substitution Laws Can Lower Drug Outlays under Medicaid," *Health Affairs* 29, no. 7 (July 2010): 1383–90.

51. National Conference of State Legislatures, *Condition-Specific Drug Substitution Legislation: Epilepsy*, print publication (Washington, DC: National Conference of State Legislatures, May 2009), online update March 2013, www.ncsl.org/issues-research/health/rx-substitution-by-pharmacists-state-legislation.aspx. On the pharmacoepidemiology of generic/brand switching of antiepileptic drugs, see Paul Crawford et al., "Are There Potential Problems with Generic Substitution of Antiepileptic drugs? A Review of Issues," *Seizure* 15, no. 3 (2006): 165–76; Joshua J. Gagne, Jerry Avorn, William H. Shrank, and Sebastian Schneeweiss, "Refilling and Switching of Antiepileptic Drugs and Seizure-Related Events," *Clinical Pharmacology & Thera-*

peutics 88, no. 3 (2010): 347–53; Aaron S. Kesselheim, Margaret R. Stedman, Ellen J. Bubrick, Joshua J. Gagne, Alexander S. Misono, Joy L. Lee, M. Alan Brookhart, Jerry Avorn, William Shrank, "Seizure Outcomes Following Use of Generic vs. Brand-Name Antiepileptic Drugs: A Systematic Review and Meta-analysis," *Drugs* 70, no. 5 (2010): 605–21. On the recent history of Generic Drug User Fee Amendments, see www.fda. gov/ForIndustry/UserFees/GenericDrugUserFees/default.htm.

52. Troyen A. Brennan and Thomas H. Lee, "Allergic to Generics," *Annals of Internal Medicine* 141 (2004): 126–30; Michael A. Steinman, Mary-Margaret Chren, and C. Seth Landefeld, "What's in a Name? Use of Brand versus Generic Drug Names in United States Outpatient Practice," *Journal of General Internal Medicine* 22, no. 5 (2007): 645–48; W. H. Shrank, J. N. Liberman, M. A. Fischer, J. Avorn, E. Kilabuk, A. Chang, A. S. Kesselheim, T. A. Brennan, and N. K. Choudhry, "The Consequences of Requesting 'Dispense as Written,'" *American Journal of Medicine* 124, no. 4 (April 2011): 309–17; W. H. Shrank, J. N. Liberman, M. A. Fischer, C. Girdish, T. A. Brennan, and N. K. Choudhry, "Physician Perceptions about Generic Drugs; W. H. Shrank, E. R. Cox, M. A. Fischer, J. Mehta, and N. K. Choudhry, "Patients' Perceptions of Generic Medications," both in *Health Affairs* (Millwood) 28, no. 2 (March–April 2009): 546–56; Keri Sewell et al., "Perceptions of and Barriers to Use of Generic Medications in a Rural African-American Population, Alabama, 2011," *Preventing Chronic Disease* 9 (2012): E142; Katie Thomas, "Why the Bad Rap on Generic Drugs?," *NYT*, 5 October 2013.

53. Edwin A. Coen to William F. Haddad, 7 August 1978, box 6, f. 16, NYSGDIF.

CHAPTER 10. LIBERATING THE CAPTIVE CONSUMER

Epigraph. Consumer Reports, The New Medicine Show: Consumers Union's New Practical Guide to Some Everyday Health Problems and Health Products (Mount Vernon, NY: Consumers Union, 1989), p. 264.

1. Although health consumerism emerged as a political force earlier in the twentieth century—indeed, much earlier—the public discourse of patients as consumers was greatly amplified from the 1960s onward. This point is also argued by Nancy Tomes in "Merchants of Health: Medicine and Consumer Culture in the United States, 1900–1940," *Journal of American History* 88, no. 2 (2001): 519–47; and in her forthcoming book *Impatient Consumers: Consumer Culture and the Making of Modern American Medicine.*

2. E. F. Trapp to John Gardner, 9 August 1967; E. F. Trapp to John Goddard, 15 August 1967; E. F. Trapp to John Goddard, 30 August 1967; Jeanne M. Mangels to E. F. Trapp, 31 August 1967; all from FDAR, accession 1938–1974, box 1967, 500.62.

3. Edna M. Lovering to E. F. Trapp, 17 August 1967, FDAR, accession 1938–1974, box 1967, 500.62.

4. E. F. Trapp to Edna M. Lovering, 20 November 1967, FDAR, accession 1938–1974, box 1967, 500.62.

5. Lizabeth Cohen has narrated the history of three "waves" of the consumer movement in *A Consumers' Republic: The Politics of Mass Consumption in Postwar America* (New York: Vintage Books, 2003), which compares the consumer movement

to other movement politics over the course of the twentieth century, including the civil rights movement and the women's movement. Other historians, most notably Lawrence Glickman in *Buying Power: A History of Consumer Activism in America* (University of Chicago Press: 2009), have complicated this periodic history of the self-conscious consumer *movement* as only one oscillation in the more dynamic chronology of consumer *activism*. As historians of medicine have demonstrated, consumer activism in the sphere of health care has been central to several social movements from health feminism to the health wing of the Black Panthers; see, e.g., Wendy Kline, *Bodies of Knowledge: Sexuality, Reproduction, and Women's Health in the Second Wave* (Chicago: University of Chicago Press, 2010); Alondra Nelson, *Body and Soul: The Black Panther Party and the Fight against Medical Discrimination* (Minneapolis: University of Minnesota Press, 2011). For a recent account of the role of pharmaceutical consumerism in global consumer activism in the 1970s, see Matthew Hilton, *Prosperity for All: Consumer Activism in an Era of Globalization* (Ithaca, NY: Cornell University Press, 2009).

6. Statement of Mildred E. Brady, 11 May 1960, *Administered Prices* vol. 21, p. 11532.

7. *Administered Prices*, vol. 14, p. 7839. The political derivation of this dispute is best described in Richard McFadyen, "Estes Kefauver and the Drug Industry" (PhD diss., Emory University, 1973). For earlier accounts, see Estes Kefauver with Irene Till, *In a Few Hands: Monopoly Power in America* (New York: Pantheon Books, 1965); Richard Harris, *The Real Voice: The First Fully Documented Account of Congress at Work* (New York: Macmillan, 1964). For more recent treatments, see Daniel Carpenter, *Reputation and Power: Organizational Image and Drug Regulation at the FDA* (Princeton, NJ: Princeton University Press, 2010); and Dominique Tobbell, *Pills, Power, and Policy: The Struggle for Drug Reform in Cold War America*, Milbank Series on Health and the Public (Berkeley: University of California Press, 2012).

8. As Kefauver's team demonstrated, the prices paid by such consumers could vary as much as twentyfold among "wonder drugs" of the era, from cortisone to antibiotics to tranquilizers to antidiabetic agents. See Richard McFadyen, "Estes Kefauver and the Drug Industry" (Ph.D. diss., Emory University, 1973); see also *Administered Prices*, vols. 4, 15–17, and 20.

9. Statement of Mildred E. Brady, 11 May 1960, *Administered Prices*, vol. 21, p. 11532.

10. Brady continued,

> The electric appliance owner's lot is more difficult but wood cookstoves and iceboxes are available, even if they are not too handy to live with. Furthermore, since the rates of the utility are controlled by governmental regulations, the consumer can join with his neighbors to protest excessive charges; and his protest, if soundly based and persistently noisy, can be expected to have an influence, however small, on what he pays the gas and electric companies. For the recalcitrant taxpayer, the road is still rougher, but we all know the story of Henry David Thoreau. Not that CU is endorsing this or any other tax evasion strategy. I refer to it simply to remind us here that even the tax rebel Thoreau suffered only jail when he chose not to submit to a payment he felt was unfairly levied. But what could a fever-ridden Thoreau have done had it been the price of his physician's prescription that he felt was an unjust exercise of economic power?

Statement of Mildred E. Brady, 11 May 1960, *Administered Prices*, vol. 21, p. 11532. On the sociology of the sick role, see Talcott Parsons, *The Social System* (Glencoe, IL: Free Press, 1951); Samuel William Bloom, *The Word as Scalpel: A History of Medical Sociology* (Oxford: Oxford University Press, 2002); John C. Burnham, "The Death of the Sick Role," *Social History of Medicine* 25, no. 4 (2012): 761–76.

11. Quotation from *Consumer Reports, The Medicine Show: Some Plain Truths about Popular Remedies for Common Ailments* (New York: Simon & Schuster, 1961), p. 181, also cited in Nancy Tomes, "The Great American Medicine Show Revisited," *Bulletin of the History of Medicine* 79, no. 4 (2005): 627–63., p. 651. Several of the CU's founding members—including Arthur Kallett and Harold Aaron—had provided some of the original impetus behind Kefauver's investigation.

12. Arthur Kallet and F. J. Schlink, *100,000,000 Guinea Pigs: Dangers in Everyday Foods, Drugs, and Cosmetics* (New York: Vanguard Press, 1932). It remains a minor irony that this work was mobilized to press for legislation that became the Food, Drugs, and Cosmetics Act of 1938, which indirectly led to the creation of the prescription-only drug (and therefore increased the "captivity" of the medical consumer).

13. Harold Aaron, *Health and Bad Medicine* (New York: Robert McBride, 1940), p. viii.

14. Initial quotation from Joan Cook, "Harried Shoppers Rely on a Research Group," *NYT*, 30 Nov 1959, 25, as cited in Glickman, *Buying Power*, p. 258. Remaining quotations from *Consumer Reports, Medicine Show*, p. 3. The volume contains many critiques of medicalization and marketing regarding "the mouth odor fallacy" (ch. 27) and "dandruff, shampooing, and baldness" (ch. 28). For a cultural history of the promotion of these conditions around specific consumer products, see Roland Marchand, *Advertising the American Dream: Making Way for Modernity, 1900–1940* (Berkeley: University of California Press, 1986).

15. *Consumer Reports, Medicine Show*, p. 181. Also cited in Tomes, "Great American Medicine Show Revisited," p. 651.

16. Jeremy Greene and David Herzberg, "Hidden in Plain Sight: The Popular Promotion of Prescription Drugs in the 20th Century," *American Journal of Public Health* 100, no. 5 (2010): 793–803. See also David Herzberg, *Happy Pills in America from Miltown to Prozac* (Baltimore: Johns Hopkins University Press, 2010).

17. Note that the attention to potential cost savings by purchasing drugs by *generic name* in 1961 preceded the existence of a self-defined *generic drug* or *generic manufacturer*. *Consumer Reports, Medicine Show*, p. 221.

18. "Even where doctors prescribe by generic name," the 1963 edition warned, "consumers can't shop around for savings on patented drugs, for which the price to be charged by each licensee is fixed by the patent holder." *Consumer Reports, The Medicine Show: Some Plain Truths about Popular Remedies for Common Ailments*, 2nd ed. (New York: Simon & Schuster, 1963), p. 195.

19. Byrd, Cong. Rec., April 26, 1967, p. 10894.

20. Statement of Durward Hall, "A Bad Prescription for Drugs," Cong. Rec., 9 August 1967, p. 21980; "Has Drug Industry Met Its Nader?," *BusinessWeek*, 10 June, 1967, pp. 104–110. For other prominent reviews, see Carl M. Cobb, "How to Save Money . . . when Buying Drugs," *Boston Globe*, 25 April, 1967, A1; "Get Well Cheaper the Hard-Name Way," *New Republic*, 6 May 1967, p. 7.

21. Interview with Richard Burack, *NYT*, 8 May 1967. "Patients (who are 'captive consumers') have a right to know for what services they are paying." Richard Burack, *Handbook of Prescription Drugs*, 1st ed. (New York: Pantheon Books, 1967), p. 33.

22. Not until 1964 did the AMA's Council on Drugs recommended that prescription bottles display what they contain to the consumer, and these recommendations were not written into state laws in many cases until the 1970s. An editorial accompanying the Council on Drugs proposal in *JAMA* noted that opiates and barbiturates should not be named for patients and that all physicians knew that "some patients were better off not knowing what they were taking." J. H. Hoch, "Labeling Prescriptions with Names of Ingredients," *Journal of the South Carolina Medical Association* 60 (1964): 231–32, quotation p. 232.

23. Burack, *Handbook of Prescription Drugs*, p. 7, emphasis mine. When the second edition was published in 1970, there still was no nationwide requirement for generic drug names (or drug names at all) to appear on prescription bottles.

24. As George Nichols, Burack's chief at Cambridge Hospital, introduced in a foreword to the first edition, "the list of drugs in this book, for all its deceiving brevity, subserves many purposes. It suggests to the medical student a framework on which to build his knowledge of clinical pharmacology. It provides the practicing physician with a much-needed guide for simpler and more effective therapy. It offers the patient understanding of what his physician is about. Finally, it provides information about costs and suggestions for controlling them which are of potential use to all interested in seeing an end to unbridled escalation in the expense of medical care. Burack, *Handbook of Prescription Drugs*, pp. xiii–xiv.

25. "No single patient of mine who has been treated with a generic-equivalent drug has experienced anything but the effect which could be expected. I have been unable to observe in patients any difference whatsoever between the efficacy of generics and that of brand-name drugs, nor has there ever been a suggestion that the former are any more likely to cause untoward side-effects." Burack, *Handbook of Prescription Drugs*, p. 72.

26. Wesley R. Wells to John R. Graham, 17 May, 1968, f, "Consumer Response," RBP.

27. CU, review of *The Handbook of Prescription Drugs*, by Richard Burack, *Consumer Reports*, January 1968, pp. 47–48.

28. Ibid., p. 48.

29. Ibid., pp. 47–48.

30. Alfred Gilman to Sen. Gaylord Nelson, 11 July 1967, in *The New Handbook of Prescription Drugs*, by Richard Burack (New York: Pantheon Books, 1970), p. 63. As Nelson's aide Ben Gordon later wrote to Burack:

> We'll handle Gilman at the proper time and at our convenience. In the meantime, we are making a damn good record. We really clobbered Squibb and put a lot of documents into the record which would amaze you. They will be available to you. Perhaps I told you this before—I don't recall, but I called Gilman. He admitted that Stetler asked him to write the letter and gave him the names of the senators, etc. He also told me that he is a consultant to three companies and gets paid by them. I told him that this statement will have to be read by him before the Subcommittee. He didn't want to come but he is resigned to the fact that he has to. I told him it will be in the next two

or 3 months. I also asked him what he meant by "legalized piracy" and think that we've got him by the balls, and at the proper time we'll start squeezing. But don't be in a rush and keep cool.

Ben Gordon to Richard Burack, 26 October, 1967, RBP.

31. "Book Club Selection Boosts Generic Drugs," *Drug News Weekly*, 13 March 1967, clipping found in f 5, News Clippings & Letters about the "Handbook," RBP.

32. The fourth version was still a financial success, selling nearly a quarter of a million copies in the first few months. Phyllis Benjamin to Richard Burack, 12 January 1977, f 4, "Handbook Rx Drugs," RBP.

33. Richard Burack and Fred J. Fox, *New Handbook of Prescription Drugs* (New York: Ballantine, 1976), quotations pp. xviii, xx–xxi, respectively.

34. Ibid., p. xxi, emphasis mine.

35. Jeremy Greene and Scott Podolsky, "Keeping Modern in Medicine: Pharmaceutical Promotion and Physician Education in Postwar America," *Bulletin of the History of Medicine* 83, no. 2 (2009): 331–77.

36. Testimony of Mildred Brady, *Administered Prices*, vol. 21, p. 11536.

37. Chauncy Leake, *Administered Prices*, part 18, p. 10430, as cited in Greene and Podolsky, "Keeping Modern in Medicine."

38. Paul Rand Dixon, *Administered Prices*, part 14, p. 8144, as cited in Greene and Podolsky, "Keeping Modern in Medicine."

39. "The Fond du Lac Study," in United States Senate, Select Committee on Small Business, Subcommittee on Monopoly, *DIAA*, part 1, 1961, pp. 698–806, quotation p. 700. Subsequent observational studies would note this "considerable hiatus between the American medical stereotype and the individual physician's human cognitive limitations"; see Robert R. Rehder, "Communication and Opinion Formation in a Medical Community: The Significance of the Detail Man," *Journal of the Academy of Management* 8 (1965): 282–89, as cited in Greene and Podolsky, "Keeping Modern in Medicine."

40. Institute for Motivational Research, *Research Study on Pharmaceutical Advertising* (Croton-on-Hudson, NY: Pharmaceutical Advertising Club, 1955), p. 39, as cited in Greene and Podolsky, "Keeping Modern in Medicine."

41. Statement of Seymour Blackman, executive secretary of Premo Pharmaceutical Laboratories, *Administered Prices*, vol. 18, p. 8241.

42. E.g., Morton Mintz, *The Therapeutic Nightmare*, 5th ed. (New York: Houghton Mifflin: 1965).

43. James Long, *Essential Guide to Prescription Drugs* (New York: Harper and Row, 1977), p. xiii.

44. Ibid., p. xi.

45. Ibid., p. xiv. For more on the history of the *Pill Book*, currently in its 15th edition, see Lewis Grossman, "FDA and the Rise of the Empowered Consumer" (paper presented at the "FDA in the Twenty-First Century," Petrie Flom Center, Cambridge, MA, 3–4 May 2013).

46. M. Laurence Lieberman, *The Essential Guide to Generic Drugs* (Harper & Row, 1986). "It is difficult," Lieberman concluded in an op-ed in the *New York Times*, "to generalize about the worth of generic versus brand-name pharmaceuticals." M. Laurence Lieberman, "Elusive Equivalence," *NYT*, 17 December 1985, A26.

47. Lieberman, *The Essential Guide to Generic Drugs*, p. 154, emphasis in original; see also p. 126 on dicyclomine.

48. Ibid. p. 141.

49. *Consumer Reports, New Medicine Show*, p. 264.

CHAPTER 11. GENERIC CONSUMPTION IN THE CLINIC, PHARMACY, AND SUPERMARKET

Epigraph. "Generic Groceries: Cheaper, but What's in Them?," *Changing Times* 32, no. 12 (1978): 36–39, quotation pp. 36–37.

1. Ibid.

2. See, for example, Sergio Sismondo, "Ghosts in the Machine: Publication Planning in the Medical Sciences," *Social Studies of Science* 39 (2009): 949–52; Kalman Applebaum, "Getting to Yes: Corporate Power and the Creation of a Psychopharmaceutical Blockbuster," *Culture, Medicine and Psychiatry* 33 (2009): 185–215. Robert Proctor, *Golden Holocaust: Origins of the Cigarette Catastrophe and the Case for Abolition* (Berkeley: University of California Press, 2012); Marion Nestle, *Food Politics: How the Food Industry Influences Nutrition and Health* (Berkeley: University of California Press, 2002).

3. Estelle Brodman, "The Physician As Consumer of Medical Literature," *Bulletin of the New York Academy of Medicine* 61, no. 3 (1985) 266–74, quotation p. 272. A series of studies found that physicians (and "physician's wives") were far more likely to have "unnecessary" appendectomies and hysterectomies than the public at large. John P. Bunker and Byorn William Brown Jr., "The Physician-Patient As an Informed Consumer of Surgical Services," *NEJM* 290, no. 19 (1974): 1051–55.

4. Nancy Tomes, "The Great American Medicine Show Revisited," *Bulletin of the History of Medicine* 79, no. 4 (2005): 627–63, p. 634; on the prescription as a linking device, see Jeremy Greene and Elizabeth Watkins, eds., *Prescribed: Writing, Filling, Using, and Abusing the Prescription in Modern America* (Baltimore: Johns Hopkins University Press, 2012).

5. As consumer historian Lawrence Glickman has noted, though originally coined in the 1920s, "the word 'consumerism' is reminted in the 1960s not by advocates in the consumer movement but by its opponents in the business world, who aimed to belittle it as a subset of their main enemy, modern liberalism." Lawrence Glickman, *Buying Power: A History of Consumer Activism in America* (Chicago: University of Chicago Press: 2009), p. 16. Michael Pertschik, chief counsel for the Senate Commerce Committee (later head of the FTC) predicted in April of 1969 that the role of consumerism in the federal government "will flourish more with Nixon as president than they did under LBJ." Weinberger quotation from *FDCR*, 19 January 1970, p. T&G5. See also *FDCR*, 7 April 1969, p. 6; *FDCR*, 20 October 1969, pp. 11–12.

6. *FDCR*, 20 October 1969, p. 12, emphasis mine.

7. *FDCR*, 2 November 1970, p. 5.

8. Quotations from *FDCR*, 18 May 1970, p. 7. *FDCR*, 23 Nov 1970, p. 11; *FDCR*, 14 Dec 1970, p. 4. The rhetorical strategy of divorcing the consumer from consumer advocates was used broadly by conservative groups and free-market advocates. See Glickman, *Buying Power*, esp. pp. 277–95.

9. "Medical Consumerism," *Archives of Internal Medicine* 128 (1971): 469–71, quotation pp. 469–70.

10. *FDCR*, 4 Oct 1971, p. 15.

11. *FDCR*, 15 April 1974, p. A-1; *FDCR*, 26 April 1971, p. 25.

12. Edward G. Feldmann to William S. Apple, 27 August, 1975, "Correspondence in Favor of Substitution," box 18.11, f 33.46.1.1, APhAA.

13. Minutes, Drug Research Board, exploratory meeting on anti-substitution problems in drugs, 21 June 1974, box 5, f 9, NYSGDIF, p. 8.

14. William C. Cray, *The Pharmaceutical Manufacturers Association: The First 30 Years* (Washington, DC: Pharmaceutical Manufacturers Association, 1989), p. 221.

15. Dominique Tobbell, *Pills, Policy, and Policy: The Struggle for Drug Reform in Cold War America*, Milbank Series on Health and the Public (Berkeley: University of California Press, 2012); see also Tobbell, "Eroding the Physicians Control over Therapy," in Greene and Watkins, *Prescribed*, pp. 66–90.

16. Robert C. Drizen, "*Oksenholt v. Lederle Laboratories*: The Physician As Consumer," *Northwestern University Law Review* 79, no. 1 (1984): 460–83, quotation p. 470. For impacts of the *Oksenholt* decision on later cases of medical devices, see e.g., Phelps v. Sherwood Medical Industries, 836 F.2d 296 (7th Cir. 1987), http://law.justia.com/cases/federal/appellate-courts/F2/836/296/420080/.

17. *FDCR*, 15 February 1971, p. T&G6. See also *FDCR*, 15 March 1971, p. 6; *FDCR*, 11 June 1973, p. 12.

18. Nelson Rockefeller quotation from *FDCR*, 15 March 1971, p. 6; see also *FDCR*, 15 February 1971, p. T&G6.

19. *FDCR*, 21 January 1974, p. T&G1. A year earlier, when the APhA had looked into revoking the licenses of Revco pharmacists because Revco advertised drug prices to consumers, the Department of Justice threatened to prosecute the APhA for violation of antitrust law. *FDCR*, 1 January 1973, p. 15.

20. *FDCR*, 7 February 1977, p. T&G2.

21. Ibid. Giant had for the past two years been building up its own private drug quality control laboratories with chemists and pharmacologists able to conduct chromatography, disintegration, and dissolution testing on all generic products to guarantee conformity with compendial standards.

22. Ibid.

23. Quotation from *FDCR*, 14 February 1977, p. A-1, continues: "The agency supports this innovation on the part of the private sector and believes it will help to control health care costs by fostering competition and educating the consumer." See also *FDCR*, 7 February 1977, pp. T&G2-3; *Wood-Gundy Progress Report*, 7 April 1977, box 7, f 24, NYSGDIF.

24. Fantle quotations from *FDCR*, 7 February 1977, p. T&G3; *FDCR*, 14 February 1977, p. A-1. Giant's 1970s listings of generic drugs are echoed in Walmart's twenty-first-century listing of four-dollar generics. Niteesh K. Choudhry and William H. Shrank, "Four-Dollar Generics—Increased Accessibility, Impaired Quality Assurance," *NEJM* 363, no. 20 (2010): 1885–87.

25. Stetler demanded that the pharmacy chains disclose the names of the firms making the drugs as well to avoid confusion. Quotation from *FDCR*, 21 February 1977; *Wood-Gundy Progress Report*, quotation p. 4.

26. Barrie G. James, *The Marketing of Generic Drugs: A Guide to Counterstrategies for the Technology Intensive Pharmaceutical Companies* (London: Associated Business Press, 1980), p. 1. On the popularity of segmentation in postwar marketing and market research, see Lizabeth Cohen, *The Politics of Mass Consumption in Postwar America* (New York: Vintage, 2003), ch. 7.

27. Lester A. Neidell, Louis E. Boone, and James W. Cagley, "Consumer Responses to Generic Products," *Journal of the Academy of Marketing Science* 12, no. 4 (1985):161–76, quotation p. 162; D. Cudabeck, "French Chain's Unbranded Products—Loved by Consumers, Hated by Admen," *Advertising Age*, June 28, 1976; C. G. Burck, "Plain Labels Challenge the Supermarket Establishments," *Fortune*, 26 March 1979, pp. 70–76.

28. Quotation from Robert O. Aders and Roger L. Jenkins, "Outlook for the Retail Food Industry," *Survey of Business* 16 (1980): 23–26, quoted in Martha R. McEnally and Jon M. Hawes, "The Market for Generic Brand Grocery Products: A Review and Extension," *Journal of Marketing* 48 (1984): 75–83, p. 75. By 1982, a literature of more than four hundred articles (mostly in trade journals) discussed the marketing of generic brand grocery products. As McEnally and Hawes claimed, by 1984, "the introduction and subsequent marketing of generic brand grocery products has made an important and lasting impression on marketing thought and practice." See also Neidell et al., "Consumer Responses to Generic Products."

29. *FDCR*, 5 March 1979, p. T&G 5.

30. J. Barry Mason and William O. Bearden, "Generic Drugs: Consumer, Pharmacist, and Physician Perceptions of the Issue," *Journal of Consumer Affairs* 14, no. 1 (1980): 193–205, quotation p. 194. See also *What Is a Generic Drug?* (Montpelier, VT: Vermont Public Interest Research Group, 1969), box 4, f 5, NYSGDIF; "Patents on 117 of Top 200 Rx Items to Lapse by '84," *American Drug Merchandising* 167 (1973): 19.

31. Neidell et al., "Consumer Responses to Generic Products," p. 174. Other studies include A. J. Faria, "Generics, the New Marketing Revolution," *Akron Business and Economic Review*, Winter 1979, pp. 33–38; P. E. Murphy and G. R. Laczniak, "Generic Supermarket Items: A Product and Consumer Analysis," *Journal of Retailing*, Summer 1979, pp. 3–14; J. Zbniewski and W. H. Heller, "Rich Shopper, Poor Shopper, They're All Trying Generics," *Progressive Grocer*, March 1979, pp. 92–106. By 1984, a systematic review of marketing research into of the demographic profiles of generic consumers found that generic consumers tended to be middle aged rather than young or old, come from larger rather than smaller households and from middle-income (rather than low- or high-income) populations, and to be characterized by a higher level of education. McEnally and Hawes, "Market for Generic Brand Grocery Products."

32. McEnally and Hawes, "Market for Generic Brand Grocery Products," p. 75.

33. Harold Hopkins, "Food Marketing without Frills," *FDA Consumer* 12, no. 9 (1978): 6–9, p. 8.

34. "Generic Groceries," pp. 36–37.

35. C. G. Burck, "Plain Labels Challenge the Supermarket Establishments," *Fortune* 26 March 1979, pp. 70–76.

36. Hopkins, "Food Marketing without Frills," p. 8. See also McEnally and Haws, "Market for Generic Brand Grocery Products," p. 75.

37. "Generic Groceries," p. 37. See also Isadore Barmash, "Inflation Puts the Squeeze on Supermarkets, Too," *NYT*, 4 June 1978, pp. F1 and F9. On the broader

problem of "imitation food," see Suzanne White Junod, "Chemistry and Controversy: Regulating the Use of Chemicals in Foods, 1883–1959" (PhD diss., Emory University, 1994).

38. Esther Peterson, having since left Giant Foods to once again become the special assistant for Consumer Affairs in Jimmy Carter's White House, suggested the popularity of generic goods stemmed from "the desperate feeling on the part of people that they must get away from frills that are costing them extra." "Generic Groceries," p. 38.

39. Hopkins, "Food Marketing without Frills," pp. 6–7.

40. *Food & Drug Letter* 59 (1978): 4.

41. Hopkins, "Food Marketing without Frills," pp. 6–7.

42. *FDCR*, January 7 1985, quotation p. 9.

43. In 1986, Lederle, one of the leading generic manufacturers, contracted the public opinion firm Hill & Knowlton and the Gallup polling corporation to determine current attitudes toward quality in generic products among "sophisticated consumers." *FDCR*, 18 December 1989, p. TG13. The "No Logo" movement that became associated with antiglobalization politics in the 1990s offers one possible counternarrative regarding the later resurgence of generic consumerism. Naomi Klein, *No Logo: No Space, No Choice, No Jobs* (New York: Picador, 2000).

44. Kenneth J. Arrow, "Uncertainty and the Welfare Economics of Health Care," *American Economic Review* 53, no. 2 (1963): 1–9; William D. Savedoff, "Kenneth Arrow and the Birth of Health Economics," *Bulletin of the World Health Organization* 82, no. 2 (2004): 139–40; Peter Temin, *Taking Your Medicine: Drug Regulation in the United States* (Cambridge: Harvard University Press, 1980); Paul Starr, *The Social Transformation of American Medicine* (New York: Basic Books, 1983).

CHAPTER 12. SCIENCE AND POLITICS OF THE "ME-TOO" DRUG

Epigraph. Statement of Walter Modell, *DIAA*, v. 1, p. 320.

1. Joshua Gagne and Niteesh Choudhry, "How Many Me-Too Drugs Is Too Many?," *JAMA* 305, no. 7 (2011): 711–12; P. C. Austin, M. M. Mamdani, and D. N. Juurlink, "How Many "Me-Too" Drugs Are Enough? The Case of Physician Preferences for Specific Statins," *Annals of Pharmacotherapy* 40, no. 6 (June 2006): 1047–51.

2. To confuse matters, many actors in 1960s and 1970s and even early 1980s also used the term *me-too drug* to designate chemically identical generic drugs being produced by a generic manufacturer. "A me-too, as the phrase is being used," in the 1968 report of the NAS-NRC DES, was an old molecule, "a drug which was marketed . . . on the basis that one or more pioneering manufacturers had processed the same or a similar product through the FDA procedure." *FDCR*, 5 February 1968, p. 12.

3. United States Task Force on Prescription Drugs, *The Drug Makers* (Washington, DC: Government Printing Office, 1968), p. 21, emphasis mine.

4. Edith A. Nutescu, Hayley Y. Park, Surrey M. Walton, Juan C. Blackburn, Jamie M. Finley, Richard K. Lewis, and Glen T. Schumock, "Factors That Influence Prescribing within a Therapeutic Drug Class," *Journal of Evaluation in Clinical Practice* 119, no. 4 (2005): 357–65.

5. Marcia Angell, "Excess in the Pharmaceutical Industry," *Canadian Medical Association Journal* 171, no. 12 (2004): 1451. Marcia Angell, *The Truth about the Drug Companies: How They Deceive Us and What to Do about It* (New York: Random House, 2004). "New Drugs, Big Dollars," *Consumer Reports*, November 2007, www.consumer reports.org/health/doctors-hospitals/medical-ripoffs/new-drugs-big-dollars/medical -ripoffs-ov3_1.htm. On more recent critiques of me-too drugs, especially in light of in- novation studies and the emerging discipline of comparative effectiveness research, see J. DiMasi and C. Paquette, "The Economics of Follow-on Drug Research and Development Trends in Entry Rates and the Timing of Development," supplement, *PharmacoEconomics* 22 (2004): 1–14; B. Pekarsky, "Should Financial Incentives Be Used to Differentially Reward 'Me-Too' and Innovative Drugs?," *PharmacoEconom- ics* 28, no. 1 (2010): 1–17; F. M. Clement, A. Harris, J. J. Li, K. Yong, K. M. Lee, and B. J. Manns, "Using Effectiveness and Cost-Effectiveness to Make Drug Coverage Deci- sions: A Comparison of Britain, Australia, and Canada," *JAMA* 302, no. 13 (7 October 2009): 1437–43.

6. Estes Kefauver, introduction to *DIAA*, vol. 1, p. 29. Kefauver did not invent the term *me-too*, nor was he the first to apply it to the world of pharmaceuticals (for ear- lier usage, see Arthur A. Cottew, "Mycolysine," *Medical World*, 28, no. 6 (1910): 262. But the hearings for S. 1552 brought an issue previously confined to professional publica- tion to a broader level of popular and political visibility.

7. Statement of Estes Kefauver, *DIAA*, vol. 1, p. 29.

8. Statement of Louis Lasagna, *DIAA*, vol. 1, quotation p. 302; Statement of Walter Modell, *DIAA*, vol. 1, quotation p. 320. Note that concern for physicians as victims of a predatory pharmaceutical marketing apparatus can be found in the 1870s, if not ear- lier; see Joseph M. Gabriel, "Restricting the Sale of 'Deadly Poisons': Drug Regulation and Narratives of Suffering in the Gilded Age," *Journal of the Gilded Age and Progres- sive Era* 9, no. 3 (2010): 145–69.

9. Modell acknowledged that at times such modifications, even if they did not change therapeutic efficacy, could produce a drug that might have a differential safety or even idiosyncratic adverse effect profile. "In such cases a molecular mod- ification, which under other circumstances might not be as effective, becomes the drug of choice." Statement of Walter Modell, *DIAA*, vol. 1, quotation p. 321. Statement of Louis Goodman, *DIAA*, vol.1, quotation p. 219.

10. Statement of Louis Goodman, *DIAA*, vol. 1, quotations pp. 221–22. Goodman elaborated an argument of subpopulation difference.

> What if a congeneric compound was definitely superior only in a small number of pa- tients, would it pass muster? Let me cite an example. I will refer later to an anticonvul- sant drug, called Trimethadione, extremely useful in children with petit mal epilepsy. Shortly after it came on the market in the mid-1940s, a close chemical congener was introduced, namely, Paramethadione, in which a methyl group of the parent drug is replaced by an ethyl group. The two drugs are almost as alike as two peas in a pod, with regard to action, use, and toxicity. In retrospect, at the time Paramethadione was released, it is doubtful whether, under the proposed legislation, it could have merited patent protection. If the company for this reason had decided not to market it, the medical profession and hundreds of children with petit mal would have been deprived

of a drug which, as the years and experience have advanced, proved to be a very useful agent. For reasons we do not understand, some cases respond well to Paramethadione that do not respond to Trimethadione, and some patients who have severe toxic reactions to the latter agent can readily tolerate the former drug.

11. Statement of Vannevar Bush, *DIAA*, vol. 4, quotation p. 2202.

12. Statement of John Christian Krantz, *DIAA*, vol. 4, p. 2370. "The treatment of a patient with a drug is a highly personalized relationship." Krantz elaborated, "The drug must tailor-fit the patient. It must provide him with maximal relief and minimal untoward side effects. Atropine may relieve my gastrointestinal distress, but not yours. By molecular manipulation, methylhopmatropine results. This may be the ideal drug for you. If it is, you deserve it, and many other anticholinergics should be and are available for your biochemical individuality."

13. John C. Krantz, ed., *Fighting Disease with Drugs: The Story of Pharmacy, a Symposium* (Baltimore: Williams & Wilkins, 1931); John C. Krantz, *Recollections: A Medical Scientist Remembers* (Baltimore: Schneidereth & Sons, 1978).

14. Statement of William Apple, *DIAA*, vol. 5, quotation p. 2760; Statement of Abraham Ribicoff, *DIAA*, vol. 5, quotation p. 2581; Statement of Eugene Beesley, *DIAA*, vol. 4, quotation p. 2003. See also G. F. Roll, "Molecular Modification: A Classic Research Method," 17 July 1961; John Blair to E. Winslow Turner, "Patent and Antitrust Part of Drug Hearings," box 6, f "Patent hearings: memoranda, etc., Aug 1961," SOAM.

15. Statement of Estes Kefauver, *DIAA*, vol. 4, p. 2317. Kefauver commended the PMA for "what I regard as the best presentation by an industry of its case since that offered by the U.S. Steel Corporation just before World War II" and noted that the PMA agreed with many provisions of S. 1552: the empowerment of the FDA to enforce claims of efficacy of a product, the extension of FDA's inspection powers, and the ability of HEW to establish usable generic names of a product.

16. *FDCR*, 16 October 1967, quotation p. 7.

17. *FDCR*, 3 March 1969, p. 11.

18. *FDCR*, 1 September 1969, quotation p. 12; *FDCR*, 31 March 1969, quotation p. 8.

19. Testimony of Walter Modell, *Competitive Problems*, vol. 1, pp. 295–96.

20. Statement of Leonard Scheele, *Competitive Problems*, vol. 6, p. 2291.

21. Benoit Majerus, "Magic Bullet to the Head?," in *Histories of the Therapeutic Revolution*, ed. Jeremy Greene, Elizabeth Watkins, and Flurin Condreau (forthcoming).

22. "A root problem of course, is that congeners with no inherent advantages over the older drug from which they are copied can be and are successfully marketed through high-power promotion." Richard Burack to Robert E. Jones, 21 February 1968, f "Consumer Response," RBP.

23. *FDCR*, 30 July 1973, p. 7.

24. *FDCR*, 9 July 1979, p. T&G4.

25. *FDCR*, 4 August 1969, p. 10.

26. Hans G. Engel, "Is There a Place for Me Too Drugs?," *Private Practice*, October 1976, pp. D1–D2, quotation p. D1.

27. In his history of clinical pharmacy, Frank Ascione credits the origins of the therapeutic interchange concept to an effort by hospitals, managed care organizations, and third-party payers to "go beyond generic product selection" in efforts to

control cost increases. Frank J. Ascinoe, Duane M. Kirking, Caroline A. Gaither, and Lynda S. Welage, "Historical Overview of Generic Medication Policy," *JAPhA* (2001): 567–77.

28. Paul L. Doering, William C. McCormick, Deobrah L. Klapp, and Wayne L. Russell, "Therapeutic Substitution and the Hospital Formulary System," *American Journal of Hospital Pharmacy* 38 (1981): 1949–51, quotation pp. 1949–50.

29. E. P. Abraham, "A Glimpse at the Early History of the Cephalosporins," *Reviews of Infectious Diseases* 1, no. 1 (1979): 99–105, quotation p. 105; J. M. T. Hamilton-Miller and W. Brumfitt, "Whither the Cephalosporins?," *Journal of Infectious Diseases* 130, no. 1 (1974): 81–84; Walter Sneader, *Drug Prototypes and Their Exploitation* (New York: John Wiley & Sons, 1996), p. 388.

30. Sneader, *Drug Prototypes and Their Exploitation*, p. 388.

31. T. H. Maugh, "A New Wave of Antibiotics Builds," *Science* 214, no. 4526 (1981): 1225–28; Michael W. Noel and James Paxinos, "Cephalosporins: Use Review and Cost Analysis," *American Journal of Hospital Pharmacy* 35 (1978): 933–35; P. N. Johnson and L. P. Jeffrey, "Restricted Cephalosporin Use in Teaching Hospitals," *American Journal of Hospital Pharmacy* 38, no. 4 (1981): 513–17.

32. P. L. Doering, W. C. McCormick, D. K. Klapp, and W. L. Russell, "State Regulatory Positions concerning Therapeutic Substitution in Hospitals," *American Journal of Hospital Pharmacy* 38 (1981): 1900–1903; Paul L. Doering, Deborah L. Klapp, William C. McCormick, and Wayne L. Russell, "Therapeutic Substitution Practices in Short-Term Hospitals," *American Journal of Hospital Pharmacy* 39 (1982): 1028–32. William W. McCloskey, Philip N. Johnson, and Louis P. Jeffrey, "Cephalosporin-Use Restrictions in Teaching Hospitals," *American Journal of Hospital Pharmacy* 41 (1984): 2359–52; Norman T. Suzuki and William B. Breuninger, "Automatic Interchange Policy for Cefotetan and Cefoxitin," *American Journal of Hospital Pharmacy* 45 (1988): 1864; N. T. Suzuki, "Automatic Interchange Policy for First-Generation I.V. Cephalosporins," *American Journal of Hospital Pharmacy* 44 (1987): 505. See also Richard Segal, Lorelei L. Grines, and Dev S. Pahak, "Opinions of Pharmacy, Medicine, and Pharmaceutical Industry Leaders about Hypothetical Therapeutic-Interchange Legislation," *American Journal of Hospital Pharmacy* 45 (1988): 570–77.

33. Paul L. Doering, Wayne L. Russell, Deborah L. Klapp, and William C. McCormick, "Therapeutic Substitution: Has Its Time Arrived?," *Hospital Formulary* 19 (1984): 36–40, quotation p. 38.

34. Harry Schwartz, "American Substitution Scene: Bitter Turf Battle," *Private Practice*, September 1982, pp. 54–60, quotation p. 60; *FDCR*, 16 December 1985, quotation p. 16.

35. Schwartz, "American Substitution Scene," p. 60. See also Julie Fairman, "The Right to Write," in *Prescribed: Writing, Filling, Using, and Abusing the Prescription in Modern America*, ed. Elizabeth Watkins and Jeremy Greene (Baltimore: Johns Hopkins University Press, 2012).

36. Judith Nelson, "What Ever Happened to the Pharmacy Class of '47?" *Private Practice*, September 1982, pp. 61–66, quotation p. 66.

37. *FDCR*, 14 April 1986, quotation p. 10. The PMA announced in December of 1985 that joint APhA-PMA Committee on Therapeutic Substitution and Rx Sampling would try to maintain dialogue on issues before they threaten pharmacy/industry re-

lations. *FDCR*, 16 December 1985, pp. 15–16. The issue came to a head in 1985 in Iowa, as state legislature passed a bill okaying therapeutic substitution, but it was vetoed by the governor in response to lobbying from PMA. By June of 1986, the APhA-PMA committee issued a tempered statement that "therapeutic substitution in which pharmacists and physicians interrelate on behalf of the patient is appropriate, but ... therapeutic substitution, in which pharmacists decide without the physician, is not appropriate." *FDCR*, 14 July 1986, p. TG6.

38. *FDCR*, 14 December 1987, p. TG6.

39. David S. Jones and Jeremy A. Greene, "The Contributions of Prevention and Treatment to the Decline in Cardiovascular Mortality: Lessons from a Forty-Year Debate," *Health Affairs* 31, no. 10 (2012): 2250–58.

40. Brian B. Hoffman, *Adrenaline* (Cambridge, MA: Harvard University Press), 2013. Joseph Gabriel, *Medical Monopoly* (Chicago: University of Chicago Press, 2014).

41. R. P. Ahlquist, "A Study of the Adrenotropic Receptors," *American Journal of Physics* 153 (1948): 586–600; Vivian Quirke, "Putting Theory into Practice: James Black, Receptor Theory, and the Development of Beta-Blockers at ICI, 1958–1978," *Medical History* 50, no. 1 (2006): 69–92; C. E. Powell and I. H. Slater, "Blocking of Inhibitory Adrenergic Receptors by a Dichloro Analog of Isoproterenol," *Journal of Pharmacology and Experimental Therapeutics* 122 (1958): 480–88; N. C. Moran and M. E. Perkins, "Adrenergic Blockade of the Mammalian Heart by a Dichloro Analogue of Isoproterenol," *Journal of Pharmacology and Experimental Therapeutics* 124 (1958): 223–37.

42. On the therapeutic proliferation of beta-blockers, see Rein Vos, *Drugs Looking for Diseases*: *Innovative Drug Research and the Development of Beta Blockers and the Calcium Antagonists* (Dordrecht: Kluwer Academic, 1991); William H. Frishman, "Beta-Adrenoreceptor Antagonists: New Drugs and New Indications," *NEJM* 305, no. 9 (1981): 500–506.

43. Frishman, "Beta-Adrenoreceptor Antagonists," p. 501.

44. A. M. Lands, A. Arnold, J. P. McAuliff, F. P. Luduena, and T. G. Brown Jr., "Differentiation of Receptor Systems Activated by Sympathomimetic Amines," *Nature* 214 (1967): 597–98. This hairsplitting work would find an analogue in Black's other project—at SKF—distinguishing histamine-1 (H1) from histamine-2 (H2) receptors, which coalesced around the discovery of the first H2-blocker, the powerful antiulcer drug Tagamet (cimetidine), arguably the world's first blockbuster drug. Quirke, "Putting Theory into Practice," pp. 88–90.

45. William H. Frishman, "Clinical Differences between Beta-Adrenergic Blocking Agents: Implications for Therapeutic Substitution," *Journal of the American Heart Association* May 1987, 1190–98, quotation p. 1190.

46. Ibid.

47. Richard A. Levy and Dorothy L. Smith, "Clinical Differences among Nonsteroidal Antiinflammatory Drugs: Implications for Therapeutic Substitution in Ambulatory Patients," *Annals of Pharmacotherapy* 23 (1989): 76–85. Richard A. Levy, "Clinical Aspects of Therapeutic Substitution," *PharmacoEconomics*, 1st ser., 1 (1992): 41–44.

48. John G. Ballin, "Therapeutic Substitution—Usurpation of the Physician's Prerogative," *JAMA* 257, no. 4 (1987): 528–29, quotation p. 528. Gerald J. Mossinghoff, "Opposition to Therapeutic Interchange," *American Journal of Hospital Pharmacy* 45

(1988): 1065. "State Chapters React to Drug Companies' Positions on Therapeutic Interchange," *American Journal of Hospital Pharmacy* 44 (1987): 2665, 2669.

49. Donald C. McLeod, "Therapeutic Drug Interchange: The Battle Heats Up," *Drug Intelligence and Clinical Pharmacy* 22 (1988):716–18, quotation p. 716.

50. Ibid., p. 717.

51. Adam Hedgecoe and Paul Martin, "The Drugs Don't Work: Expectations and the Shaping of Pharmacogenetics," *Social Studies of Science* 33, no. 3 (2003): 327–64.

52. Darryl Rich, "Experience with a Two-Tiered Therapeutic Interchange Policy," *American Journal of Hospital Pharmacy* 46 (1989): 1792–98.

53. A. Zoloth, J. L. Yon, and R. Woolf, "Effective Decision-Making in a Changing Healthcare Environment: A P&T Committee Interview," *Hospital Formulary* 24, no. 2 (1989): 85–87, 90, 93, 96, 98, quotation p. 93.

54. Even as more and more states—including Washington and Wisconsin—were investigating the legal basis of therapeutic substitution by the mid to late 1980s, their efforts were being surpassed by private insurers seeking means to contain costs through substitution practices. By the end of the 1980s, federal legislation was making it more difficult for states to pursue therapeutic substitution just as a new arena for substitution was coming into focus: managed care formularies. But the ability of states to act in this arena was swiftly contested in state legislatures and on a federal level. At the end of the 1980s, Sen. David Pryor (D-AR) proposed a national P&T committee to help create a consensus on which drug classes were and were not exchangeable. Pryor's Pharmaceutical Access and Prudent Purchasing Act was intended to adjudicate, pharmaceutically speaking, what minimal "essential benefits" a state should provide—but it was strongly opposed by industry. Pryor's plan for a national P&T would precipitate the opposite, under the Omnibus Budget Reconciliation Act of 1990, when Sen. Pryor reached an agreement with prominent US pharmaceutical manufacturers to prohibit Medicaid formularies in exchange for rebate programs to Medicaid. William J. Moore and Robert J. Newman, "Drug Formulary Restrictions as a Cost-Containment Policy in Medicaid Programs," *Journal of Law and Economics* 36, no. 1 (1992): 71–97; Bryan L. Wasler, Dennis Ross-Degnan, and Stephen B. Soumerai, "Do Open Formularies Increase Access to Clinically Useful Drugs?," *Health Affairs* 15, no. 3 (1996):95–109.

CHAPTER 13. PREFERRED DRUGS, PUBLIC AND PRIVATE

Epigraph. Lawrence W. Abrams, "The Role of Pharmacy Benefit Managers in Formulary Design: Service Providers or Fiduciaries?," *Journal of Managed Care Pharmacy* 10, no. 4 (2004): 359–60.

1. These changes to managed care structures happened swiftly as the programs ballooned in the 1980s. On the emergence of for-profit managed care in the early 1980s, see Beatrix Hoffmann, *Health Care for Some: Rights and Rationing in the United States since 1930* (Chicago: University of Chicago Press, 2012). Only one-third of HMOs were for profit in 1985, but one year later there were more for-profit than nonprofit HMOs; by the end of the 1990s HMOs were overwhelmingly a for-profit sector. This section relies on the analysis by Bradford H. Gray, "The Rise and Decline of the HMO:

A Chapter in US Health Policy History," in *History and Health Policy: Putting the Past Back In*, ed. Rosemary A. Stevens, Charles E. Rosenberg, and Lawton R. Burns (New Brunswick: Rutgers University Press, 2006), pp. 309–37.

2. William C. Cray, *The Pharmaceutical Manufacturers Association: The First 30 Years* (Washington, DC: PMA, 1989), p. 337.

3. Gerald J. Mossinghoff, "Opposition to Therapeutic Interchange," *American Journal of Hospital Pharmacy* 45 (1988): 1065.

4. One key focus of the managed care backlash was the "gag rules" prohibiting physicians from discussing incentive/substitution structures with patients. Thomas Bodenheimer, "The HMO Backlash—Righteous or Reactionary?," *NEJM* 335, no. 21 (1996): 1601–4; Robin Toner, "Rx Redux: Fevered Issue, Second Opinion," *NYT*, 10 October 1999.

5. David S. Hilzenrath, "Art Imitates Life when It Comes to Frustration with HMOs," *Washington Post*, 10 February, 1998, p. C01; Nancy-Ann DeParle, "As Good As It Gets? The Future of Medicare+Choice," *Journal of Health Politics, Policy and Law* 27, no. 3 (2002): 495–512; National Health Policy Forum, *Managed Care: As Good As It Gets?* (site visit report, George Washington University, Washington, DC, January 8–11, 2002).

6. The three firms were PCS Health Systems, Inc., Merck-Medco Managed Care, and Diversified Pharmaceutical Services. Sheila R. Shulman, "Pharmacy Benefit Management Companies (PBMs): Why Should We Be Interested?," *PharmacoEconomics*, 1st ser., 14 (1998):49–56.

7. Quotation from Robert F. Atlas, "Wrangling Prescription Drug Benefits: A Conversation with Express Scripts' Barrett Toan," *Health Affairs* W5 (2005): 191–98; James G. Dickinson, "Pharmacy Turmoil Follows Merck-Medco News," *Medical Marketing and Media* 28, no. 11 (1993): 66–68.

8. "A highly refined formulary," he continued, "limits the possibility of generic substitution and therapeutic interchange. As such, a highly refined formulary can drive up the costs of a drug benefit considerably." Abrams, "The Role of Pharmacy Benefit Managers in Formulary Design: Service Providers or Fiduciaries?"

9. Michael Gray, "PBMs: Can't Live with 'Em, Can't Live without 'Em," *Medical Marketing and Media* 30, no. 9 (1995): 48–52, quotation pp. 50–51.

10. Ibid., p. 52.

11. Shulman, "Pharmacy Benefit Management Companies"; see also Fed. Reg. 236 (5 January 1998).

12. Federal Trade Commission, "Merck Settles FTC Charges That Its Acquisition of Medco Could Cause Higher Prices and Reduced Quality for Prescription Drugs," 27 August 1998 (press release), www.ftc.gov/opa/1998/08/merck.htm.

13. "PBMs Switch Brand Name Drugs for Deals," *Medical Marketing and Media* 36, no. 10 (2001): 18–29; Milt Freudenheim, "With Ties Lingering, Medco Leaves Merck," *NYT*, 20 August 2003, p. C2; Abrams, "The Role of Pharmacy Benefit Managers in Formulary Design," 359–60; Milt Freudenheim, "Documents Detail Big Payments by Drug Makers to Sway Sales," *NYT*, 13 March 2003, p. C1; Allison Dabbs Garrett and Robert Garis, "PBMs: Leveling the Playing Field," *Valparaiso University Law Review* 42, no. 1 (2007): 33–80. Mark Lowry, "Pharmacy Coalition Endorses PBM Reform Legislation," *Drug Topics*, 18 September 2013. http://drugtopics.modernmedicine.com /drug-topics/news/pharmacy-coalition-endorses-pbm-reform-legislation.

14. This section relies upon the analysis Dan Fox provides in *The Convergence of Science and Governance: Research, Health Policy, and American States* (Berkeley: University of California Press, 2010), quotation p. 76; see also Daniel M. Fox, "Evidence of Evidence-Based Health Policy: The Politics of Systematic Reviews in Coverage Decisions," *Health Affairs* 24, no. 1 (2005):114–22; Reforming States Group, "State Initiatives on Prescription Drugs: Creating a More Functional Market," *Health Affairs* 22, no. 4 (2003):128–36, quotation cited in Fox, *Convergence of Science and Governance*, p. 76; Michele Mello, David M. Studdert, and Troyen A. Brennan, "The Pharmaceutical Industry versus Medicaid—Limits on State Initiatives to Control Prescription-Drug Costs," *NEJM* 350, no. 6 (2004): 208–13.

15. Jeremy A. Greene, "Swimming Upstream: Comparative Effectiveness Research in the US," *PharmacoEconomics* 27, no. 12 (2009): 979–82.

16. For earlier use of systematic reviews in evidence-based policy making, see Andrew Oxman and Daniel Fox, eds., *Informing Judgment: Case Studies of Health Policy and Research in Six Countries* (New York: Milbank Memorial Fund, 2001).

17. Fox, *Convergence of Science and Governance*, pp. 78–83; Robert Pear and James Dao, "States' Tactics Aim to Reduce Drug Spending," *NYT*, 21 November 2004.

18. Fox, *Convergence of Science and Governance*, pp. 81–82.

19. Mark Gibson, "Making the Best Use of Limited Resources for Drug Evaluations" (paper presented at a Cochrane Collaborative Colloquium, Sao Paolo, 26 October 2007), as cited in Fox, *Convergence of Science and Governance*, p. 88–89. "Major Black Groups Call OMB Medicaid Drug Plan Dangerous," *New Pittsburgh Courier*, 11 August 1990; Richard A. Levy, "Ethnic and Racial Differences in Responses to Medicines: Preserving Individualized Therapy in Managed Pharmaceutical Programmes," *Pharmaceutical Medicine* 7 (1993): 139–65, p. 139; Steven Epstein, *Inclusion: The Politics of Difference in Medical Research* (Chicago: University of Chicago Press, 2007), esp. pp. 71–73, 138; Adam Hedgecoe, *The Politics of Personalised Medicine: Pharmacogenetics in the Clinic* (Cambridge: Cambridge University Press, 2004); David S. Jones, How Personalized Medicine Became Genetic, and Racial: Werner Kalow and the Formation of Pharmacogenetics," *Journal of the History of Medicine and Allied Sciences* 68, no. 1 (2013): 1–48.

20. Valentine J. Burroughs, Randall W. Maxey, and Richard A Levy, "Racial and Ethnic Differences in Response to Medicines: Toward Individualized Pharmaceutical Treatment," *Journal of the National Medical Association* 94 (2002): 1–26, quotation p. 2; see also George Curry, "Guarding against Generic Racism in Medicine," *Los Angeles Sentinel*, 3 October 2002; Valentine Burroughs, Randall Maxey, Lavera Crawley, and Richard Levy, *Cultural and Genetic Diversity in America: The Need for Individualized Pharmaceutical Treatment* (Washington, DC: NPC and NMA, 2004). On the broader articulation of suspicion of generic drugs by African American physician groups, see Anne Pollock, *Medicating Race: Heart Disease and Durable Preoccupations with Difference* (Durham, NC: Duke University Press, 2012).

21. Other critiques of inclusion/difference in the articulation of therapeutic similarity came from mental health advocates, the National Alliance on Mental Illness, the American Psychiatric Association, along with the AMA. Tom Toolen, "States Misuse Evidence-Based Medicine," *Medical Herald*, April 2005, p. 1, as cited in Fox, *Convergence of Science and Governance*, pp. 92, 94.

22. SmithKline Beecham immediately threatened to cease all business with Maine wholesalers. Carl J. Seiden, "Caution: Politics Ahead," *Medical Marketing and Media* 35, no. 10 (2000): 114–24. Shawna Lydon Woodward, "Will Price Control Legislation Satisfactorily Address the Issue of High Prescription Drug Prices? Several States Are Waiting in the Balance for *PhRMA v. Concannon*," *Seattle University Law Review* 26 (2002): 169–95. J. Robert Pear, "Justices Voice Skepticism on Taking Drug-Cost Case," *NYT*, 23 January 2003, A18; Bill Wechsler, "Maine Court Victory Major Setback for Pharma," *Pharmaceutical Executive* 23, no. 7 (2003): 20.

23. I.e., Idaho Medicaid presented a savings of $340,000 after the first 6 months of the DERP review of anticonvulsants, with a report of positive outcomes and no negative outcomes. Fox, *Convergence of Science and Governance*, pp. 102–3.

24. Kalipso Chalkidou, Sean Tunis, Ruth Lopert, Lise Rochaix, Peter T. Sawicki, Mona Nasser, and Bertrand Xerri, "Comparative Effectiveness Research and Evidence-Based Health Policy: Experience from Four Countries," *Milbank Quarterly* 87, no. 2(2009): 339–67. Andrew Oxman and Daniel Fox, eds., *Informing Judgment: Case Studies of Health Policy and Research in Six Countries* (New York: Milbank Memorial Fund, 2001). Institute of Medicine, *Initial National Priorities for Comparative Effectiveness Research* (Washington, DC: National Academies Press, 2009). Jerry Avorn, "Debate about Funding Comparative-Effectiveness Research," *NEJM* 360, no. 19 (2009): 1927–29; Gail R. Wilensky, "The Policies and Politics of Creating a Comparative Clinical Effectiveness Research Center" (online only), *Health Affairs* 28 no. 4 (2009): w719-29.

25. On the opposition of reason and rationality in Cold War science, see Paul Erickson, Judy L. Kein, Lorraine Daston, Rebecca Lemov, Thomas Stern, and Michael D. Gordin, *How Reason Almost Lost Its Mind* (Chicago: University of Chicago Press, 2013).

CHAPTER 14. THE GLOBAL GENERIC

Epigraph. IMS Health, *Brand Renewal: Maximizing Lifecycle Value in an Ever More Generic World*, 2007, www.imshealth.com/deployedfiles/ims/Global/Content /Solutions/Healthcare%20Measurement/Pharmaceutical%20Measurement/Brand _Renewal.pdf.

1. IMS Health, *The Global Use of Medicines: Outlook through 2016*, 2012, www .imshealth.com/deployedfiles/ims/Global/Content/Insights/IMS%20Institute%20for %20Healthcare%20Informatics/Global%20Use%20of%20Meds%202011/Medicines _Outlook_Through_2016_Report.pdf.

2. IMS Health, *Brand Renewal*.

3. Núria Homedes and Antonio Ugalde, "Multisource Drug Policies in Latin America," *Bulletin of the World Health Organization* 64 (2005): 66–67; On the historical relevance of counterfeit drug markets in sub-Saharan Africa, see Kristin Peterson, *Speculative Markets: Drug Circuits and Derivative Life in Nigeria* (Durham, NC: Duke University Press, 2014).

4. For an emerging comparative ethnography of generic drugs across multiple sites in Latin America, including Mexico, Brazil, and Argentina, see Cori Hayden, "A Generic Solution? Pharmaceuticals and the Politics of the Similar in Mexico," *Current*

Anthropology 48, no. 4 (2007): 475–95; and Hayden, "No Patent, No Generic: Pharmaceutical Access and the Politics of the Copy," in *Making and Unmaking Intellectual Property: Creative Production in Legal and Cultural Perspective*, ed. Mario Biagioli, Peter Jaszi, and Martha Woodmansee (Chicago: University of Chicago Press, 2011), pp. 285–304. For ethnographic accounts of generic drugs in India, see Stefan Ecks and Soumita Basu, "The Unlicensed Lives of Antidepressants in India: Generic Drugs, Unqualified Practitioners, and Floating Prescriptions," *Transcultural Psychiatry* 46, no. 1 (2009): 86–106; also Kaushik Sunder Rajan, "Pharmaceutical Crises and Questions of Value: Terrains and Logics of Global Therapeutic Politics," *South Atlantic Quarterly* 111, no. 2 (2012): 321–46. On the heterogeneity of technology transfer in the developing world more generally, see Sanjaya Lall, *Developing Countries as Exporters of Technology* (London: Macmillan, 1982).

5. Halfdan Mahler, address to the Twenty-Eighth World Health Assembly, Geneva, 15 May 1975, A28/11, 1975, WHOA. On the shifting postcolonial politics of the World Health Assembly in the 1970s, see Nitsan Chorev, *The World Health Organization between North and South* (Ithaca, NY: Cornell University Press, 2012).

6. Portions of this section adapted form Jeremy A. Greene, "Making Medicines Essential: The Emergent Centrality of Pharmaceuticals in Global Health," *BioSocieties* 6, no. 1 (2011): 10–33.

7. Daniel Azarnoff to Expert Committee on the Selection of Essential Drugs, 19 July 1977, E 19 81 1, f 1, Geneva, CH, WHOA, p. 2.

8. *SCRIP World Pharmaceutical News* 259 (1977): 23.

9. S. Michael Peretz, "An Industry View of Restricted Drug Formularies," *Journal of Social and Administrative Pharmacy* 1, no. 3 (1983): 130–33.

10. Egli to Mahler, untitled letter, 1 December 1977, E 19 81 1, f 1, WHOA.

11. Ibid.

12. The question of technology transfer in the 1970s helped foment a broad-based critique of multinational corporations through UN bodies such as the United Nations Centre on Transnational Corporations (UNCTC), though most UN recommendations in this area ultimately depended on cooperation with multinational corporations. See, e.g., UNCTC, *Transnational Corporations in the Pharmaceutical Industry of Developing Countries* (New York: United Nations, 1984). On the LCPFP, see M. Stork, W. B. Wanandi, A. S. Arambulo, *Guidelines and Recommendations for the Establishment of a Low Cost Pharmaceutical Formulation Plant (LCPFP) in Developing Countries* (Geneva: WHO, 1980); Rosalyn King, *The Provision of Pharmaceuticals in Selected Primary Health Care Projects in Africa: Report of a Survey*, USAID, contract no. Afr/0135-C-1092, November 1981, p. 73, http://pdf.usaid.gov/pdf_docs/PNAAY435.pdf; "Drug Formulation in Developing Countries," *Bulletin of the World Health Organization* 59, no. 4 (1981): 531; Najmi Kanji et al., *Drugs Policies in Developing Countries* (London: Zed Books, 1992).

13. Barrie G. James, *The Marketing of Generic Drugs: A Guide to Counterstrategies for the Technology-Intensive Pharmaceutical Companies* (London: Associated Business Press, 1980), p. 84. While Servipharm AG had no direct association with the WHO's essential drugs program, "the product line follows the WHO's general drug selections and includes antibiotics, antiparasitics and analgesics prices to compete with generic competitors. Ciba-Geigy's move into the branded generic market in the

Third World parallels their acquisition of Tutag, a branded and commodity-generic manufacturer and marketer in the US."

14. *FDCR*, 21 September 1987, pp. 6–7.

15. William C. Cray, *The Pharmaceutical Manufacturers Association: The First 30 Years* (Washington, DC: PMA, 1989), p. 292.

16. Meir Perez Pugatch, *The International Political Economy of Intellectual Property Rights* (Edward Elgar, 2004), pp. 66–69. A similar argument would be leveled in a PMA complaint against generic manufacturers in Thailand a few years later. What would happen to American consumers, the PMA asked in its brief, if "Thai generics of uneven quality are exported to other countries in the region?" The PMA warned that such products would quickly be repackaged by counterfeiters "so that they appear to be a product of the U.S. and [sold] into the diversion market in this country." Quotations from *FDCR*, 11 February 1991, p. T&G3.

17. Cray, *Pharmaceutical Manufacturers Association*, p. 311.

18. Quotation from Cray, *Pharmaceutical Manufacturers Association*, p. 312.

19. Maurice Cassier and Marilena Correa, "Patents, Innovation and Public Health: Brazilian Public-Sector Laboratories' Experience in Copying AIDS Drugs," *RECIIS* 1, no. 1 (2003): 83–90; Jean-Paul Gaudilliere, "How Pharmaceuticals Became Patentable in the Twentieth Century," *History and Technology*, 24, no. 2 (2008): 208.

20. United Nations Centre on Transnational Corporations, *Transnational Corporations in the Pharmaceutical Industry of Developing Countries*, p. 90. The form of bilateral intellectual property action regarding pharmaceuticals that the PMA pushed forward in Brazil would become increasingly common over the late 1980s and 1990s, most famously illustrated by the tasking of Vice President Al Gore as emissary to South Africa in 1998 to try to shut down the importation of generic versions of US-patented antiretroviral drugs as the South African HIV/AIDS epidemic was reaching catastrophic proportions. See "Ethics and HIV Treatments: Al Gore under Fire," *Annals of Oncology* 10 (1999): 1261–67; Cassier and Correa, "Patents, Innovation and Public Health"; Pugatch, *International Political Economy of Intellectual Property Rights*, p. 68.

21. *FDCR*, 6 March 1989, p. 6.

22. *FDCR*, 19 October 1992, pp. 8–10. On the ramifications for NAFTA for the emerging generics sector in Mexican pharmaceuticals, see Cori Hayden, "A Generic Solution? Pharmaceuticals and the Politics of the Similar in Mexico," *Current Anthropology* 48, no. 4 (2007): 475–95; on the relation of NAFTA and the WTO in the differential articulation of pharmaceutical intellectual property between Mexico and Brazil in the 1990s and 2000s, see Kenneth C. Shadlen, "The Politics of Patents and Drugs in Brazil and Mexico: The Industrial Bases of Health Policies," *Comparative Politics* 10 (2009): 41–58.

23. Joao Biehl, *Will to Live: AIDS Therapies and the Politics of Survival* (Princeton, NJ: Princeton University Press, 2009); Cassier and Correa, "Patents, Innovation and Public Health."

24. Cassier and Correa, "Patents, Innovation and Public Health."

25. These drugs were all off patent and generically available: ampicillin, ranitidine, ketoconazole, furosemide, as well as a newer antiasthma agent, albuterol. Andréa D. Bertoldi, Aluísio J. D. Barros, and Pedro C. Hallal, "Generic Drugs in Brazil: Known

by Many, Used by Few," *Cas: Saúde pública, Rio de Janeiro* 21, no. 6 (2005): 1808–15; Hayden "No Patent, No Generic," in Biagioli, Jaszi, and Woodmansee, *Making and Unmaking Intellectual Property*, pp. 285–304.

26. A subsequent US International Trade Commission report on Indian generic pharmaceuticals defined the phrase *plain vanilla generics* as a technical term, meaning "commodity generics that are 'off-patent' in the regulated markets [which offer little or no innovative value over the innovator's product]," William Greene, "The Emergence of India's Pharmaceutical Industry and Implications for the U.S. Generic Drug Market," United States International Trade Commission, Office of Economics working paper 2007-05-A, p. iii.

27. Ibid., p. 1.

28. For more on the general development of the Indian generics sector, see Kaushik Sunder Rajan, *Biocapital: The Constitution of Postgenomic Life* (Durham, NC: Duke University Press, 2008); Kaushik Sunder Rajan, "Pharmaceutical Crises and Questions of Value: Terrains and Logics of Global Therapeutic Politics," *South Atlantic Quarterly* 111, no. 2 (2012): 321–46; Stefan Ecks and Soumita Basu, "The Unlicensed Lives of Antidepressants in India: Generic Drugs, Unqualified Practitioners, and Floating Prescriptions," *Transcultural Psychiatry* 46, no. 1 (2009): 86–106.

29. The Patent Act, paired with the Drug Price Control Order of 1970, allowed for a kind of market exclusivity for reverse engineering innovation. Greene, "The Emergence of India's Pharmaceutical Industry," p. 3.

30. Note the similarity between this list of nations and the list named by Mossinghoff in 1985 as "pirates" of US-based innovative intellectual property. Sanjaya Lall, "Developing Countries as Exporters of Technology: A Preliminary Assessment," in *International Resource Allocation and Economic Development*, ed. H. Giersch (Tubingen: J. C. B. Mohr, 1979); Sanjaya Lall, *Developing Countries as Exporters of Technology: A First Look at the Indian Experience* (London: Macmillan, 1982). Lall makes frequent reference to Joseph Schumpeter's models of saltatory technological progress in his predictions to explain the heterogeneity of technology-based industry within the global South. On comparative statistics of Indian vs. US pharmaceutical production and consumption, see UNCTC, *Transnational Corporations in the Pharmaceutical Industry of Developing Countries*, p. 100; Greene "Emergence of India's Pharmaceutical Industry," pp. 4–5, 16–17; Nitin Shukla and Tanushree Sangal, "Generic Drug Industry in India: The Counterfeit Spin," *Journal of Intellectual Property Rights* 14 (2009): 236–40.

31. Greene "Emergence of India's Pharmaceutical Industry," pp. 4–5.

32. Germán Velásquez, "The Access to Drugs between the New Rules of International Trade and the Right to Health" (paper presented at the "Drugs, Standards, and the Practices of Globalization" conference, Paris, 9 December 2010).

33. Ibid, p. 19. See also Vikas Bajaj, "In India, a Developing Case of Innovation Envy," *NYT*, 9 December 2009. Rachel Zimmerman and Jesse Pesta, "Drug Industry, AIDS Community is Jolted by Cipla AIDS-Drug Offer," *WSJ*, 8 February 2001. Donald G. McNeil, "India Alters Law on Drug Patents," *NYT*, 24 March 2005. Divya Rajagopal, "Cipla Launches Four in One Drug for HIV Patients," *Economic Times*, 14 August 2012, http://articles.economictimes.indiatimes.com/2012–08-14/news/33201146_1_line -treatment-cipla-tenofovir-and-emtricitabine.

34. UNCTC, *Transnational Corporations in the Pharmaceutical Industry*, pp. 99, 101.

35. Greene "The Emergence of India's Pharmaceutical Industry, pp. 6–7.

36. Shukla and Sangal, "Generic Drug Industry in India," p. 238. On the relations of the two forms of piracy, see Adrian Johns, *Piracy: Intellectual Property Wars from Gutenberg to Gates* (Chicago: University of Chicago Press, 2010).

37. *FDCR*, 20 August 1990, p. 12. See also Joanna Breitstein, "I Pray for the Welfare of Your Company," *Pharmaceutical Executive* 26, no. 10 (2006): 49–57. It should be noted that Indian generic companies, like Ranbaxy and Dr. Reddy's were also busily acquiring generic manufacturers on Europe and North America. Dr. Reddy's purchased the German firm Betapharm Arzneimittel for $572 million—a purchase that overnight made Dr. Reddy's Germany's fourth-largest generic drug manufacturer. Ranbaxy, in turn, purchased manufacturing capacity in Europe with the acquisition of the Romanian company Terapia and the French firm RPG Aventis.

38. Jennifer Bayot, "Teva to Acquire Ivax, Another Maker of Generic Drugs," *NYT*, 26 July 2005, C5.

39. Robert Koenig, "Giant Merger Creates Biotech Power," *Science* 271, no. 5255 (1996): 1490; "When Giants Unite," *Pharmaceutical Executive* 17, no. 2 (1997): 46–62. See also Walter Armstrong, "The Book on Daniel Vasella," *Pharmaceutical Executive* 29, no. 10 (2009): 46–58; Breitstein, "I Pray for the Welfare of Your Company," p. 57; Natasha Singer, "That Pill You Took? It May Be Teva's," *NYT*, 8 May 2010; Yoram Gabison, "The Rise and Fall of Teva: How Israel's Global Pharma Star Lost Its Vitality," *Ha'aretz*, 26 October 2013, p. 1; Dror Reich, "CEO of Teva, Israel's Largest Company, Steps Down in Shock Resignation," *Ha'aretz*, 30 October 2013, p. 1; Eytan Avriel, "Teva's Woes Extend beyond CEO's Sudden Departure," *Ha'aretz*, 31 October 2013, p. 3.

40. Breitstein, "I Pray for the Welfare of Your Company," p. 52.

41. Singer, "That Pill You Took?"

42. Breitstein, "I Pray for the Welfare of Your Company," pp. 57–58.

43. Armstrong, "The Book on Daniel Vasella," quotation p. 54. See also Stan Bernard, "Big Pharma's Most Feared Competitor," *Pharmaceutical Executive* 6 (2010): 30–31.

CONCLUSION. THE CRISIS OF SIMILARITY

Epigraph. Ludwik Fleck, *Genesis and Development of a Scientific Fact*, ed. Fred Bradley and Robert K. Merton (Chicago: University of Chicago Press, 1979), p. 20.

1. In February 2012, the FDA issued a draft guidance on its biosimilars pathway, but at time of press the pathway has yet to be fully elucidated or realized. "Teva Receives EU Marketing Authorization for TevaGrastim," *Bloomberg News*, 16 September 2008; "Teva Announces FDA Grants Approval for Tbo-filgrastim for the Treatment of Chemotherapy-Induced Neutropenia" (press release), 30 August 2012, www.tevapharm.com/Media/News/Pages/2012/1730014.aspx.

2. "Biosimilars," *Health Affairs*, 10 October 2013, http://healthaffairs.org/blog/2013/10/10/health-policy-brief-biosimilars/.

3. Angela Creager, "Biotechnology and Blood: Edwin Cohn's Plasma Fractionation Project, 1940–1953," in *Private Science: Biotechnology and the Rise of the Molecular Sciences*, ed. Arnold Thackray (Philadelphia: University of Pennsylvania Press, 1998), pp. 39–62; Alexander von Schwering, Heiko Stoff, and Betina Wahrig, *Biologics: A History*

of Therapeutic Agents Made from Living Organisms in the Twentieth Century (London: Pickering & Chatto, 2013). For alternate histories of biotech avant la lettre outside the realm of therapeutics, see Jack Ralph Kloppenberg Jr., *First the Seed: The Political Economy of Plant Biotechnology* (Madison: University of Wisconsin Press, 2005). For more recent histories of biotech therapeutics see Sally Hughes, *Genentech: The Beginnings of Biotech* (Chicago: University of Chicago Press, 2011); Nicholas Rasmussen, *Gene Jockeys: Life Science and the Rise of Biotech Enterprise* (Baltimore: Johns Hopkins University Press, 2014).

4. As of 2013, the US insulin market was limited to three manufacturers: Eli Lilly, Novo Nordisk, and Sanofi-Aventis, all of whom sold distinct and incommensurable insulin products under brand name. Lilly sells insulin lispro under the brand name Humalog and recombinant insulin under the brand name Humulin; Sanofi-Aventis sells insulin glulisine under the brand name Apidra and insulin glargine under the brand name Lantus; Novo Nordisc sells insulin aspart under the brand name Novolog. None of these can be considered to be generic drugs. *FDA Orange Book of Approved Drug Products with Therapeutic Equivalence Evaluations*, accessed 1 November 2013, www.accessdata.fda.gov/scripts/Cder/ob/default.cfm; David M. Dudzinski, "Reflections on Historical, Scientific, and Legal Issues Relevant to Designing Approval Pathways for Generic Versions of Recombinant Protein-Based Therapeutics and Monoclonal Antibodies," *Food & Drug Law Journal* 60 (2005): 143–260; Michael Bliss, *The Discovery of Insulin* (Chicago: University of Chicago Press, 1982).

5. On the administration of insulin patents in relation to the standardization of different insulin formulations, see Bliss, *Discovery of Insulin*; Christiane Sindig, "Making the Unit of Insulin: Standards, Clinical Work and Industry," *Bulletin of the History of Medicine* 76 (2002): 231–70; Maurice Cassier and Christiane Sindig, " 'Patenting in the Public Interest': Administration of Insulin Patents by the University of Toronto," *History and Technology* 24, no. 2 (2008): 153–72; Ulrike Thoms, "The German Pharmaceutical Industry and the Standardization of Insulin Before the Second World War," in von Schwering, Stoff, and Wahrig, *Biologics*, pp. 151–72.

6. For a longer view of the interplay of patent and trademark law and living systems, from *Ex Parte Latimer* to *Parke, Davis & Co. v. H. K. Mulford Co.* to *Diamond v. Chakrabarty* to *Association for Molecular Pathology v. Myriad Genetics*, see Daniel J. Kevles, "New Blood, New Fruits: Protections for Breeders and Originators," in *Making and Unmaking Intellectual Property: Creative Production in Legal and Cultural Perspective*, ed. Mario Biagioli, Peter Jaszi, and Martha Woodmansee (Chicago: University of Chicago Press, 2011); Christopher Beauchamp, "Patenting Nature: A Problem of History," 16, no. 257 (2013): 257–312.

7. Dudzinski, "Issues Relevant to Designing Approval Pathways"; Rasmussen, *Gene Jockeys*, esp. ch. 2.

8. *FDCR*, 14 May 1990, p. T&G10.

9. Brian Reid, "US Biotech Prepares to Fight Generic Biologicals," *Nature Biotechnology* 20 (2002): 322. Huub Schellekens and Jean-Charles Ryff, " 'Biogenerics': The Off-Patent Biotech Products," *Trends in Pharmacological Sciences* 23, no. 3 (2002): 119–21.

10. Schellekens and Ryff, " 'Biogenerics,' " p. 120. See also Jeffrey Fox, "Lawsuits Anticipated on Generic Biologicals Front," *Nature Biotechnology* 21, no. 7 (2003): 721–22;

Valerie Junod, "Drug Marketing Exclusivity under United States and European Union Law," *Food and Drug Law Journal* 59 (2004): 1–56.

11. Bruce Manheim Jr., Patricia Granahan, and Kenneth Dow, " 'Follow-on Biologics': Ensuring Continued Innovation in the Biotechnology Industry," *Health Affairs* 25, no. 2 (2006): 394–404; Rasmussen, *Gene Jockies*, esp. ch. 5.

12. John Yoo, "Taking Issues in the Approval of Generic Biologics," *Food and Drug Law Journal* 60 (2005): 33–43.

13. Jeffrey Fox, "Democrats Prioritize Pricing, Generics and Drug Safety," *Nature Biotechnology* 25, no. 2 (2007): 150–51; Gregory Davis et al., "Recommendations regarding Technical Standards for Follow-on Biologics: Comparability, Similarity, Interchangeability," *Current Medical Research and Opinion* 25, no. 7 (2009): 1655–61; Robert Pear, "In House, Many Spoke with One Voice: Lobbyists," *NYT*, 14 November 2009. Investigative journalists in the United Kingdom likewise revealed that extensive lobbying from Amgen had helped produce a blanket 2008 parliamentary advisory against approval of biosimilar products. Nigel Hawkes, "There's No Biological Alternative, Says Parliament," *BMJ* 336 (15 March 2008): 588.

14. A. B. Engelberg, A. S. Kesselheim, and J. A. Avorn, "Balancing Innovation, Access, and Profits for Biologics," *NEJM* 361, no. 20 (2009): 1917–19; H. Grabowski, G. Loing, and R. Mortimer, "Implementation of the Biosimilar Pathway: Economic and Policy Issues," *Seton Hall Law Review* 41, no. 2 (2011): 511–57.

15. Henry Miller, "Why an Abbreviated FDA Pathway for Biosimilars Is Overhyped," *Nature Biotechnology* 29, no. 9 (2011): 794–95. Megerlin et al. offer a slightly more optimistic interpretation of European data; see Francis Megerlin, Ruth Lopert, Ken Taymor, and Jean-Hughues Trouvin, *Health Affairs* 32, no. 10 (2013): 1803–10.

16. John Hubbard, senior vice president of worldwide development operations for Pfizer USA, has called biobetters "the market opportunity" for Pfizer's future development. Generics and Biosimilars Initiative, "Biobetters Rather than Biosimilars," http://gabionline.net/Biosimilars/General/Biobetters-rather-than-biosimilars; for more discussion of biobetters and "supergenerics," see Cori Hayden, "Distinctively Similar: A Generic Problem," *UC Davis Law Journal* 47, no. 2 (2013): 601–32.

17. Mark McCamish and Gillian Woollett, "The State of the Art in the Development of Biosimilars," *Clinical Pharmacology and Therapeutics* 91, no. 3 (2012): 405–16; Federal Trade Commission, *Authorized Generic Drugs: Short-Term Effects and Long-Term Impact* (Washington, DC: FTC).

18. Jessica DeMartino, "Biosimilars: Approval and Acceptance?," supplement, *Journal of the National Comprehensive Cancer Network* 9, no. 3 (2011): S-6–S-9. BIO and PhRMA likewise advocated for uniquely naming biologics at the time of market entry to reduce physician and patient confusion and aid pharmacovigilance efforts. "BIO and PhRMA insist on unique names for biosimilars," *Biosimilar News*, 4 July 2012.

19. Robert Ulin, *Vintages and Traditions: An Ethnohistory of Southwest French Wine Cooperatives* (Washington, DC, Smithsonian Institute Press, 1996). Ulin's analysis of how the appellation system itself has skewed our historical understanding of wine toward the makers of *grand crus* (at the expense of the cooperatives that produce table wine) resonates on several levels with the present analysis. I thank Sophia Roosth for referring me to Ulin's work.

20. All cites from Duff Wilson, "Drug Firms Face Billions in Losses in '11 as Patents End," *NYT*, 6 March 2011.

21. Jack W. Scannell, Alex Blanckley, Helen Boldon, and Brian Warrington, "Diagnosing the Decline in Pharmaceutical R&D Efficiency," *Nature Reviews: Drug Discovery* 11, no 3 (2012): 191–200, p. 193.

22. Other modifications of Moore's Law in biotech abound, such as Carlson's Law, which relates to alarming speed by which protein and genome sequencing has dropped in price. Chris Kelty, in his discussion of Moore's Law in *Two Bits: The Cultural Significance of Free Software* (Durham, NC: Duke University Press, 2008), discusses these curves in the broader context of fears that the acceleration of technological development has escaped human control.

23. Ibid. This paragraph based on Scannell et al., "Diagnosing the Decline in Pharmaceutical R&D Efficiency."

24. Scannell et al., "Diagnosing the Decline in Pharmaceutical R&D Efficiency."

25. IMS Health, *Generic Medicines: Essential Contributors to the Long-Term Health of Society*, 2010, www.imshealth.com/imshealth/Global/Content/Document/Market_Measurement_TL/Generic_Medicines_GA.pdf.

26. Several scholars have studies the role of promissory futures in the contemporary functioning of biomedical institutions; see Mike Fortun, "Medicated Speculations in the Genomics Futures Markets," *New Genetics and Society* 20 (2001): 139–56; Nik Brown, "Hope against Hype: Accountability in Biopasts, Presents and Futures," *Science Studies* 16, no. 2 (2003): 3–21; Charis Thompson, *Making Parents: The Ontological Choreography of Reproductive Technologies* (Cambridge: MIT Press, 2005); Michael Morrison, "Promissory Futures and Possible Pasts: The Dynamic of Contemporary Expectations in Regenerative Medicine," *BioSocieties* 7, no. 1 (2012): 3–22.

27. William Apple, address at PMA Annual Meeting, Boca Raton, FL, April 1960. Also cited by Dominique Tobbell in *Pills, Power, and Policy: The Struggle for Drug Reform in Cold War America and Its Consequences*, Milbank Series on Health and the Public (University of California Press, 2012).

28. These examples represent a few of the unanticipated consequences of the incentive structure for generic drug firms under current health policy structures. After a decade of litigation, the FTC has only in 2013 been recognized to have authority to block "pay for delay" transactions in which innovator firms pay large amounts of funds to early generic competitors to *not* bring their drugs to market, thereby maintaining market exclusivity. Conversely, in 2012, the Supreme Court ruled that, unlike brand-name pharmaceutical manufacturers, generic manufacturers were *not* liable when patient were injured because of unlabeled adverse effects from their products, meaning that generic firms continue to have little reason to monitor the long-term safety of their drug products. Finally, the number of generic drugs *not* being produced in adequate numbers to supply American hospitals has been increasing over the past decade, jumping from two hundred to more than three hundred between 2012 and 2013. Aaron S. Kesselheim and Nathan Shiu, "*FTC v. Actavis:* The Supreme Court Issues a Reversal on Reverse Payments," *Health Affairs*, 21 June 2013, http://healthaffairs.org/blog/2013/06/21/ftc-v-actavis-the-supreme-court-issues-a-reversal-on-reverse-payments/; Aaron S. Kesselheim, Jerry Avorn, and Jeremy A. Greene, "Risk, Responsibility, and Generic Drugs," *NEJM* 367, no. 18 (2012): 1679–81; Margaret Clapp, Michael A. Rie, and Phillip L. Zweig, "How a Cabal Keeps Generic Scarce," *NYT*, 2 September 2013.

INDEX

Abbott, 80, 123, 236, 316n40

Abbreviated New Drug Application (ANDA), 66, 71–72, 73, 74, 75, 83, 85, 86, 87, 300n19, 300–301n29

Abraham, Edward P., 221

Abrams, Lawrence, 231

absorption, 45, 94, 98–102, 104, 111–13, 117, 118, 121, 188

Academy of Pharmaceutical Sciences, 142

Accum, Frederick, 57

acetaminophen, 33, 42

adrenalin, 22, 27, 224–25, 283n3

Adriani, John, 63, 64, 115, 116

adulterated medicines, 11, 49, 55–57, 60, 62, 63, 64–65, 71

advertising: of branded generics, 81; of brand vs. generic names, 42–43, 140; of generic drugs, 77–78; by Giant Foods, 199–200; in *JAMA*, 28, 31, 43; of look-alike drugs, 132; physician prescribing practices and, 185–86; by Purepac, 78–80; of quality, 77–79, 126–27, 198, 199–201, 202. *See also* marketing

Ahlquist, R. P., 225

albuterol, 15, 33

alchemy, 14

Allegra, 259

allergy drugs, 93, 188–89, 212, 229, 259

ALLHAT study, 237

American Association of Retired Persons (AARP), 147–48, 153, 165, 167, 206

American Druggist, 77, 79, 127

American Drug Index, 40

American Drug Manufacturers Association, 28

American Enterprise Institute, 241

American Epilepsy Foundation, 169

American Journal of Pharmacy, 142

American Medical Association (AMA), 17; AMA-USP 1961 joint committee on nomenclature, 45; concerns about federal oversight of generic drug naming, 44–45, 47, 49–50; Council on Drugs, 31–32, 45, 63, 68, 115, 246, 304n1, 321n22; Council on Pharmacy and Chemistry, 27, 28, 31, 178, 285n18; Department of Drugs, 227; Fond du Lac Study, 185, 192; *New and Nonofficial Drugs,* 27; role in drug naming, 27, 28, 31–32, 33, 35, 37–38, 40, 47, 49–50; Seal of

Acceptance program, 28, 31, 41, 287n26; on therapeutic substitution, 223–24, 227–28

American Pharmaceutical Association (APhA), 27, 32, 39, 46, 64, 140, 141–43, 144, 195, 216, 310n25; AARP and, 147–48; on brand substitution, 140–42, 147–50; Policy Committee on Professional Affairs, 222

American Pharmaceutical Manufacturers Association, 28

American Society of Hospital Pharmacists, 98

Amgen, 264, 265

amphetamines, 41, 60, 61, 71, 72, 73, 82, 99

ampicillin, 74–75, 80, 143, 148, 223

Anderson, Sharon, 125

Angell, Marcia, 213

antibiotics, 2, 12, 128; ampicillin, 74–75, 80, 143, 148, 223; cephalosporins, 221–24, 228–29, 230; chloramphenicol, 73, 74, 75, 93, 110–14, 116, 118, 119, 128, 304n1, 305n7; erythromycin, 123, 316n40; penicillins, 42, 62, 69, 74, 80, 82, 128, 148, 221, 223, 306n20; tetracycline, 74, 119, 148, 156

antidepressants, 2, 229

antihypertensives, 237

antipsychotics, 237

antiretroviral production: in Brazil, 251–53, 336n20; in India, 254, 255–56, 257, 337n33

antisubstitution laws, 11, 56 57, 141–43, 310n19; repeal of, 147–49, 151, 155, 195. *See also* substitution, generic

Apple, William, 64, 141, 142, 168, 195, 216, 297n31, 310n17

Archambalt, George F., 140

Argentina, 250, 254

Arrow, Kenneth, 208

As Good As It Gets (film), 232–33

aspirin, 22, 26, 173, 283n3, 285n13, 308n48

atorvastatin, 133, 138, 236, 270

Aurobindo (India), 257

authorized generics, 267

Ayerst, 74, 80, 124–25

Azarnoff, Daniel, 246

Azilect, 259

azithromycin, 259

Baer, William J., 236

Ballin, John, 227